BRITISH COLUMBIA: ITS RESOURCES AND PEOPLE

BRITISH COLUMBIA: ITS RESOURCES AND PEOPLE

edited by

CHARLES N. FORWARD

Western Geographical Series Volume 22

Department of Geography, University of Victoria
Victoria, British Columbia
Canada

1987 University of Victoria

Western Geographical Series, Volume 22

editorial address

Harold D. Foster, Ph.D.
Department of Geography
University of Victoria
Victoria, British Columbia
Canada

Publication of the Western Geographical Series has been generously supported by the Leon and Thea Koerner Foundation, the Social Science Federation of Canada, the National Centre for Atmospheric Research, the International Geographical Union Congress, the University of Victoria and the Natural Sciences and Engineering Research Council of Canada.

BRITISH COLUMBIA: ITS RESOURCES AND PEOPLE

(Western geographical series; ISSN 0315-2022; v. 22)
Includes bibliographical references.
ISBN 0-919838-12-X

1. British Columbia — Description and travel — 1981- . I. Forward, Charles N., 1927- . II. University of Victoria (B.C.), Dept. of Geography. III. Series.
FC3817.4.B75 1987 917.11 C86-091358-9
F1087.B75 1987

ACKNOWLEDGEMENTS

Many members of the Department of Geography, University of Victoria, contributed to the successful completion of this volume. When the project was initiated all of the authors were faculty members in the department and their co-operation and understanding during the editorial process was greatly appreciated. Special thanks are due to the members of the Technical Services division under the direction of Ian Norie: Diane Macdonald, who undertook the tasks of both typesetting and layout; and Ole Heggen and Ken Josephson, who did the cartography and photographic production. Finally, the efforts of Debbie Penner, Jennifer Roy and Donna Schmidt, who typed the initial manuscript drafts, are acknowledged with thanks.

University of Victoria
Victoria, British Columbia

Charles N. Forward
Editor

PREFACE

The production of this book was a co-operative effort by a number of faculty and staff members of the Department of Geography, University of Victoria who set out to produce a comprehensive, geographical appreciation of the westernmost province in the 1980s. It follows in the footsteps of the pioneering volume published as part of the Canadian contribution to the success of the International Geographical Union congress held in Montreal in 1972 (Robinson, J.L. (ed.), *British Columbia*. Toronto: University of Toronto Press, Studies in Canadian Geography, 1972). No claim is made that it is fully balanced and complete in coverage, but it does embrace a wide range of topics in physical and human geography. Several things may be identified that recur throughout the book. First, evidence of the distinctiveness of British Columbia within Canada appears frequently, as the various aspects of the province are examined. Second, the unity that has evolved within a diversity of landscapes and activities is apparent in many aspects of British Columbia's geographic character, indicating that the province has reached a certain degree of maturity in the transformation of its vast realm from an unknown wilderness to an integrated regional economy. Third, the future prospects of British Columbia within various aspects of the economy or social organization are commented on in many of the chapters. In the first two chapters the historical development and regional character of British Columbia are examined, with the goal of providing a general background for the more detailed treatment of special topics that follows. The physical environment is considered in chapters three to five, leading to the examination of a number of specific resource topics in chapters six to thirteen. Finally, several aspects of population, ethnic groups, urban areas and tourism are investigated in chapters fourteen to eighteen.

Charles N. Forward
Victoria, British Columbia

PLATE A A view near Greenwood in the southern Interior. *K. Josephson Photo* ▶

PLATE B Tofino on the west coast of Vancouver Island. *B.C. Government Photo* ▶

TABLE OF CONTENTS

LIST OF PLATES

LIST OF FIGURES

LIST OF TABLES

PLATE 1 Rodeo at Fort St. John. *B.C. Government Photo*▶

1 EVOLUTION OF REGIONAL CHARACTER[1]

Charles N. Forward
University of Victoria

British Columbia is a most distinctive province within the nation. It is generally considered a separate major region of Canada, whereas the adjacent western provinces of Alberta, Saskatchewan and Manitoba are usually lumped together as the Prairie Provinces. Its dominantly mountainous character, its extensive coastline and its general orientation to the Pacific rather than inland set it apart from interior Canada. Historically, European occupance and development of British Columbia came from the Pacific via initial settlements on Vancouver Island and the southwest corner of the mainland. For nearly half a century this coastal-focussed settlement area west of the mountain barriers expanded in isolation from the rest of Canada before significant development occurred on the Prairies. The first transcontinental railway in the 1880s finally provided a transport connection, but did not change the well established coastal orientation of economic activity which has continued to the present day. Despite the clear-cut distinctiveness of British Columbia relative to the rest of Canada, there exists considerable diversity within the province and a number of distinctive geographical regions have developed. Diversity of terrain is inevitable in a dominantly mountainous country where there are bound to be numerous valleys, and in British Columbia there are extensive areas of plateau and plain as well. Tremendous variation in climate, vegetation and soils occurs both vertically, owing to the great ranges of elevation from sea level to high mountain peaks, and latitudinally, with the province spanning more than 11 degrees (48°19′N to 60°00′N), not to mention the longitudinal variations from marine to continental conditions. The development of the province from the southwestern focus, first eastward and then northward, over a long period of time, and the utilization of different resource combinations in various areas, produced a diversity of human occupance that augmented this physical diversity. Although it is difficult to identify specific regional boundaries, an attempt is made to define and characterize the major regions that have evolved.

MAJOR GEOGRAPHICAL REGIONS

A total of 11 major regions have been identified, all of which could be subdivided into a number of minor regions at a more detailed scale of analysis, which is beyond the scope of this chapter (Figure 1,1). Only the major regional divisions have been recognized, in order to characterize diversity within a limited and clearly distinctive set of regions. The discussion of each region takes the form of a brief historical and geographical thumbnail sketch. They are discussed from south to north, roughly in accordance with the areal expansion of economic development. Maps of the development of railway and highway transportation networks are presented to illustrate the many references to these systems throughout the text (Figures 2,1 and 3,1). Sources of information include a number of provincial publications on regions, the *Bulletin Area*[2] series, the *Economic Survey*[3] series, the *Development Possibilities*[4] series, the *Regional Index*[5] and *Holland*[6] on landforms; the federally published *Area Profiles*[7]; the *Atlas of British Columbia*[8]; and several books, Barker[9], Ormsby[10], Robinson[11], and Robinson and Hardwick[12].

Vancouver Island

Consisting of a mountainous backbone flanked by a rugged, deeply indented coast on the west and a narrow coastal lowland on the east, Vancouver Island offered a varied terrain for human occupance. The moderate, marine climate of high precipitation nurtured luxuriant coniferous forests of large trees, western hemlock characterizing the extensive wetter regions and Douglas fir the drier east coast. The climate also was amenable to agriculture, especially along the southeastern coast. In fact, that area achieved a widespread reputation as an attractive residential and retirement environment, owing to its moderate climate characterized by mild winters and cool, dry summers.

The Colony of Vancouver Island was established in 1849, predating the mainland Colony of British Columbia by nine years. Prior to this, what is now British Columbia was part of New Caledonia controlled by the Hudson's Bay Company which operated under a charter from the British Crown to prosecute the fur trade. The Pacific headquarters was situated at Fort Vancouver near the mouth of the Columbia River until the international boundary was extended from the Rocky Mountains to the Pacific along the forty-ninth parallel in 1846. The headquarters was shifted to Fort Victoria which had been established on the southern tip of Vancouver Island three years earlier. Although there were a number of fur trading posts scattered throughout the southern part of New Caledonia, the first

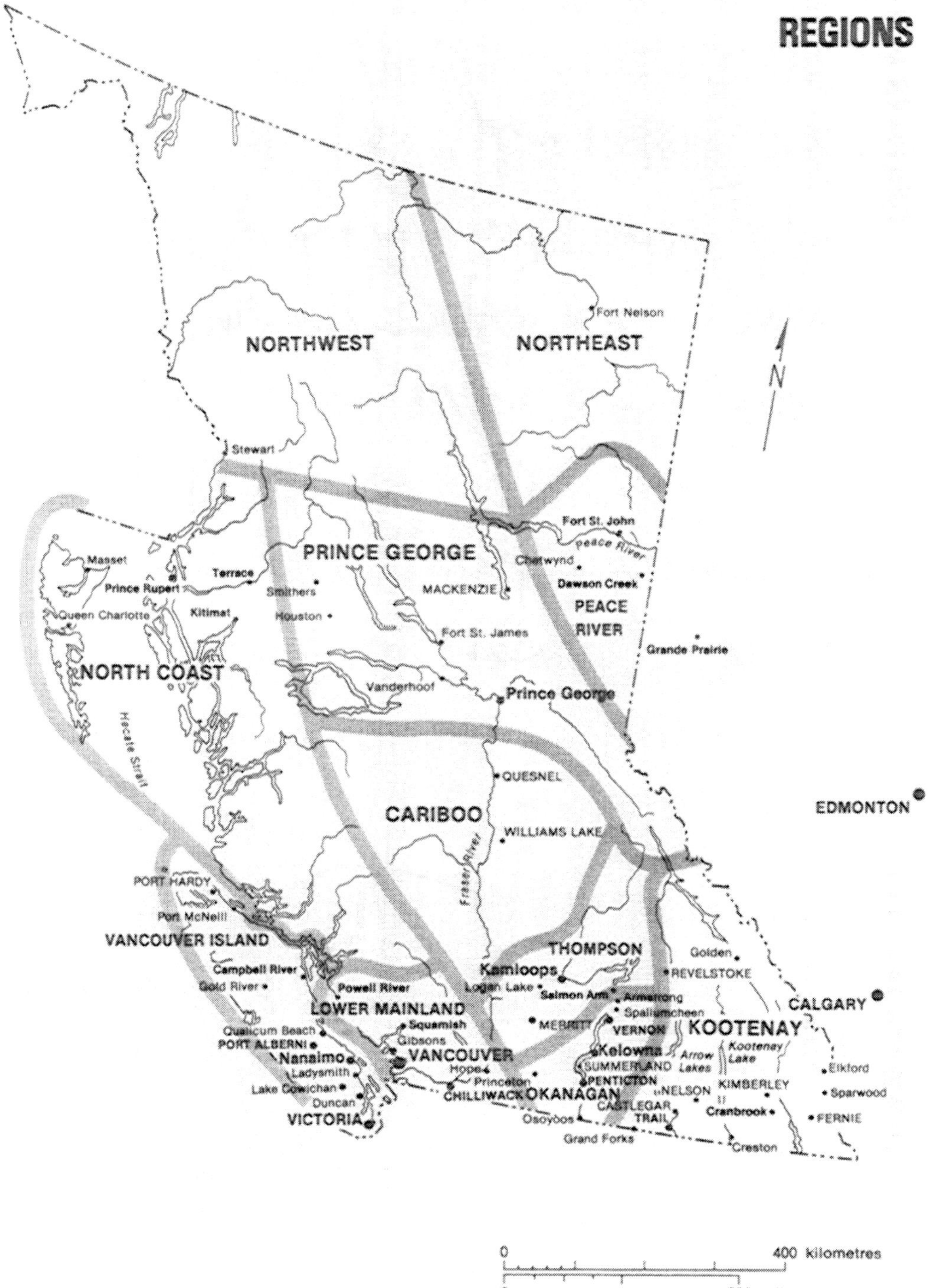

FIGURE 1.1 Geographical regions

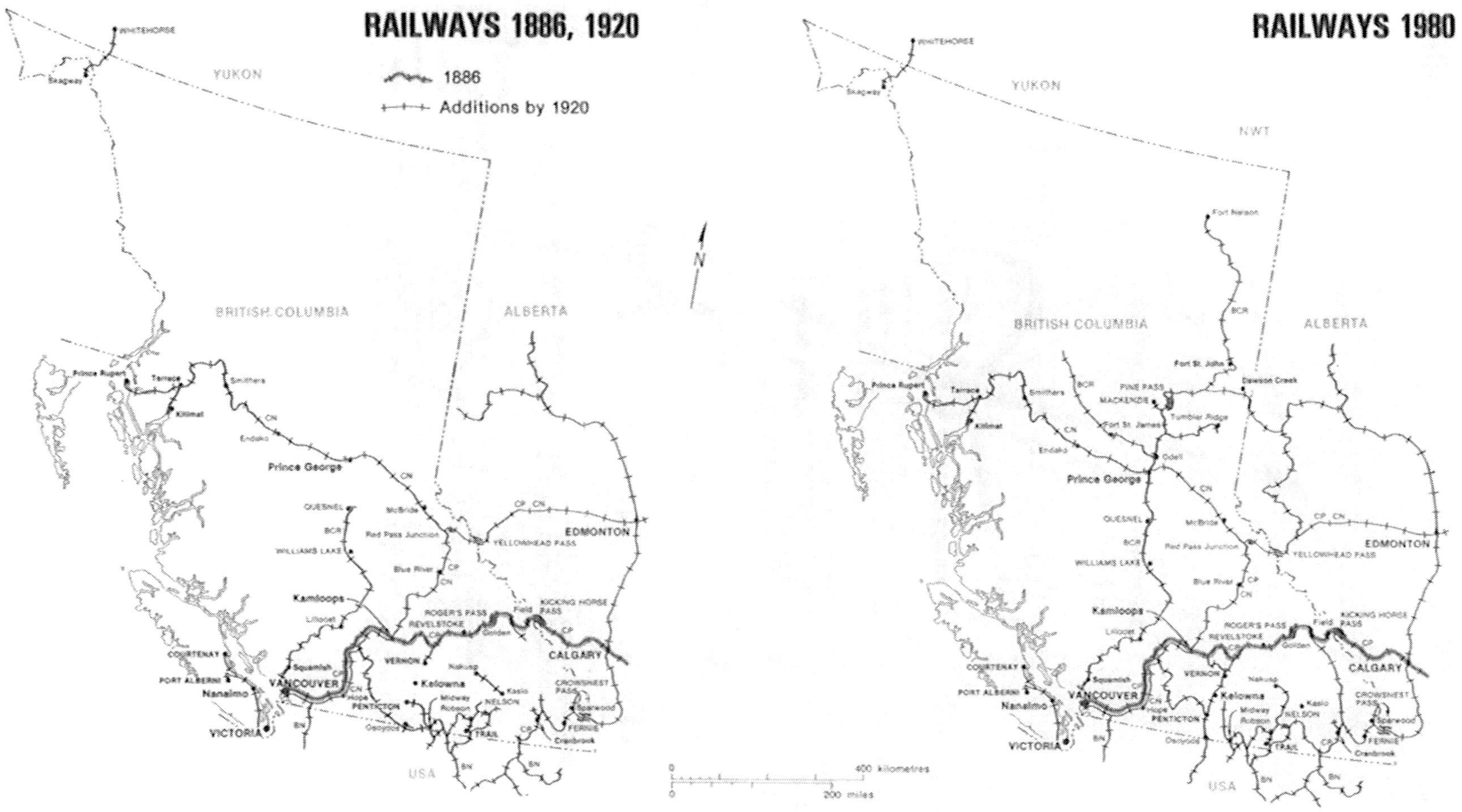

FIGURE 2,1 The railway network — 1886, 1920, 1980

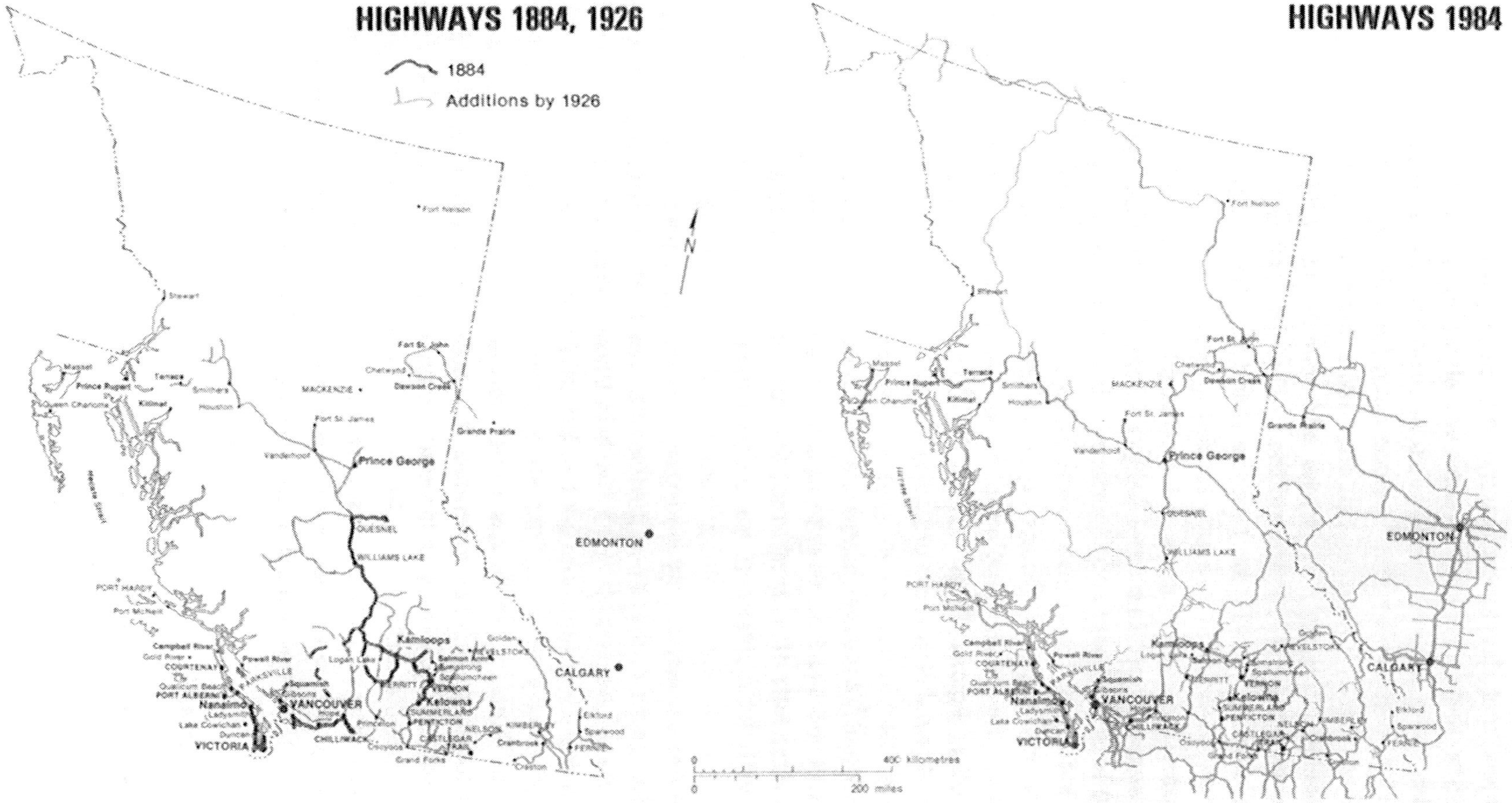

FIGURE 3,1 The highway network — 1884, 1926, 1984

settlement that developed beyond the trading post stage was Victoria. It became the capital of the separate island colony and when the two colonies were united in 1866 Victoria won the capital function over its rival, New Westminster, and became the major urban centre for the united Colony of British Columbia and, after 1871, the province. It continued as the leading commercial, financial and manufacturing centre until Vancouver usurped this role around the turn of the century.

The advance of settlement on the Island and the development of transportation occurred from south to north, with Victoria as the focal point. The completion of a wagon road preceded the railway by a few years only. The Esquimalt and Nanaimo Railway opened in 1886 and was later extended to Port Alberni and Courtenay. It was mainly the expansion of the forest industry in the rich stands of Vancouver Island that carried development to areas north of Courtenay, but even today the bulk of the population lies to the south. Population growth was steady, but not as rapid as in many other regions (Table 1,1). In 1871 the Island was the main concentration of population in the province (50 percent of the total), but by 1891 it was surpassed in population by the Lower Mainland.

At present Vancouver Island may be characterized as strongly based on the forest industry and, despite its lengthy settlement history, as an island where a recently developed area, the northern and western part, is juxtaposed with the southern tip, the oldest settlement area in British Columbia. There is some mining activity, commercial fishing and agriculture, but the bulk of the primary activity is logging and the manufacturing is dominated by pulp, paper and lumber production. An excellent highway network unifies the Island, focussing on the twin transportation nodes of Nanaimo and Victoria which are connected with metropolitan Vancouver by frequent automobile ferry services. Victoria lost the role of central place for the Island to Nanaimo which has easier access to Vancouver and a more central location, and has extended its trading hinterland to most of the Island, leaving Victoria as an isolated commercial node at the southern tip to deal with the mainland on its own. The Island's regional character is preserved largely because it is separated from the mainland by water and is an integrated, developed area adjacent to a mountainous and sparsely populated mainland, except in the vicinity of Vancouver.

Lower Mainland

An eastward extension of the Coastal Trough occupied by the Fraser River interrupts the continuity of the Coast Mountains and provides the only large area of reasonably flat land along the mainland coast of British Columbia. There on a completely land-locked natural harbor, Burrard

Table 1.1 BRITISH COLUMBIA POPULATION BY REGION AND PERCENTAGE CHANGE 1891-1981

	1891	*Change (%) 1891-1921*	*1921*	*Change (%) 1921-1951*	*1951*	*Change (%) 1951-1981*	*1981*
Vancouver Island	36,767	213	115,136	87	215,003	129	491,368
Lower Mainland	43,382	489	255,341	158	659,692	120	1,453,297
Thompson	5,740	183	16,220	106	33,426	274	124,912
Okanagan	3,360	790	29,901	136	70,608	179	196,888
Kootenay	3,405	1,528	55,436	71	94,766	64	155,272
Cariboo	5,519	163	14,511	29	18,665	245	64,347
Prince George	n.a.		9,103	232	30,229	341	133,416
North Coast	n.a.		25,338	6	27,777	141	64,633
Peace River	n.a.		1,644	679	12,801	297	50,863
Northeast	n.a.		278	394	1,373	342	4,600
Northwest	n.a.		1,674	12	1,870	172	4,874
British Columbia	98,173	434	524,582	122	1,165,210	136	2,744,470

SOURCES: *Census of Canada*, 1891, 1921, 1951, 1981.

Inlet, arose the port of Vancouver. South of the inlet the Fraser River built a large delta that provides an extensive area of flat land. The climate is similar to that of southeastern Vancouver Island, perhaps a little warmer in summer, which creates excellent conditions for agriculture.

The Lower Mainland, together with southern Vancouver Island, constitutes the heartland of British Columbia, the northern part of the Island and the rest of the province being the hinterland. This pattern has prevailed since the beginning of significant settlement on the mainland which dates back to the gold rush of 1858. New Westminster was chosen as the site for the capital of the mainland Colony of British Columbia that was established in 1858, prompted by the rush that attracted many thousands of gold seekers. Economic development consisted mainly of sawmilling on Burrard Inlet which began in the 1860s and large scale canning of salmon which came in the 1870s, as well as agriculture in the rich alluvial land of the Fraser Valley and the activities connected with servicing the placer gold mining industry. New Westminster remained the largest city until the rise of Vancouver after the arrival of the Canadian Pacific Railway in 1886, but the whole area was commercially tied to Victoria, the major port and business centre. The creation of Vancouver as the western terminus of the transcontinental railway quickly weakened Victoria's grip on provincial trade and commerce, and Vancouver soon overshadowed New Westminster as the key urban centre of the mainland. By 1900 Vancouver had become the largest city and dominant port and commercial centre of the province. Its rapid growth propelled the Lower Mainland to primacy among the regions in population, and its areal expansion eventually incorporated New Westminster, along with many other originally separate communities, into a metropolitan area that now constitutes 87 percent of the regional total (Table 1,1). Vancouver became the leading port of Canada and the leading metropolis of western Canada. Wood industries dominate its manufacturing, but fish processing, shipbuilding and a great variety of secondary industries are also important.

Besides metropolitan Vancouver, several tributary areas have been placed within the region. The term "lower mainland" refers specifically to the densely populated lowland along the Fraser from Hope to Vancouver. The flood plain of the river consists largely of prime agricultural land that, despite rapid urbanization of the zone, remains one of the leading and most productive agricultural areas of the province. Within the Coast Mountains north of Vancouver Garibaldi Provincial Park provides a major recreational area for the large urban population, particularly for skiing. The British Columbia Railway crosses the mountains along the edge of the park en route to its terminus at North Vancouver. Northwestward is a zone known as the Sunshine Coast that is tied to Vancouver by highway, with several short sections of ferry. Powell River, at the northern end of this zone, is one of the leading pulp and paper manufacturing centres of British Columbia. Among the regions the Lower Mainland functions as the focus of all of the significant transportation and communication systems and exerts tremendous dominance within the province.

Thompson

The Thompson region falls largely within the Interior Plateau, between the Coast and Cascade mountains on the west and the Columbia Mountains on the east. In contrast with the Lower Mainland and Vancouver Island, it is a relatively dry region characterized by forested and open range land and noted as cattle grazing country. Far from being flat, the plateau has a rolling surface and is deeply incised, to the extent of 800 to 1,100 metres (2,500 to 3,500 feet) below the general level, by a complex drainage network focussed on the North and South Thompson rivers which join at Kamloops to form the Thompson which, in turn, enters the Fraser River at Lytton. Extensive grassland characterizes the valley areas south and west of Kamloops, with open forest on the upper slopes and dense forest to the north.

Since the beginning of European occupance the region has been of great importance for its transportation corridors. As early as 1812 Fort Kamloops was built by the Pacific Fur Company and it later became a way station on the Hudson's Bay Company fur brigade route southward to the Columbia River via the Okanagan Valley. To the west the steep Fraser Canyon became the route of the Cariboo Wagon Road between Yale and Clinton, built in the early 1860s to provide access to the Cariboo gold fields. The building of the Canadian Pacific Railway in the 1880s along the South Thompson and Thompson rivers, then southward through the Fraser Canyon, obliterated part of the Cariboo Road, but facilitated settlement along the valleys and confirmed the significance of Kamloops which was chosen as a divisional point. Extensive cattle ranching became established in the grasslands, while logging and farming activities expanded eastward to the Shuswap Lake area and southward into the Okanagan Valley. Branch lines were established southward from Sicamous to Okanagan Lake by 1891 and from Spences Bridge to Merritt in 1906. The development of thermal coal resources in the Nicola Valley near Merritt became feasible with this rail service. The pivotal nature of Kamloops in the provincial transportation system was enhanced by the building of another transcontinental railway line, the Canadian Northern (later incorporated into the Canadian National Railways), completed in 1915. From Yellowhead Pass through the Rockies it followed the North Thompson Valley southward to Kamloops and then parallelled the Canadian Pacific to Vancouver. These major railway routes were followed more recently by modern highways, including the Trans-Canada which parallels the Canadian Pacific Railway.

Over the years forest industries and mining have become more important than agriculture, but the regional character is still strongly defined by beef cattle ranching. Widespread logging activities support a large number of sawmills, especially in the Kamloops and Shuswap Lake areas, and the pulp mill at Kamloops. Mining is concentrated in the Highland Valley west of Kamloops where nearly half a dozen copper mines and a smelter are situated. Population growth during the last thirty years has been fairly rapid (Table 1,1). A new super-highway, the Coquihalla, now under construction between Hope and Merritt, will shorten the journey from Kamloops to Vancouver and will serve to further strengthen the east-west transportation corridor role of this interior region and may foster additional economic development.

Okanagan

The region is focussed on the deep north-south trench occupied by Okanagan Lake and several smaller lakes where a high percentage of the population lives. This trench and many river valleys are incised below the Interior Plateau which is bordered by the Cascade Mountains on the southwest and the Monashee Mountains on the east. Like the valleys of the Thompson region, the Okanagan and Similkameen are relatively dry, sunny, and very warm in summer, especially south of Penticton, and are characterized by grasses and dry forest vegetation. In the southern part the bunch grass is associated with various desert shrubs. Hence, irrigation was the key to agricultural productivity.

Settlement was minimal until the latter part of the nineteenth century, despite the use of the fur brigade trail until the Columbia rivermouth fur trading base was cut off by establishment of the forty-ninth parallel boundary in 1846. Subsequently, the furs were carried westward to Fort Langley via a route between Kamloops and Hope. The gold rush of 1858 attracted some early settlers who recognized the grasslands as admirable grazing country for cattle, horses and sheep to supply the mining industry. It remained ranching country until nearly the end of the century when commercial fruit growing was introduced with the use of irrigation. The railway built in the early 1890s from the Canadian Pacific main line to Okanagan Lake where it connected with the lake steamers gave access to widespread markets for valley products. A land boom ensued during the first decade of the twentieth century and many irrigation companies were established, as fruit growing spread throughout the valley. The fruit growing industry matured during the 1920s and 1930s when additional railway and highway connections with the outside world were established, including the southern trans-provincial route of the Canadian Pacific Railway known as the Kettle Valley Line.

Today the region is noted for its wide range of fruit and truck crops. Although apples are the most important fruit produced, increasing acreages of grapes support a number of wineries and have established the Okanagan Valley as the second-ranking wine region of Canada after the Niagara Peninsula of Ontario. Losses of agricultural land to urbanization were substantial during the 1950s and 1960s, but this situation has been ameliorated considerably since the introduction of the agricultural land reserves in 1972.

Other activities in the region include the ever-present forest industry which takes a leading role, especially in the northern part around Vernon. Mining is less important today than it was earlier in the century when copper and coal were mined in the vicinity of Princeton. However, a couple of copper and molybdenum mines west of Okanagan Lake and a silver-lead mine to the east are current producers. Tourism has become an important industry and many retired people attracted by the valley environment have taken up residence. The north-south core of the region is punctuated at intervals by a number of cities, but Kelowna has emerged as the fast growing service and manufacturing centre of the valley. The Okanagan region has a most favorable image as an attractive fruit growing, recreational and residential environment.

Kootenay

A mountainous region with a pronounced north-south grain, the Kootenay (sometimes known as Kootenays because of the East Kootenay and West Kootenay regional subdivisions) is characterized by a number of major parallel valleys separated by rugged mountain chains. Great rivers with lake sections, the Columbia and its Arrow lakes, the Kootenay and its Kootenay Lake, occupy major valleys. Two basic mountain systems are separated by the striking linear feature, the Rocky Mountain Trench, along which the Columbia River flows north and the Kootenay River south. The Rocky Mountains, composed of sedimentary rocks, lie east of this trench and the Columbia Mountains, which include the Purcell, Selkirk and Monashee ranges and are largely composed of igneous rock, lie to the west. The region's reputation as a prominent North American mining area is based on its rich resources of coal occurring in the sedimentary Rockies and metallic minerals in the crystalline Columbia Mountains. Arrayed at right angles to the major eastward movement of air masses, the mountains intercept greater precipitation than is received in much of the Interior Plateau and endow the region with water flows and gradients suitable for hydro-electric power development. Although flat land is extremely limited, the climate and soils of many southern valleys are suitable for agriculture.

The long history of mining in the Kootenay and the importance of mineral products in the economy have created a mining region image comparable with the fruit growing image of the Okanagan. Placer gold mining began in the mid-1860s in the Kootenay River area around Fort Steele (originally Galbraith's Ferry) and initiated settlement. Farther north the building of the Canadian Pacific Railway through the Kicking Horse Pass during the early 1880s attracted some settlement and lumbering activities to that area. This was followed by the mining boom during the 1890s when a large number of high grade mineral deposits were brought into production, mainly silver, lead and zinc, at a number of locations in the southern Kootenay from Trail to Fernie. Toward the end of the century coal mining on a large scale got under way in the Crowsnest Pass area. Many independent railway lines were built to serve the mining activities and the Great Northern Railway extended a number of lines northward to tap the mineral exports. In an attempt to meet this competition, the Canadian

Pacific Railway extended lines southward and acquired sections of the independent lines, eventually putting together its Kettle Valley line from Crowsnest Pass to Hope. Growth of population was rapid and many towns and cities were established, but by the 1920s the boom was over and the Kootenay region has been one of slow growth since that time (Table 1,1). The lead-zinc mine at Kimberley is still in operation, after more than 90 years of mineral production, and helps to support the huge smelter and refinery at Trail. Coal production has escalated greatly during the last two decades to serve export markets, mainly in Japan.

Among other resource industries, the forest industry is the most widespread and prominent, but hydro-electric power production and agriculture are important in certain areas. With pulp mills at Castlegar and Skookumchuk, and a scattering of sawmills, especially in the Golden area of the Rocky Mountain Trench and the Castlegar-Nelson area, the forest industry is a large employer in the region. The British Columbia Hydro Mica and Revelstoke dams on the Columbia River constitute two of the three largest hydro-electric power production sites in the province and the West Kootenay Power and Light Company operates several large sites near Trail. Agriculture is important in the Grand Forks and Creston areas and in the Rocky Mountain Trench. Around Creston grain, vegetables and tree fruits are grown in addition to forage crops, while the other areas are oriented chiefly toward forage crops and beef cattle.

Development in the Kootenay region has been fragmented by the mountainous terrain and the scattered occurrence of resources. East-west transportation against the grain of the country has been achieved with great difficulty and no dominant city has emerged as in the Okanagan and Thompson regions. Nevertheless, a distinct regional character has evolved and become widely recognized.

Cariboo

A portion of the Interior Plateau between the Coast Mountains and Cariboo Mountains traversed by the Fraser River has been known by the loosely defined regional name "Cariboo" since the gold rush period. In this context it embraces the area west of the Fraser known locally as "Chilcotin", as well as the undisputed "Cariboo" east of the river. The gold rush started on the lower Fraser River near Yale in 1858 and extended northward along the river and its tributaries until the biggest strikes of all occurred in 1861 east of Quesnel. Barkerville and other gold towns were established that attracted sizeable populations. Access to the region initially was via the Cariboo Trail from the head of Harrison Lake to Lillooet, thence northward. Mileages to road houses were measured from Lillooet and several became significant centres of population, especially 70 Mile House, 100 Mile House and 150 Mile House. The southern part of this route was largely replaced by the Cariboo Wagon Road built between Yale and Clinton.

Although much of the region is covered by dense forests, the drier southern part and the deep river valleys are characterized by grassland and open forest. Many of the disappointed gold miners and other early settlers recognized this terrain as suitable for cattle raising and turned to ranching as the gold production waned. The regional image of the Cariboo as cattle country thus was established early. The extensive forests in the region were not heavily utilized until much later, mostly after World War II. The population was substantial during the nineteenth century, comparable with that of the Thompson region, but the Cariboo was one of slow growth until the 1950s (Table 1,1).

Industrial development was hampered by inadequate transportation until the building of the Pacific Great Eastern Railway (now British Columbia Railway) from Squamish on the coast through Lillooet to Quesnel by 1921. This provided a connection with Vancouver via a rail ferry from Squamish. Extensions at both ends during the 1950s, to Vancouver on the south and to Prince George and the Peace River area on the north, tied the railway to the northern line of the Canadian National Railways at Prince George and gave access to the major ports of Vancouver and Prince Rupert for export shipments. The Cariboo Highway which follows the route of the early wagon road forms the north-south stem of the modern provincial road transportation system.

Forestry has become the leading industry today and there is some mining, as well as the extensive ranching activity. Sawmilling, which was widely scattered in small settlements, has become more concentrated in Quesnel, Williams Lake and 100 Mile House. Quesnel also has a pulp mill. There are two mines near Williams Lake producing molybdenum and copper, and gold is being produced at Wells near Barkerville, in the heart of the 1860s gold district. As in the Kootenay, there is no dominant urban centre. Quesnel is the largest, but all are under 10,000 in population. The Cariboo retains its indelible ranch country image, despite the importance of other activities.

Prince George

Consisting largely of the northern part of the Interior Plateau plus a section of the Rocky Mountain Trench, this region has no widely accepted designation, as have the Kootenay and Okanagan regions. "Northern Interior", an appellation difficult to define, could have been used, but "Prince

George" was chosen because that city completely dominates the region from its strategic location at the crossing of the two rail and highway corridors that traverse the region. The rolling plateau lies between the Coast Mountains and the Rocky Mountains and is bordered on the north by a complex of mountains, including the prominent Skeena and Omineca chains. The lowest elevations are around Prince George where an extensive glacial lake produced a basin with soils suitable for agriculture. The climate of this northern part of the plateau is significantly moister than that farther south and supports dense forests of white spruce and balsam fir.

During the fur trade period this region was an important production area and one of the forts established by the North West Company in 1806, Fort St. James, became the administrative centre for the whole of New Caledonia until 1857, just before the gold rush and the establishment of the Colony of British Columbia. The site of Prince George at the confluence of the Nechako and Fraser rivers was another fur trading post, Fort George, established in 1807. While this region lay north of the Cariboo gold rush activity and remained oriented toward the fur trade much longer than southern areas, a small gold rush occurred during the early 1870s in the Germansen Landing-Manson Creek area north of Fort St. James. Widespread settlement of the region was largely delayed until the first two decades of the present century when the Grand Trunk Pacific Railway was built as a transcontinental system that entered British Columbia via Yellowhead Pass and terminated at the newly established port of Prince Rupert. Prince George became an important railway divisional point and consolidated its role as the crossroads of the north when the Pacific Great Eastern was extended northward and highways were built along both the east-west and north-south corridors. Settlement expanded chiefly along the east-west corridor from Prince George to Hazelton and since the railway was completed the Prince George region has recorded the highest rate of population growth among the major regions (Table 1,1).

Throughout the region the forest industry dominates the present economy, having greatly overshadowed other activities, such as agriculture and mining since the 1950s. The building of three pulp and paper mills at Prince George and two at Mackenzie supplemented the numerous sawmills of the region and strengthened the centralization trend in the industry that produced such a strong wood converting complex at Prince George. The agricultural potential of the Prince George-Hazelton corridor was the major attraction for early settlement and agriculture has developed into a significant industry, particularly in the Vanderhoof area. Hay, grain and other fodder crops are grown to support beef and dairy cattle. Several mines producing copper and molybdenum are operating in the Burns Lake area. These resource industries benefit from the transportation systems of the region, but the east-west and north-south routes exist mainly to serve other areas. The "crossroads" truly is an apt description of the Prince George region.

North Coast

Embracing a large portion of the Coast Mountains, along with the Queen Charlotte Islands, this region has a distinctive environment characterized by a rugged, fiorded coast, a moderate climate of high precipitation and a generally sparse population. The Queen Charlotte Islands are part of the Insular Belt of volcanic and intrusive rocks which includes Vancouver Island. Elevations are moderate compared with those of the Coast Mountains which are composed of intrusive igneous rock and rise steeply from the sea, reaching altitudes averaging 1,200 to 1,800 metres (4,000 to 6,000 feet) inland. Transportation routes penetrate to the coast only at Bella Coola, reached by a minor, mostly gravelled, highway and at Kitimat and Prince Rupert, the termini of the northern east-west rail and highway corridor. The year round heavy precipitation supports a coast forest in which western hemlock is most characteristic, but extensive areas at higher elevations in the Coast Mountains are non-forested. Although agricultural possibilities are minimal, fishing provides an important economic activity.

The intricate coastline was explored and mapped during the late eighteenth and early nineteenth centuries and the area was included within the fur trading empire. Fort Simpson was established at the mouth of the Nass River in 1832, but was soon moved westward to its present location (Port Simpson) at the entrance to Portland Canal north of Prince Rupert. The native Indian population oriented to a salmon economy was substantial, including the Haida on the Queen Charlotte Islands and the Tsimshian and Kwakiutl along the mainland coast. Even today the Indian population constitutes a high proportion of the total outside of the few industrial centres like Kitimat and Prince Rupert. The Skeena River route which became the rail and highway corridor was one of the access routes to the Omineca gold fields in the early 1870s. Steamers took prospectors up-river as far as Hazelton, the head of navigation. The canning of salmon began in the 1870s on the Skeena estuary and whaling was carried out from bases in the Queen Charlotte Islands.

The choice of a site on Kaien Island in 1905 for the terminus of the Grand Trunk Pacific Railway inaugurated the development of the port of Prince Rupert which became the largest city in the region. The railway was not completed until 1914 and the port never fulfilled the expectation of its founders, partly because the railway soon became simply a branch line of the Canadian National Railways instead of the main line of an independent and competitive transcontinental system. Other industrial developments were the establishment of several pulp and paper mills (Ocean Falls, Port Edward and Kitimat) and the building of a huge aluminum smelter and refinery at Kitimat based on a large block of inexpensive hydro-electric power generated at Kemano in the Coast Mountains. Ocean Falls, with its large newsprint mill established before World War I, was never connected with the outside world by road or rail and finally expired as a viable town when the mill closed in 1980. Until recently the region has been one of slow population growth, but port expansion at Prince Rupert promises future growth of the major city (Table 1,1).

Aside from these industrial and transportation functions, the economy of the region is largely based on logging and fishing. A large portion of the log production is sent by water transport to log converters in the south. Much of the fish processing has been centralized at Prince Rupert which also functions as the main base of the commercial fleet. Mining has never been of great importance in the region. An iron-copper mine at Tasu in the Queen Charlotte Islands which operated for a number of years closed down recently and there remains only one molybdenum mine north of Terrace. The recent expansion of port facilities at Prince Rupert to ship coal from northeastern British Columbia and grain from the Prairies in large quantities has finally justified the faith of its founders some 80 years ago. This rainy, forested land of fiords and mountains, with its northern gateway to the interior, remains one of the most distinctive regions of British Columbia.

Peace River

Physiographically, the Peace River region differs from the rest of British Columbia in lying east of the crest of the Rocky Mountains and consisting of Interior Plains and Foothills units, akin to the terrain of southwestern Alberta. The Peace River, draining a large area west of the mountains, is the only river in North America to flow through the Rocky Mountains. Despite its high latitude situation, the relatively flat Interior Plains section contains half of the cropland in British Columbia. Summers are short, but mid-summer temperatures are similar to those of central British Columbia and the duration of daylight at that latitude helps to compensate for the short summer. Besides soil resources for agriculture, the region possesses folded sedimentary, coal-bearing structures on the slopes of the Rockies, relatively flat-lying oil and gas-bearing sedimentaries in the Interior Plains, and the hydro-electric power resources of the Peace River.

As in other northern regions like Prince George, most of the settlement and development of the region was delayed until the present century, although it had been explored and occupied as part of the fur trading territory a century before. Simon Fraser established fur trading posts at Rocky Mountain House (Hudson's Hope) and Fort St. John in 1805. During the planning phase for the western railway project in the 1870s possible routes across the mountains even as far north as the Peace River were investigated and the presence of potential agricultural land was known. When the actual route of the Canadian Pacific Railway was established and little cultivable land was available there the provincial government compensated the federal government for its investment in the railway by granting control of a large area extending north and south of the river which became known as the Peace River Block. Little settlement occurred until the decade preceding World War I and the area remained an agricultural frontier until extensive filling in of available lands occurred after World War II. The railway from Edmonton reached Grande Prairie, Alberta in 1916, but was not extended to Dawson Creek until 1931. This agricultural region had developed as virtually an extension of Alberta's settlement and its trade and transportation interactions were with the Alberta heartland. It was weaned away from Alberta by British Columbia government transportation policies of the 1950s which resulted in the building of both highway and railway links between Prince George and the Peace River area.

Large scale resource development that began in the 1950s transformed the region's economy from one heavily dependent on agriculture and forestry to the more diversified economy of the present day. The rapid growth of population since 1951 reflects this development (Table 1,1). Oil and gas discoveries led to the building of pipelines to the south and the establishment of a small oil refinery at Taylor near Fort St. John. The building of the Bennett Dam on the canyon of the Peace River near Hudson's Hope in the 1960s harnessed a large block of hydro-electric power that was transmitted to markets in the south. Most recently, the development of two coal mines near Tumbler Ridge southwest of Dawson Creek has provided an additional increment of economic growth in the region. This development was made possible by the building of a branch line through the Rocky Mountains from the main line of the British Columbia Railway so that the coal could be hauled to Prince George for transfer to the Canadian National Railways and on to Prince Rupert. Despite all these changes, grain and livestock farming remain important to the region, and the prairie-like landscape sets the region apart from the rest of British Columbia.

Northeast

Composed of northern extensions of the same physiographic units as those of the Peace River region, the Northeast includes a larger band of mountainous terrain, along with the Interior Plains component. In this northern area the Rocky Mountains reach a width of 130 kilometres (80 miles), compared with a 50 kilometre (30 mile) width at the Peace River. To the west the region is bounded by the Rocky Mountain Trench, as is the Peace River region. The most striking differences from its southern neighbor are the lack of agricultural development and the sparsity of population. It is a region of boreal forests and muskeg where climatic conditions are a bit harsher than in the Peace River area, enough to push it beyond the agricultural frontier in terms of commercial farming under present conditions. Despite this reality, a considerable area of wooded land around Fort Nelson has been included within the agricultural land reserves of British Columbia.

Without the agricultural attraction, the northeast remained unsettled and undeveloped even longer than did the Peace River region, although it too was involved in the fur trade. Posts were established at Fort Nelson and Nelson Forks and the Liard River was used as a fur trading route southward from its Mackenzie River junction in the Northwest Territories. It was not until World War II that this region was intruded upon significantly by outside influences and permanent settlement. Early in the war the Northwest Staging Route was established between southern Canada and the Yukon, necessitating the building of a string of rudimentary airports, among which were Beatton River, Fort Nelson and Smith River within this region. This was followed by the construction of the Alaska Highway by American military personnel in 1942 from Dawson Creek to Fairbanks, traversing the northeast through Fort Nelson. Since then the servicing of highway traffic has been a major activity of the settlements scattered along this key transportation link. An additional transport connection was added with the extension of the British Columbia Railway from Fort St. John to Fort Nelson in 1971.

Industrial development of the region consists mainly of natural gas production from a number of fields discovered since exploration began in the 1950s and sawmilling activity concentrated mainly at Fort Nelson. A network of natural gas pipelines connects the various fields with the trunk lines in the Peace River area. Oil pipelines have not yet been added because discoveries have consisted chiefly of natural gas, although this could change in the future. Fort Nelson is the petroleum industry focus, as it is for most activities in the region. Accounting for nearly 80 percent of the regional population, the dominance of Fort Nelson is overwhelming (Table 1,1).

Northwest

The Northwest consists of a complex of mountains and plateaus lying between the formidable Coast and St. Elias mountains on the west and the Rocky Mountains on the east. The major mountain ranges are the Cassiar, Omineca and Skeena, and the plateaus include the Yukon and Stikine. It is a vast wilderness region of sparse population that is even less developed than the Northeast. Boreal forests in which spruce predominate cover valley and plateau surfaces, but extensive areas at higher elevations are classed as alpine tundra. Like the Northeast, it is a non-agricultural region.

The Northwest has a long history of European acquaintance that spanned the fur trade and gold mining periods so characteristic of other regions in British Columbia. However, exploitation of the fur resources was hampered by Russian, and after 1867, American, ownership of the Alaska Panhandle which presented difficulties of access. Agreements with the Russians allowed the Hudson's Bay Company to establish posts at the mouths of the Stikine and Taku rivers and for a short time there was a post at Dease Lake. A minor gold rush on the Stikine River in the 1860s resulted in the establishment of the Stikeen Territory that soon after was incorporated into the Colony of British Columbia. An ambitious attempt to establish a telegraph line to Europe via Asia began in 1865, with the cutting of a right of way through the Northwest. The project was abandoned the next year when a trans-Atlantic cable was opened. This right of way was used by the Canadian government in 1901 for a telegraph line to the Yukon. In the 1870s a major gold rush occurred north of Dease Lake in an area which became known as Cassiar Mines. The Klondike gold rush of 1898 attracted so many prospectors that some fanned out into British Columbia, discovering gold east of Atlin Lake. By 1900 Atlin had become a flourishing city. This area is still producing gold today.

Access to the region has long been a problem. For many years the Telegraph Trail was the only route crossing the region. With the opening of the Alaska Highway, roads were extended south to Atlin and Cassiar and in the 1960s a road from Cassiar to Stewart was constructed. In the late 1970s the British Columbia government was intent on opening up the region to resource development and an extension of the British Columbia Railway from Fort St. James to Dease Lake was begun, with the ultimate goal being a railway line to the Yukon and perhaps on to Alaska. This extension has since been abandoned because sufficient potential freight to support the line did not materialize.

Mining is the major economic support of the region, though potential hydro-electric power and forest resources exist. Copper mining near

Stewart and asbestos mining at Cassiar are supplemented by gold and silver mining in the Atlin, Cassiar and Omineca Mountain areas. The Stikine and Iskut rivers have been considered by British Columbia Hydro for Power development, but there is little likelihood of any major project, in view of the limited potential demand. The vastness of the territory, the weak transport network and the sparse economic development has produced few important towns and none that is central to the area like Fort Nelson. Stewart, the only port outlet on Canadian soil, is the largest centre, followed by Cassiar, Atlin and Lower Post, all peripherally situated.

CONCLUSION

Through a combination of physical diversity and a process of economic and social development over a period of nearly 200 years, a distinct set of major geographical regions has evolved in British Columbia. Some are more distinctive and widely recognized than others and some are more clearly defined and bounded than others. However, all are now integrated by transportation systems into a functioning whole, although the northern regions were effectively tied to the rest of British Columbia only within the last 30 years. The most widely recognized regions probably are Vancouver Island, Lower Mainland, Okanagan, Kootenay, Cariboo and Peace River. Among the others Thompson and Prince George probably have the weakest tradition of popular usage, but they exist as distinct regional entities that may acquire clearer identities and boundaries in the future.

REFERENCES

1. The research assistance of Diana Hocking who compiled the information for the maps and the table is gratefully acknowledged.

2. BRITISH COLUMBIA, *Bulletin Area*. Victoria: Department of Lands, Forests and Water Resources, Lands Service: No. 1 Kootenay, 1959; No. 2 Okanagan, 1964; No. 3 Lower Coast, 1959; No. 4 Vancouver Island, 1974; No. 5 Quesnel-Lillooet, 1959; No. 6 Kamloops, 1960; No. 7 Fraser-Fort George, 1964; No. 8 Prince Rupert-Smithers, 1959; No. 9 Atlin, 1961; No. 10 Peace River, 1962.

3. BRITISH COLUMBIA, *Economic Survey*. Victoria: Department of Industrial Development, Trade and Commerce, Economics and Statistics Branch: Hope and District, 1959; Chilliwack and District, 1960; Kamloops Region, 1961; Kelowna Region, 1962; East and West Kootenays, 1963; Fernie and District, 1963; Grand Forks-Greenwood Region, 1965; Peace River-Liard Region, 1966; South Coastal Region, 1967; Kimberley Area, 1967; Cranbrook Region, 1968; Cariboo-Chilcotin Region, 1969; Central Kootenay Region, 1970; Bulkley-Nechako Region, 1970; Columbia-Shuswap Region, 1970; Okanagan-Shuswap, 1971; Lillooet-Nicola Region, 1972; Skeena-Queen Charlotte Region, 1973; Lillooet-Thompson Region, 1982; Northwest Region, 1982.

4. BRITISH COLUMBIA, *Summary Report of Development Possibilities*. Victoria: Department of Economic Development: North East Region, 1975; Kootenay Region, 1976; Central Region, 1976; Mid-Coast, Northern Vancouver Island and Queen Charlotte Islands Region, 1976.

5. BRITISH COLUMBIA, *British Columbia Regional Index*. Victoria: Ministry of Economic Development, 1978.

6. HOLLAND, S.S., *Landforms of British Columbia*. Victoria: Department of Mines and Petroleum Resources, Bulletin No. 48, 1964.

7. CANADA, *Area Profiles*. Ottawa: Department of Manpower and Immigration, Vol. 1, Atlantic and Pacific Regions, 1968.

8. FARLEY, A.L., *Atlas of British Columbia*. Vancouver: University of British Columbia Press, 1979.

9. BARKER, M.L., *Natural Resources of British Columbia and the Yukon*. Vancouver: Douglas, David and Charles, 1977.

10. ORMSBY, M.A., *British Columbia: A History*. Vancouver: Macmillan, 1958.

11. ROBINSON, J.L. (ed.), *British Columbia*. Toronto: University of Toronto Press, Studies in Canadian Geography, 1972.

12. ROBINSON, J.L. and HARDWICK, W.G., *British Columbia: One Hundred Years of Geographical Change*. Vancouver: Talonbooks, 1973.

PLATE 2 Mount Waddington in the Coast Mountains. *B.C. Government Photo*▶

PLATE 3 The Peace River near Hudson's Hope. *B.C. Government Photo*▶

2 LANDMARK MAPS IN BRITISH COLUMBIA'S PAST[1]

Patricia P. Gilmartin
University of South Carolina

European exploration of British Columbia began in the late eighteenth century. By the late nineteenth century, most of the province's major geographical features had been identified and mapped, though not in complete detail and accuracy, of course. The maps of this era played an integral role in the surveying, settlement, and development of the region. As the nature and location of known geographic elements were recorded, maps also paved the way for further expansion into unexplored areas until, gradually, the voids on the map were filled in.

Progress in cartography does not necessarily proceed at a steady pace, however. There are sometimes precipitating factors, such as a significant person, event, or scientific advance, that, in a single, dramatic step, raise the standard of cartographic excellence to a new level which is not improved upon for years afterward. It is the premise of this paper that among British Columbia's early historical maps a few can be identified as truly landmark maps, in that they advanced or summarized existing geographic knowledge with a new thoroughness and quality that was not surpassed for decades. Although the maps are incomplete and inaccurate by today's standards, they can still be admired in an historical context. And from them can be traced the geographical and cartographical history of British Columbia.

In order to appreciate this history more fully, one might ask what motivated the men who first explored and mapped the wilderness that was to become Canada's westernmost province. The answer can be summarized in two words: politics and economics. Politically, in order to maintain sovereignty against competition from Spain, Russia, and later, the United States, the British government undertook major exploration and mapping projects in Canada. And in 1871 when British Columbia became a province of the Dominion of Canada, the political terms of union included commitments from the federal government to survey, inventory, and map much of the province. Economically, the region's vast natural resources — furs, gold and other minerals, fish, lumber, land — all required cartographic records to locate, allocate, and administer, and were diligently mapped and re-mapped over the years. These historic political and economic impulses form a common thread that weaves through the cartographic annals of British Columbia.

CHARTING THE COAST

What is now British Columbia was first seen by Europeans from the decks of small Spanish sailing vessels that ventured northward along the Pacific coast from Spanish America. Members of the earliest expeditions, one led by Perez in 1774 and the other by Bruno Hezeta in 1775, had only a sketchy view of the rugged promontories and islands which characterize the coastline. As a result of these voyages, however, the first known chart of British Columbia's Pacific margin was published. The *Carta Reducida* of 1775 by Bodega y Quadra, one of Hezeta's officers, showed a rudimentary outline of the northwest-trending coast, a few islands, and some indications of possible inlets on the mainland.[2]

The first English-sponsored voyage along the northwest coast of America followed soon after the Spanish ships. Captain James Cook's third and final expedition took him to the Oregon coast and north to Bering Strait in 1778 and 1779. Adverse weather conditions along the coast of British Columbia prevented Cook and his men from carrying out detailed observations, however, and the charts of this area which ensued were very similar to the earlier Bodega chart.

British, Spanish, and even American ships continued to probe the Pacific coast and islands of British Columbia for more than a decade, but without significant cartographic advances. Then in 1792 the British Admiralty assigned Captain George Vancouver to undertake the task of surveying the Pacific coast from 30°N to 60°N and exploring any navigable rivers that might provide access to eastern settlements and trading posts.[3]

While carrying out this mission, the Vancouver expedition encountered two Spanish vessels headed by Galiano and Valdes which also were engaged in their own exploration from the Strait of Juan de Fuca northward. The two parties agreed to co-operate and to exchange information which was later incorporated into their respective charts. The British expedition was larger and better equipped than the Spanish, but benefited, nevertheless, from explorations carried out by Galiano and Valdes. Vancouver acknowledged Spanish contributions and retained on his charts many of their names for places that he had not explored personally. As a result, these Spanish names still add an exotic flavor to an otherwise Anglo-Saxon and Indian menu of modern coastal toponymy.

Vancouver's surveys of America's west coast occupied the summers of 1792 through 1794. The resulting charts were published as a folio atlas in 1798. Figure 1,2 shows a portion of a much more extensive chart from the atlas which summarizes the labors of those three summers and depicts the entire coast from Baja California to the Gulf of Alaska. The segment reproduced here extends from just below the Columbia River northward to approximately the outlet of the Stikine River, encompassing most of today's Washington and British Columbia coastline. The track of Vancouver's ship, the *Discovery*, and her armed tender, the *Chatham*, is also illustrated.

The chart shows extensive coastal detail, including major inlets, landmarks, navigation hazards, innumerable islands, and such topographic information as the surveyors could ascertain from their vantage points on the ships and in longboats. Somewhat more attention was given to the mainland than to islands because one of the mission's major assignments was to search for a navigable inland waterway.

Mountains were symbolized primarily with a pictorial style of representation, although on some of the major islands a rather unsophisticated version of hachuring was used. The contrast between these two methods of terrain representation is evident in a comparison between the mountains of Vancouver Island and those on the mainland opposite. Vancouver Island, which on this chart is equitably named "Quadra and Vancouver Island", was shown for the first time to be an island rather than part of the mainland. Interestingly, Vancouver and his men failed to recognize the area of the Fraser River delta as the outlet of a major river, and so the river does not appear on his charts.

In spite of this uncharacteristic oversight, the charts of Vancouver represent an extraordinary cartographic achievement and surely deserve to be called a landmark in the annals of British Columbia mapping. The detail and accuracy of the charts are all the more remarkable, considering the complexity of the coastline, the dangers of unknown currents and submerged hazards, the vagaries of west coast weather, and the limitations of eighteenth century equipment. Vancouver's work on the coast was not surpassed for more than 50 years until the surveys of George Henry Richards from 1857 to 1863.

Again sponsored by the British Admiralty, Richards conducted hydrographic surveys of the newly-defined international boundary in the Strait of Juan de Fuca, the coast of Vancouver Island, and the mainland to 51°N. His work was published in two volumes as the *Vancouver Island Pilot*.[4]

The responsibility for coastal charting shifted to the Government of Canada when British Columbia joined Confederation in 1871. The Canadian Hydrographic Service is still engaged today in producing, improving, and updating the charts for the longest national coastline in the world.[5]

FIGURE 1,2 The coast of British Columbia from a 1798 chart by George Vancouver. ▶

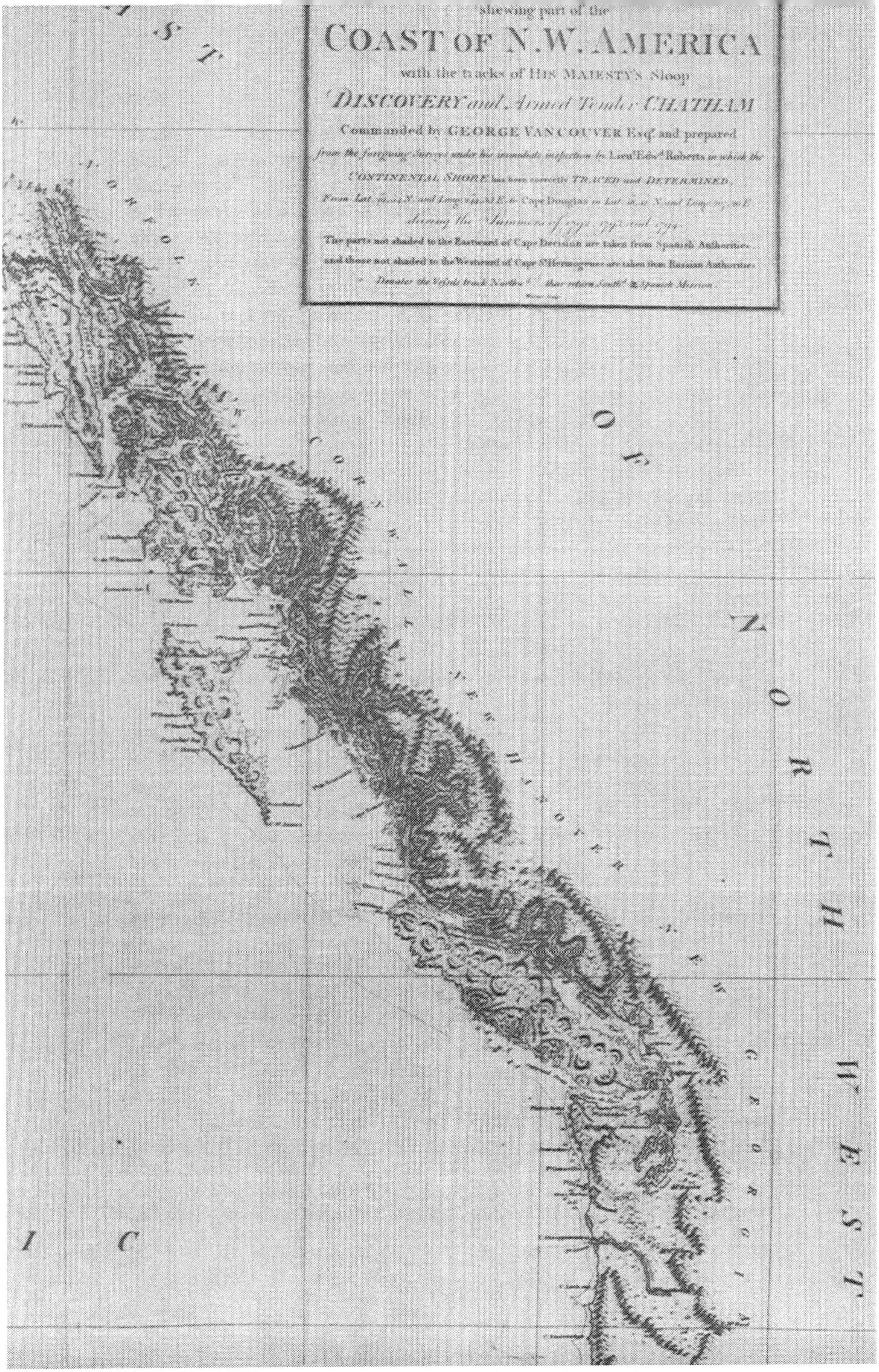
shewing part of the
COAST OF N.W. AMERICA
with the tracks of HIS MAJESTY'S Sloop
DISCOVERY and Armed Tender CHATHAM
Commanded by GEORGE VANCOUVER Esq.r and prepared
from the foregoing Surveys under his immediate inspection by Lieu.t Edw.d Roberts in which the
CONTINENTAL SHORE has been correctly TRACED and DETERMINED,
during the Summers of 1792, 1793 and 1794.
The parts not shaded to the Eastward of Cape Decision are taken from Spanish Authorities
and those not shaded to the Westward of Cape St Hermogenes are taken from Russian Authorities.
OF
NORTH
WEST
NEW CORNWALL
NEW HANOVER
NEW GEORGIA

BRITISH COLUMBIA'S INTERIOR UNFOLDS

At the same time that British and Spanish ships were reconnoitering the coastline of British Columbia, other explorers were pressing toward the same coast overland from the shores of the Great Lakes and Hudson Bay. Most of these men represented great fur trading companies, such as the North West Company and the Hudson's Bay Company which competed vigorously to open fresh trading areas in the West. In addition, they hoped to discover a viable route to the Pacific in order to facilitate shipping to and from their western trading posts.

The fur trade industry existed physically as a chain of outposts stretched across the country, linked by a sequence of lakes and rivers which served as the prime means of travel, trade, and communication between posts. The earliest maps of the interior were produced as an outgrowth of the fur trade and, as a result, concentrated heavily on hydrographic features.

Alexander Mackenzie was the first white man to reach the Pacific coast of Canada from inland. Coincidentally, he arrived at Dean Channel near Bella Coola in July, 1793, only about a month after Vancouver's anchorage there.[6] Mackenzie's arduous trek for the North West Company took him hundreds of miles across wild and rugged territory, but secured the fur trade of the region for his sponsors and opened the doors to further exploration from the East. His traverse covered only a narrow east-west strip of the wilderness, but it included part of the drainage systems of several important rivers, the Peace, Parsnip, upper Fraser (which he mistook for the Columbia), West Road, Dean, and Bella Coola.

Information from Mackenzie's explorations and the Vancouver voyages soon found its way onto commercial maps of the day, appearing first on the 1802 edition of the map of North America by the eminent English cartographer, Aaron Arrowsmith. Indeed, the achievements of these two explorers comprised all that was known then of today's British Columbia. Arrowsmith's map summarized it as an "H"-shape: a detailed coastline based on Vancouver's charts, a schematic rendering of the Rocky Mountain chain, and, as a crosspiece, Mackenzie's route to the Pacific.

The northwestern portion of Arrowsmith's map of 1802 is shown in Figure 2,2. It is characterized by high quality engraving, as were all of the Arrowsmith company's maps, with meticulous attention to linework and lettering. Landforms are symbolized with a combination of hachure lines and diagrammatic renderings. Clearly shown are Mackenzie's route across British Columbia and various points of interest along the way, such as Indian villages, sites of portages around river rapids, and, of course,

FIGURE 2,2 The northwestern portion of Arrowsmith's 1802 map of North America. ▶

DOG RIB'D
INDIANS
Great Bear L.
Current very Strong
Remarkable Rock with an Encampment upon it
INLAND IND.
Marten Lake
Very Hilly
Anawed L.
BEAVER
IND.
Horn Mountain
STRONG BOWS
IND.
River of the Mountains
SLAVE LAKE
Slave Fort
CHEPAWAYANS
Buffaloe and Moose
River of the Mountains
Buffaloe River
Caribeuf Mountains
Salt Springs
Peace Pt
Peace River
Falls
Clear
Fort Chepawyan
ATHAPESCOW LAKE
High Rocks
Whitefish L.
New House
Unjigah or Peace River
Peace Riv. Fort
Athapescow or Elk River
Forks
Fort
ROCKY MOUNTAINS
ROCKY MOUNTAIN IND.
Finlay's Branch
Rapids
Knisteneaux War Road
Lesser Slave Lake
East Branch
Elk River
K
Red Deer L.
Beaver L.
Athapescow Riv.
Pembina Riv.
Principal Branch
NASCUD IND.
NAGILER IND.
ATNAH IND.
SLOUACUS IND.
Salmon R.
Edmonton Ho.
Buckingham Ho.
Moose Lake
BLOOD IND.
BLACKFOOT IND.
Beaver Hills
Return 1793
FALL IND.
Great quantity of Coals in this Creek
Red Deer River
Devils Nose
Woods
No Woods
Bow Hills
the Nose
South Branch
Pyramid
Askow or Bad R.
A Spitcheevee
No Woods
Stone River
CATTANAHOWES

prominent landmark features. The only other entry west of the mountains shows the territory of the Long Hair and Blue Mud Indian groups and the Great Lake River in the southeast corner of the province. This river is probably today's Elk River which drains, eventually, into Kootenay Lake.[7] At that time, however, it was thought to flow into the Oregon (Columbia) River, as is indicated on Arrowsmith's map. Mackenzie also supposed that the Fraser River, whose upper reaches he travelled for some distance, was part of the Columbia River system, and this misconception is reflected in his own maps and the one shown here. Like Vancouver's charts, then, this map of 1802 shows no major rivers emptying into the Strait of Georgia.

The map contains numerous references to Indian territories and settlements, as might be expected, since the inland explorers relied heavily on native guides and geographical accounts. The somewhat greater detail seen in Figure 2,2 for the prairie regions east of the Rocky Mountains can also be attributed to assistance from the Indians and earlier expeditions by Mackenzie and other trading company explorers.

Aaron Arrowsmith's 1802 map of North America summarizes eloquently all that was then known of the trackless territory west of the Rocky Mountains. With the knowledge gathered by the Mackenzie expedition to the Pacific coast, cartographers were able, at least, to show with reasonable accuracy the breadth and character of the topography beyond the barrier of the mountains.

Mackenzie's trek to the coast opened the way for numerous other traders, explorers and surveyors, most of whom also were backed by fur trading companies. Men like Simon Fraser, David Thompson, David Stuart, John McLeod, Robert Campbell, Samuel Black, Archibald McDonald, and many others contributed to the geographical knowledge of British Columbia during the first half of the nineteenth century. This continuous accumulation of information was duly recorded on maps of the day, with the Arrowsmith family remaining the preeminent publishers of maps of North America. By 1850 the major topographic and hydrographic features of British Columbia's southern half had been explored, at least superficially. It is worth noting, however, that the main purpose of exploration at this time was still primarily economic — extension of the overland fur trade.

The mid-century period was pivotal in British Columbia's economic and political history. When the international boundary was established in 1846 at 49°N latitude the Hudson's Bay Company lost the vast, resource-rich Oregon Territory to the United States. In anticipation of this boundary settlement, the company had shifted its fur trading headquarters to Fort Victoria at the southern tip of Vancouver Island. A short time later the Company was granted Vancouver Island for the purpose of establishing a Crown colony. Colonization created a need for topographic, cadastral, and other administrative maps, and the discovery of coal deposits on the Island led to further exploration and mapping requirements.

The task of surveying the Hudson's Bay Company holdings in the vicinity of Fort Victoria was assigned eventually to a young Irishman, Joseph Pemberton. A skilled and conscientious engineer, Pemberton surveyed and mapped a significant proportion of southern Vancouver Island, including the townsite of Victoria, the adjacent Saanich Peninsula and Sooke areas, and extensive tracts in the Cowichan Valley and around Nanaimo. A portion of one of his maps entitled *The Southeastern Districts of Vancouver Island* is shown in Figure 3,2. This map was published in London by John Arrowsmith in 1855. In its entirety, the map depicts the Saanich Peninsula and the Island as far west as Sooke and north to Cowichan Bay. Most of the detail is concentrated on the southernmost land areas, however. Prominent topographic and hydrographic features are shown and identified by name, as are various land districts and the site of Victoria. This map represents the first detailed and accurate survey of southern Vancouver Island, and Pemberton's engineering and drafting skills served his employer and the future province well. His excellent work was an outstanding contribution to the cartographic chronicles of British Columbia. So highly was Pemberton regarded that after leaving the Company's employ he was appointed the first Surveyor General of the Colony in 1860.

Another event that occurred around mid-century had even more far-reaching effects on the history of British Columbia, as it is reflected in her maps. That event was, of course, the discovery of gold in 1858 in the Fraser and Thompson river gravels. Tens of thousands of gold-seekers flooded into those areas which formerly had accommodated only a few remote trading posts. The consequences of this virtual overnight boom were tremendous in every respect. Cartographically, it led to demands for a different kind of map. The fur trader's maps of river and lake routes were not adequate for the needs of prospectors, settlers, and the ancillary services and administrative functions which accompanied them.

In response to the general chaos resulting from the gold rush, James Douglas, Governor of the Colony of British Columbia, requested assistance from London, and late in 1858 a detachment of Royal Engineers began to arrive. In addition to their military duties, they were instructed to locate and survey public lands and roads and to select a site for the capital of the colony.[8]

Within the brief span of four years from 1859 to 1863 the Royal Engineers produced a number of large-scale cadastral surveys and town plans, and several smaller scale route and regional maps, some of which were

FIGURE 3,2 A portion of Joseph Pemberton's map of southeastern ▶ Vancouver Island published in 1855.

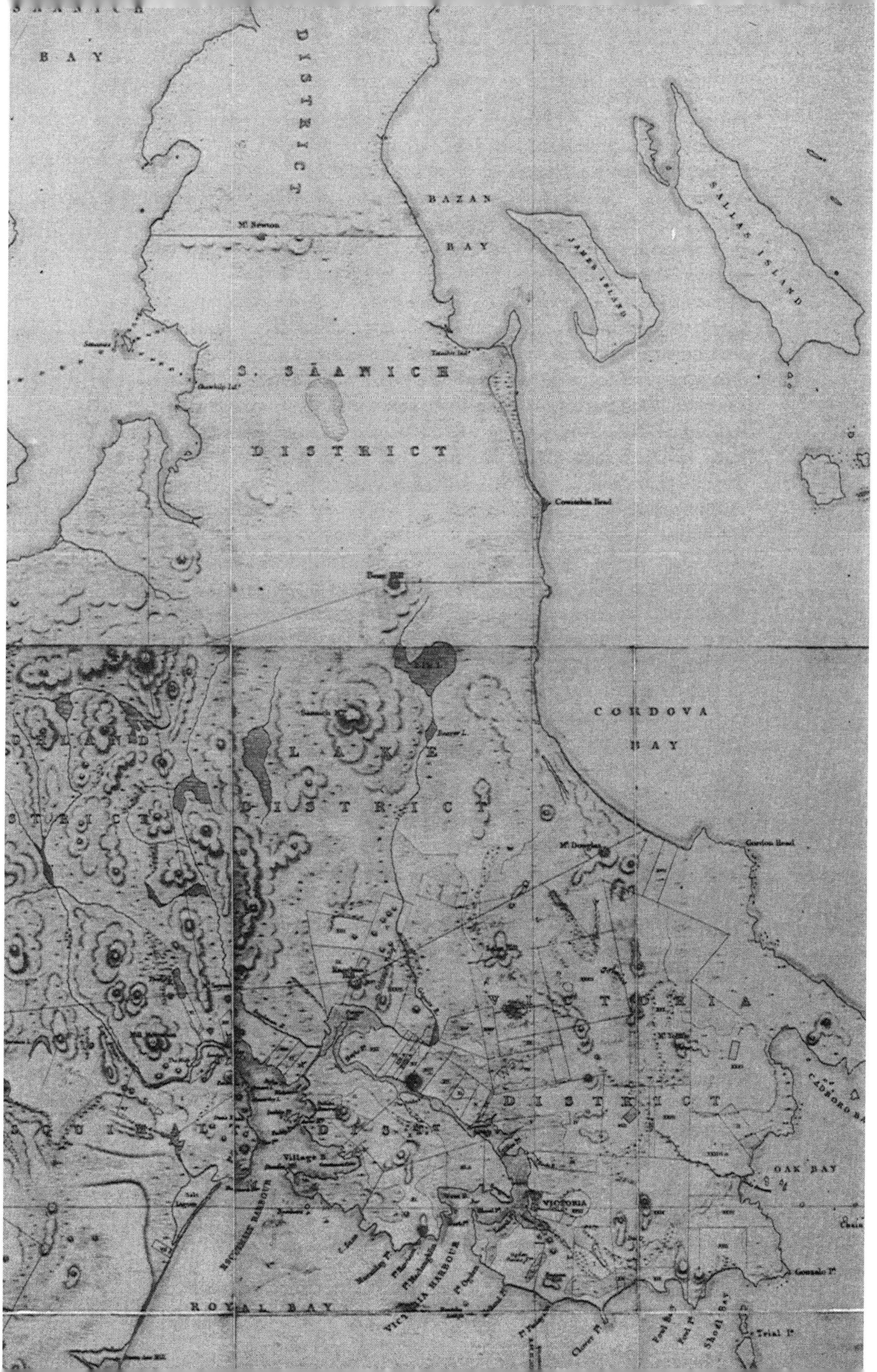
BAY
DISTRICT
BAZAN
BAY
JAMES ISLAND
SALLAS ISLAND
Mt. Newton
S. SAANICH
DISTRICT
CORDOVA
BAY
LAKE
DISTRICT
Mt. Douglas
Gordon Head
VICTORIA
DISTRICT
OAK BAY
VICTORIA
ESQUIMALT HARBOUR
VICTORIA HARBOUR
ROYAL BAY
Trial I.

printed on their own lithographic press.[9] Evidently, they planned, but did not publish, a general map of the colony.[10]

The cartographic contributions of the Royal Engineers can hardly be exaggerated. They achieved a new standard of detail and accuracy for the areas which they mapped and established an examplar for the later district map series of southern British Columbia.[11] A portion of a representative map by the Royal Engineers is shown in Figure 4,2. Published in 1861 and entitled *British Columbia: New Westminster to Lillooet*, the map shows the southwest mainland from the Strait of Georgia inland to Lillooet, southeast Vancouver Island, and the Gulf Islands at a scale of 10 miles to the inch. The content of the map is limited to areas actually surveyed by the Royal Navy and Engineers: coastal outlines, some mainland hydrography, travel routes from New Westminster to Lillooet via the Fraser River and Harrison Lake, and such settlements as then existed. Other regions were left empty. Topographic detail is minimal, consisting only of hachure lines indicating relative relief along mainland lakes and rivers. As the map's title implies, the primary purpose of the map was specifically to show access between New Westminster, the colony's capital at that time, and Lillooet. In addition, however, it provided a very detailed outline of the coast and islands of the straits of Georgia and Juan de Fuca. The international boundary along the forty-ninth parallel is designated also. It is interesting to note that many of the place names in use today were already established in 1861, as evidenced by the toponymy on this map.

The cartography of the Royal Engineers reveals a commitment to accuracy on their part, both in geographic content and in cartographic production. They summarized effectively the Engineers' excellent accomplishments in surveying and developing British Columbia's lower mainland and, thus, facilitated further settlement in the area.

By the end of the colonial period in 1871 other pieces of the British Columbia puzzle had fallen into place. The southern Rockies and upper Columbia-Kootenay region were explored by expeditions headed by Captain John Palliser (1858-59) and Walter Moberly (1864-66). The Cariboo gold rush in 1861 led to several surveys designed to facilitate access to the gold fields. The Yellowhead Pass area and the North Thompson River country were investigated by Milton and Cheadle. And a route from Bute Inlet through the Coast Mountains to the gold areas was pioneered by Alfred Waddington, though his planned toll road was never completed. Subsequent gold discoveries on the Stikine, Peace, Parsnip, Finlay, and Omineca rivers drew prospectors ever farther northward and revealed further details of British Columbia's interior.

FIGURE 4,2 A portion of a Royal Engineers map of southwestern ▶ British Columbia published in 1861.

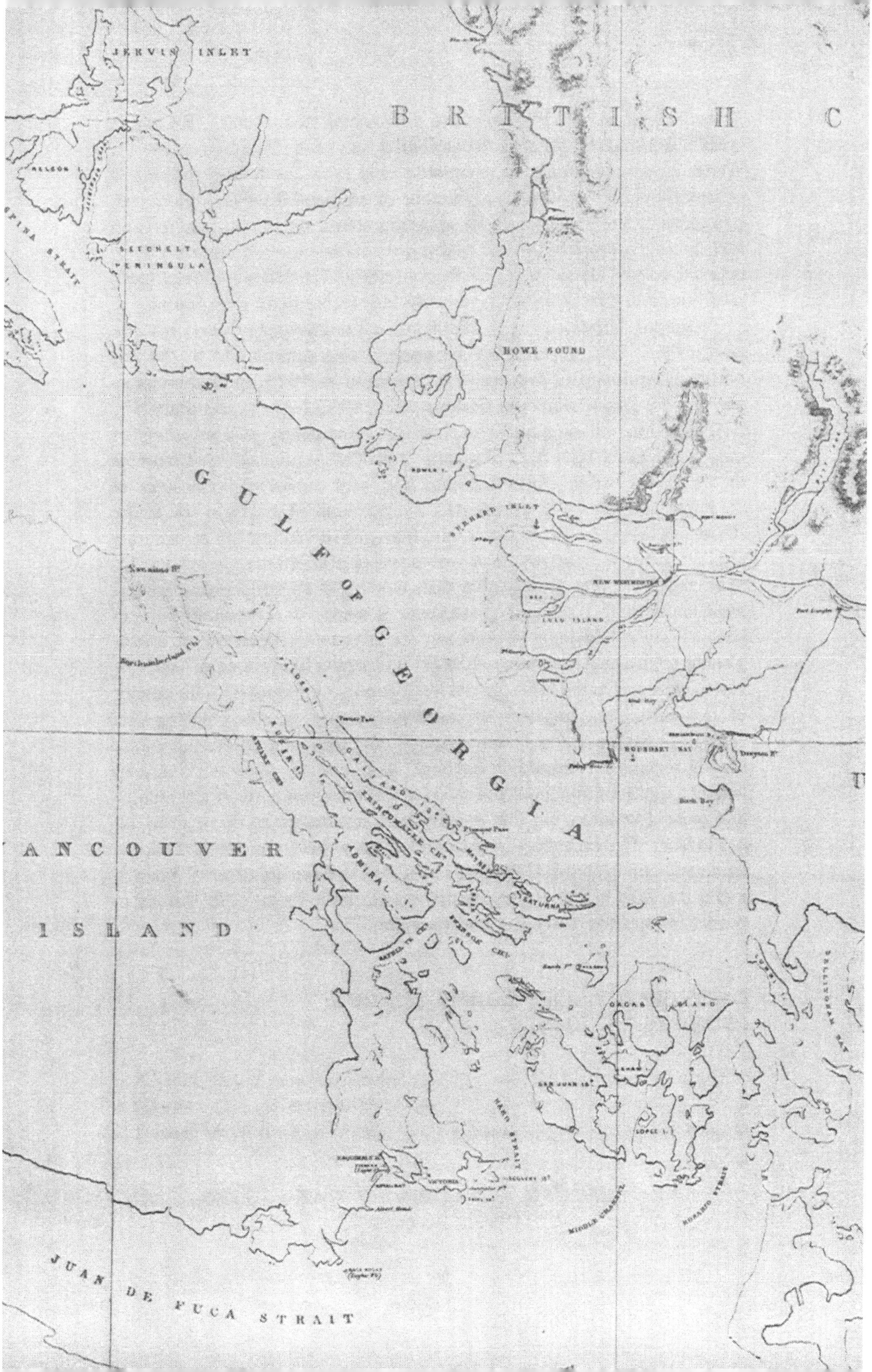
JERVIS INLET
B R I T I S H C
HOWE SOUND
G U L F OF G E O R G I A
ANCOUVER
I S L A N D
JUAN DE FUCA STRAIT

The British Columbia that became a province of the Dominion of Canada in 1871 was a vastly different region than it had been just 20 years earlier. Within a wilderness formerly populated only by a few trading posts and native villages had sprung up a number of permanent white settlements, connected by roads in some cases, agricultural and logging endeavors, and various mining ventures. Much of this growth can be attributed directly or indirectly to gold fever. Although the various gold rushes were individually relatively brief, their consequences were far-reaching and long-lasting.

Geographic knowledge also had increased enormously during the colonial period. This accumulated information was summarized in the first official general map of the province produced in 1871 under the direction of J.W. Trutch, Surveyor General of British Columbia, and drafted by J.B. Launders, an ex-member of the Royal Engineers. A portion of the map is shown in Figure 5,2. The map is entitled *Map of British Columbia to the 56th Parallel, North Latitude* and was published at a scale of 25 miles to the inch. It represented the culmination of the work of the Royal Engineers, the Royal Navy, and many other pioneers in as accurate and handsome a presentation as was possible at that time. From a technical standpoint, the cartographic skills evident in its appearance are still most impressive. Topographic character is symbolized qualitatively with hachure lines. Additional information about terrain and elevation is supplied through descriptive notes, such as "undulating forest plateau" and elevation measures in the form of spot heights and areal ranges. The interior of Vancouver Island had been explored by this time, and the resulting data were incorporated in the Trutch map. By contrast, the Queen Charlotte Islands remained relatively undetailed. Copious, meticulously-rendered lettering appears throughout the map, identifying hundreds of geographic and cultural features, but this is difficult to appreciate on a reduced reproduction. This exceptional work certainly deserves to be recognized as a landmark map in British Columbia's past, for as Pearson stated, "it bench marks the close of exploratory mapping and the full entry into the era of precise instrumental surveying and mapping."[12]

CARTOGRAPHY AFTER CONFEDERATION: FROM PRIVATE TO PUBLIC

The cartographic implications of Confederation were substantial. The federal government became directly responsible for certain major mapping projects in the province and also contributed financial support for provincial

FIGURE 5,2 A portion of the J.W. Trutch map published in 1871. ▶

ventures. Under the terms of Confederation, the central government was obliged, among other things, to construct a railroad across British Columbia, linking it with the rest of the continent. This necessitated extensive detailed surveying and mapping of potential routes for the line. Indeed, the railroad project eventually involved 75,800 kilometres (47,000 miles) of route exploration and 19,400 kilometres (12,000 miles) of actual survey on the southern mainland and Vancouver Island.[13] The federal government also undertook scientific exploration and mapping in the new province. The Geological Survey of Canada prepared studies of vegetation, soils, topography, hydrology, and, of course, geology and minerals, thus providing new information about unexplored areas and deeper insights into more familiar regions. Bolstered by funds from the federal treasury, mapping by the provincial government expanded to meet new needs in administration, public works, and resource development.[14] The transcontinental railroad accessed eastern markets for British Columbia's wealth of natural resources, most of which required maps for inventory and development purposes.

The railroad ran both ways, of course, and brought westward waves of entrepreneurs and settlers. As the population and economy expanded, the need for cadastral surveys and maps became critical and was addressed by both provincial and federal agencies. Even so, settlers' demands for surveyed land (at 10 shillings an acre) soon exceeded the amount of surveyed land available. A preemption system was instituted which allowed settlers to claim and occupy land before it was surveyed and paid for. Cadastral mapping capabilities were pressed beyond their capacity, both in surveying land that had been preempted and in meeting demands for maps which showed lands still unclaimed.

In these resource and cadastral maps one sees the beginnings of special purpose or "thematic" mapping for the province. A number of regional map series after the style of the Royal Engineers' district maps were initiated following Confederation. These were useful in providing general information at a large scale to the public and to public administrators. And, of course, the smaller scale general map of British Columbia continued to be improved throughout the late nineteenth century.

When British Columbia became part of Canada, only 70 years after Aaron Arrowsmith could put only two mountain ranges and one cross-country route on his map of the region, most of the major features of the province were known and mapped, at least approximately. This represents a remarkable achievement in light of the province's vast territory and complex, often inhospitable, terrain and weather. Today's maps are compiled by computers in governmental mapping agencies from data gathered by high technology instrumentation. It is easy to forget the early cartographic contributions by men who made up in courage and determination what they lacked in equipment and support. The map of the province was filled in gradually by individual explorers and surveyors like Mackenzie and Pemberton, small groups of men such as the Royal Engineers and Vancouver and his crew, and innumerable others whose knowledge was gathered and summarized by Trutch in his outstanding map. Thanks to these men, and others like them, the job of completing and refining the map of British Columbia was made much easier for those who followed.

REFERENCES

1. The author wishes to acknowledge the assistance of the staff of the Map Library of the Provincial Archives of British Columbia, Victoria, and Brian Turnbull, Curator, University Map Collection, University of Victoria.

2. FARLEY, A.L., *The Historical Cartography of British Columbia*. Madison: University of Wisconsin, Department of Geography, unpublished Ph.D. dissertation, 1960, p. 48.

3. LAMB, W.K., "Maps Relating to the Vancouver Expedition", *ACML Bulletin*, 28, 1978, pp. 8-18.

4. FARLEY, *op. cit.*, p. 276.

5. NICHOLSON, N.L. and SEBERT, L.M., *The Maps of Canada*. Folkestone, Kent: William Dawson and Sons, 1981, p. 148.

6. ANDREWS, G.S., "Surveys and Mapping in British Columbia Resources Development", *Transactions of the Seventh British Columbia Natural Resources Conference*. Victoria: British Columbia Natural Resources Conference, 1954, p. 98.

7. FARLEY, *op. cit.*, p. 143.

8. WOODWARD, F., "The Royal Engineers' Mapping of British Columbia: 1845-1906", *WAML Bulletin*, 7, 1976, pp. 28-42.

9. SPITTLE, J.D., "Early Printed Maps of British Columbia: A Preliminary Report on the Maps, Plans and Charts Printed and Published in the Crown Colony Between 1858 and 1871", *ACML Proceedings*, 7, 1973, p. 20.

10. FARLEY, *op. cit.*, p. 248.

11. *Ibid.*

12. PEARSON, D.F., "An Historical Outline of Mapping in British Columbia", *The Canadian Cartographer*, 11, 1974, p. 119.

13. ANDREWS, *op. cit.*, p. 102.

14. PEARSON, *op. cit.*, p. 120.

MAP REFERENCES

Figure 1,2 VANCOUVER, George, *A chart showing part of the coast of Northwest America with the tracks of His Majesty's sloop* Discovery *and armed tender Chatham*. Copied from a facsimile *Atlas to Vancouver's Voyage*, London, 1798, in the Provincial Archives, Victoria, British Columbia.

Figure 2,2 ARROWSMITH, Aaron, *A map exhibiting all the new discoveries in the interior parts of North America*. London, 1795; additions to 1802. From a copy in the University Map Collection, University of Victoria, Victoria, British Columbia.

Figure 3,2 PEMBERTON, Joseph D., *The Southeastern Districts of Vancouver Island*. London, 1855. From a copy in the University Map Collection, University of Victoria, Victoria, British Columbia.

Figure 4,2 PARSONS, Robert M., *British Columbia: New Westminster to Lillooet from a general map in preparation by the Royal Engineers*. New Westminster, 1861. From a copy in the Provincial Archives, Victoria, British Columbia.

Figure 5,2 TRUTCH, Joseph, *Map of British Columbia to the 56th parallel, north latitude*. London, 1871. From a copy in the Provincial Archives, Victoria, British Columbia.

PLATE 4 Landslide near Hope. *B.C. Government Photo* ▶

3 LANDFORMS AND NATURAL HAZARDS

Harold D. Foster
University of Victoria

LANDFORMS

British Columbia lies within the Cordilleran and Interior Plains regions of Canada, two of the country's five major physiographic subdivisions.[1] The Canadian Cordillera is predominantly rugged and of high relief and includes all of the mainland and offshore islands west of the eastern margins of the Rocky Mountains. Only in the northeast does the province extend beyond the Cordillera into the Interior Plains Region. There the relief is one of plateaus, plains and prairies (Figure 1,3).

British Columbia's huge area (948,600 square kilometres [366,255 square miles]) is both geologically and geomorphologically varied. Located close to the junction of the Pacific and North American plates, two of the six large elements that make up the majority of the earth's lithosphere, British Columbia is an integral part of the circum-Pacific 'Ring of Fire'. The junction between these two plates is composed predominantly of transform faults. As a result, the Pacific and North American plates slide past one another, neither creating nor destroying the lithosphere.[2] Such movements are, however, associated with numerous shallow focus earthquakes. Two smaller plates, the Juan de Fuca and Explorer, have recently been identified in British Columbia's southern offshore waters. There the leading edge of the North American Plate overrides the smaller plates, forcing them into the mantle where they are absorbed.[3]

The variety of geological environments within British Columbia is immense. The bulk of the province is composed of folded, faulted and metamorphosed rocks, ranging in age from the Precambrian to the Recent, and in type from sedimentary to volcanic and intrusive. That part of British Columbia lying west of the Rocky Mountain Trench was involved in the Coast Range Orogeny, lasting from early Jurassic time through to post Lower Cretaceous. This was associated with major mountain building and folding and faulting on a regional scale, together with the intrusion of granitic batholithics in the Insular, Coast and Columbia mountains and elsewhere. Subsequent erosion removed the covering sedimentary rocks,

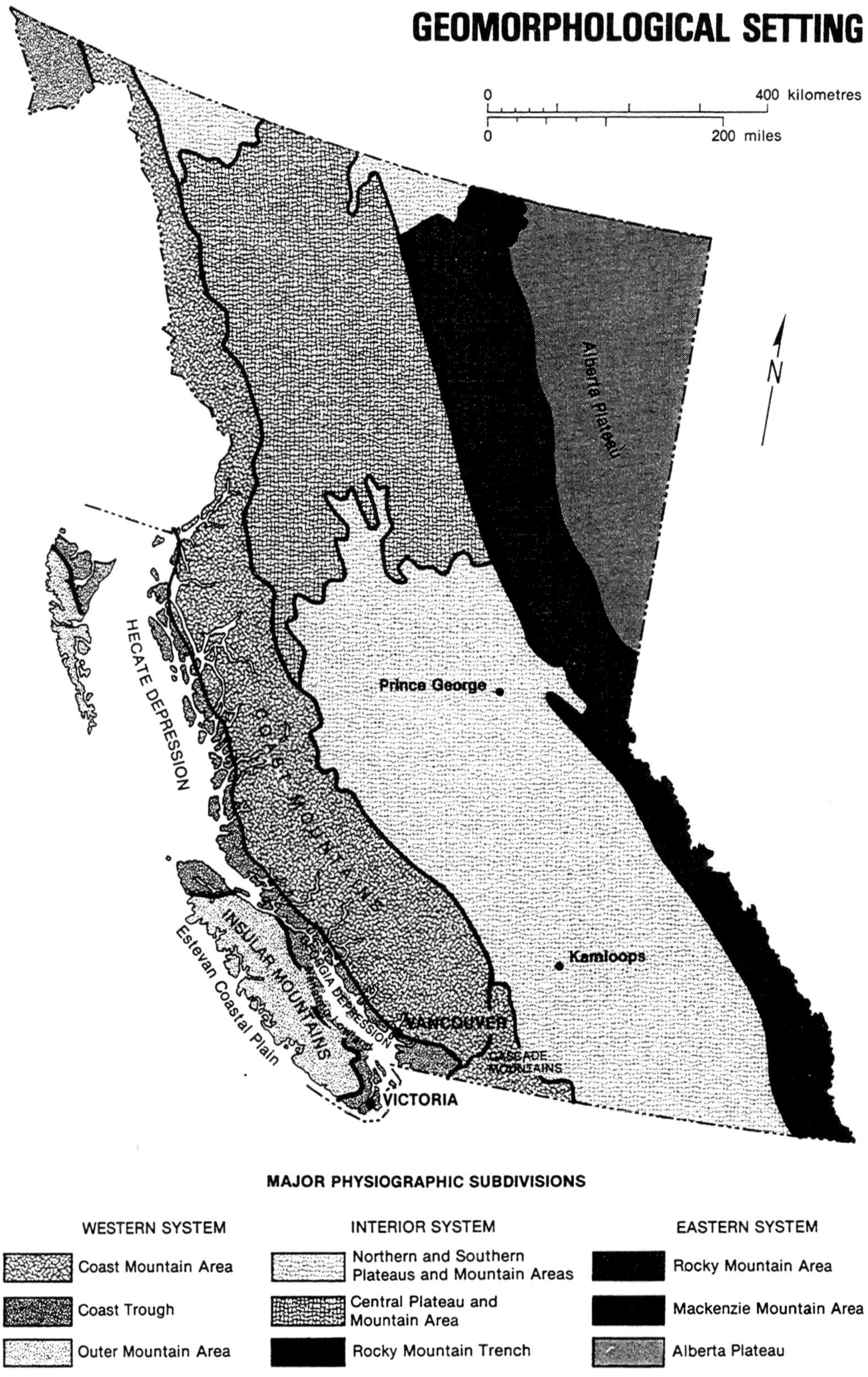

FIGURE 1,3 Major physiographic regions

exposing the underlying batholithic cores. Contemporaneous deposition of Upper Cretaceous and early Tertiary sediments also occurred in two basins. To the west lay a depression that included Vancouver Island, the Queen Charlotte Islands and the Gulf of Georgia. A second basin was located in what is now the Rocky Mountain area. The Rocky Mountains, therefore, now occupy the site of a former geosyncline, a recipient of deposition during the Palaeozoic and Mesozoic eras. The post-Paleocene Rocky Mountain Orogeny terminated this sedimentary process and culminated in the formation of the Rocky Mountains. During the Tertiary, most of the province was subjected to extensive erosion. Differential uplift and erosion in the Pliocene caused extensive rejuvenation and deep river incision, establishing the present arrangement of mountains, plateaus and plains.[4] On several occasions this regenerated topography was later subjected to extensive Pleistocene glaciations. At its maximum, during the Wisconsin, the Cordilleran Glacier Complex was a continuous and interconnected mass of valley and piedmont glaciers and an ice sheet which, although centred in British Columbia, stretched northwestward to the Aleutian Islands and southward to Mount Adams, near the Columbia River, a distance of 3,780 kilometres (2,350 miles). The shed lay between the Coast Mountains and the Rocky Mountains where the ice was at least 2,280 metres (7,500 feet) thick. Other smaller ice caps formed on Vancouver Island and on the Queen Charlottes.[5] As a result of such Pleistocene glaciations, the province is now marked by a wide range of striking glacial erosional and depositional features, including U-shaped valleys, fjords and large deposits of till and glacio-fluvial and glacio-marine sediments.

The generally high relief of British Columbia and the westerly atmospheric circulation to which it is subjected result in major variations in precipitation. This ranges from some 25 centimetres (10 inches) in parts of the province's interior to over 500 centimetres (200 inches) along the coast. Interior river basins receive most of their annual precipitation as snow during the winter months. This remains in storage until the late spring or summer when higher temperatures release it to stream flow. Mountain areas also support a multitude of glaciers and ice caps. These are found predominantly in the northwest and southwest of the Coast Mountains, one of the largest being the Llewellyn Glacier, south of Atlin Lake.

It is clear, from this brief summary of the physical geography of British Columbia, that the province has experienced a most varied geological and geomorphological history. Its active seismicity; heavy rainfall; high snowfall; steep, glaciated slopes; folded, faulted and jointed rocks; and unstable sediments are associated with a diversity of natural hazards. How these are managed provides the focus for the remainder of this chapter.

NATURAL HAZARDS

Nowhere in British Columbia is free from natural hazard risk. However, associated losses can be reduced by planning for greater safety through the adoption of numerous disaster mitigation strategies. This process of deliberate risk management normally involves the operation of complex decision-making networks, termed "safety delivery systems" by this author.[6] A heuristic model of such a system is presented as Figure 2,3. This diagram illustrates that the reduction of risk from major natural hazards almost invariably involves the co-operation of a variety of participating groups. These include international, national, regional and local governments; the scientific community; financial institutions, such as insurance companies and banks; developers; architects; realtors; the media; the owners and renters of private homes and commercial buildings; and the general public.

Regardless of the country or province involved, or indeed the hazard under consideration, the acceptable level of risk has been defined either implicitly or explicitly. The management of every safety delivery system then attempts to achieve this target by the adoption of strategies which are intended to ensure that losses, from the hazards for which it is responsible, do not rise above this threshold. Naturally, the acceptable level of risk from particular hazards varies from country to country and indeed from region to region. As a result, societies evolve in a manner that allows them to operate within specific levels of tolerance for natural and man-made events. Boundaries to what is, in effect, acceptable risk are defined either by laws or by common practice. Regulations, such as public health and building codes, usually identify the maximum event that should be guarded against and, as a result, govern the losses that must be accepted. The associated level of safety reflects a variety of factors, including the availability of capital and manpower, past experience, needs, wants and priorities.

Such safety delivery systems are not static: they evolve. Social goals may alter and, in consequence, the public may demand greater protection from earthquakes, floods, hurricanes or other hazards. Most displays of dissatisfaction with a particular safety delivery system usually take place after a disaster, which clearly demonstrates that existing standards are too low.[7] As a result, the law or associated codes and regulations may be altered and the safety delivery system involved modified to meet new objectives. In addition, change and associated improvements in security can be achieved by the provision of incentives designed to promote the adoption of innovation. These incentives which are provided to different groups within the safety delivery system may be in the form of capital, for example, grants and subsidies, income tax credits, accelerated depreciation or tax exemptions. They can also involve the provision of information

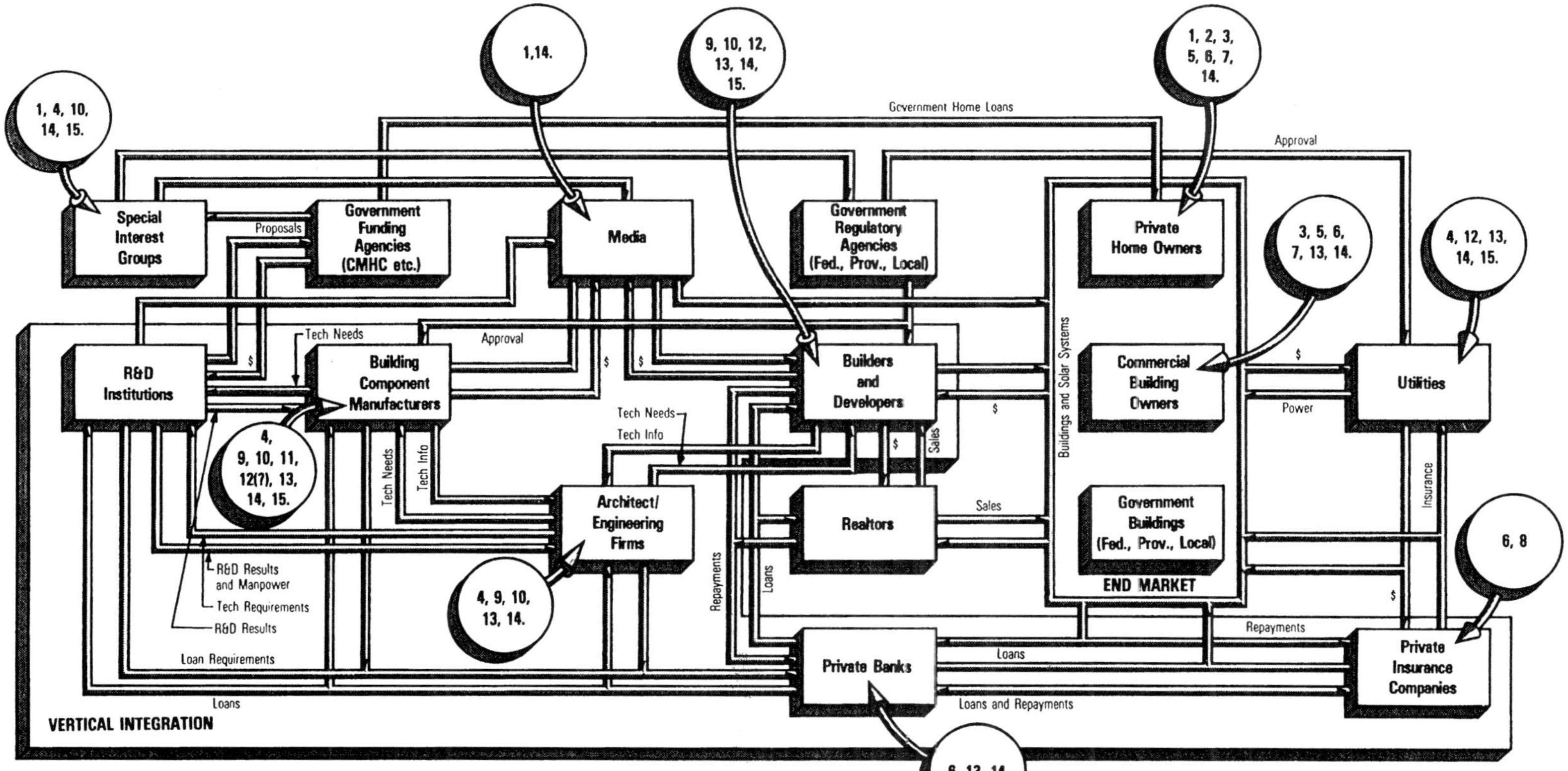

INCENTIVES

1. Grants and subsidies
2. Income tax credits
3. Income tax deductions
4. Research and development
5. Low cost loans
6. Guaranteed loans
7. Accelerated depreciation/rapid amortization
8. Government insurance and reinsurance
9. Government procurement
10. Demonstration programs
11. Government equity investment
12. Tax exempt bonds
13. Information collection
14. Information dissemination
15. Education and training programs

FIGURE 2,3 A heuristic model of a flood safety delivery system through research, development, demonstration, and education and training programs. The points where these incentives can best be used to reduce losses are also illustrated in Figure 2,3.

Regardless of the hazard involved, in theory, at least, every safety delivery system's management can choose whether or not to implement four basic groups of mitigation strategies. Typically, these involve measures to modify or remove either the hazard or the infrastructure at risk, or they seek to accommodate accepted losses with minimum social disruption. In practice, however, the choice of strategies may be limited by such factors as a lack of scientific knowledge about the hazard and its probable impact, an absence of appropriate technology, excessive cost, or legal and social objections to implementation. The selection of mitigation strategies is also influenced by the training of those involved in managing the safety delivery system and the perceptions and attitudes of various special interest groups and the general public who are served by it. In addition, some hazards strike so rarely, as, for example, meteorites, or cause so little damage that no safety delivery system exists for them. In such cases, the typical response by those adversely affected is bearing the loss.

British Columbian Safety Delivery Systems: Case Studies

A multiplicity of delivery systems have evolved in British Columbia in attempts to increase public safety and to protect individual or corporate property. Such networks include those seeking to reduce the damage caused by floods, snow avalanches, earthquakes, tsunamis and landslides. Examples of such systems are described in detail in an attempt to illustrate how they have developed and the basic principles involved in their operation.

FLOOD SAFETY NETWORK

Experience has demonstrated that floods are the most damaging of British Columbia's natural hazards. This is because access to drinking water was a major factor in the initial location of many settlements. Silt deposition on flood plains also provided fertile agricultural soils, while river transport made the clearing of the forest and trade easier propositions. As a result, many British Columbian cities were sited on river banks or lakesides. Subsequent growth, therefore, frequently took place on flood plains and is inevitably leading to escalating losses from this hazard. Although there are five rural and urban communities on Vancouver Island that are threatened by flooding, Port Alberni, Campbell River, Courtenay, Duncan and Lake Cowichan, by far the largest risks are being taken in the Fraser Valley. In 1965, for example, Sewell estimated that some 10 percent of the population in the Fraser Valley was living on land susceptible to major flood inundation.[9] This would amount to about 200,000 people as of 1985. High risk settlements on Vancouver Island have a combined population in excess of 44,000.

Over most of the province maximum river discharge occurs during the snow-melt generated, spring freshet period of May to July. The rivers of Vancouver Island, however, carry maximum flows in the winter months, as a direct result of high rainfall. Major flood years in British Columbia have been 1894, 1948, 1961, 1972, 1974, 1978, 1980 and 1981. The greatest discharges were recorded in 1894, the "flood of record" reaching a height of some 7.92 metres (25.99 feet) on the Fraser River at Mission. However, at that early date settlement was sparse and damage restricted. In contrast, in 1948 the Fraser reached 7.62 metres (25.00 feet) in the same location, causing major destruction. Almost 10 percent of the Fraser Valley was inundated. This flooding of 22,260 hectares (55,000 acres) left 200 families homeless, caused 10 fatalities and washed out 82 bridges. Together, the federal and provincial governments provided approximately 20 million dollars to rehabilitate flood victims and repair and strengthen the dyking system. This money was spent in the lower Fraser Valley, and at Kamloops, Prince George and Quesnel. Other, more recent, floods of note occurred in Kamloops in July 1972 and in the Vancouver-Squamish-Brackendale areas in 1980 and 1981. In Kamloops damage reached 8 million dollars, as 200 houses were inundated, while losses in the Squamish floods totalled 13 million and 2 million dollars in successive years. Although such damage is considerable, far larger losses can be anticipated. This great loss potential was described in the conclusions of the Fraser River Joint Advisory Board Report in 1976. That publication stated:

> Greater floods than that of 1894 can and will occur, but the specific year or years of their occurrence cannot be predicted. There is a 1-in-3 probability that the 1894 flood will be equalled or exceeded during the 60-year period from 1973 to 2032.[10]

The Report also estimated that such a flood would result in damages totalling some 500 million dollars, and render thousands of British Columbians homeless.

A major flood safety delivery system has developed in British Columbia during the past 90 years. The first legislative response to this hazard took place in 1894. The Drainage, Dyking and Irrigation Act of that year was stimulated by a major spring freshet that caused widespread inundation along the Thompson and Fraser river flood plains. Other flood-related legislation passed by the province has included the Flood Relief Act, the Fraser Valley Dyking Board Act and the Moritorium Act. All three were passed in 1948 as a direct response to the major Fraser flood of that year. The Flood Relief Act provided for a co-operative federal-provincial aid, relief and rehabilitation program in the Fraser Valley. It also established a cost sharing program in which the federal government provided 75 percent of the cost of the repair and construction of dykes and created the authority on which dyking boards rested, such as those set up in Chilliwack, Sardis and Mission. Because of the damage and the threat of subsequent flooding, dyking repairs were urgently needed in the Fraser. To address this problem, the province established the Fraser Valley Dyking Board through the 1948 Act of the same name. This body acted quickly to repair the dyking system and to increase its height. Fortunately, this was achieved prior to the 1950 freshet, which resulted in the third highest water level on record at Mission. In consequence of this prompt action no flood damage resulted. The Moratorium Act (1948) was passed to prevent hasty compensation demands and to allow government agencies time to assess the damages, evaluate claims and prepare avenues for payment to victims. Research into the flood hazard on the Fraser River culminated with the 1967 report of a Canada-British Columbia Committee, which prepared a plan for protection that would cost 33 million dollars to implement. This led to a federal-provincial agreement signed on May 24, 1968 to improve protection on the Fraser from the spring freshet. Subsequently, this was expanded to provide a comprehensive flood control program for the area. In 1976, a further 60 million dollars was made available by each level of government and the program was extended until March 31, 1984.

Numerous other pieces of legislation have also been passed which have an impact on the operation of the flood safety network. These include the Land Titles Act (1979) and the Environment Management Act (1982). The former provides that subdivisions can be prevented in areas subjected to hazards including floods, while the latter Act was designed to empower the Minister of Environment to declare an environmental emergency and expend funds without cabinet or legislative approval. The need for this power was perceived only after the 1980 Squamish floods. In addition to this provincial legislation, the municipalities also have considerable authority to influence hazard risk taking through the zoning powers granted under the 1957 Municipal Act.

The legislative diversity influencing the flood safety delivery network in British Columbia is paralleled by a relatively complex administrative framework. The Water Management Branch of the Ministry of Environment, for example, plays a major role in flood mitigation. This branch is responsible for several programs involving flood loss reduction. These include the Flood Damage Prevention Program which is essentially a warning system based on snowpack data. At regular intervals, during the winter and spring, the water equivalent of the snowpack is calculated at some 200 snow courses and, based on these data, future discharge predictions are made. Although not limited to the Fraser and Thompson river basins, it is for them that such flood predictions are most significant. Recent expansion of the system now permits flood warnings to be issued also for the Squamish River Valley.[12] In addition, the Water Management Branch has administered a Floodplain Development Control Program since 1974. This was implemented as part of the federal government's national flood damage reduction effort. The British Columbian program seeks to identify high risk areas and make recommendations on floodplain management, as well as to review subdivision applications, community plans and bylaws to see if they increase the risk of flood damage. Efforts are also made to increase public awareness of the flood hazard. As part of this program, by June 1982 some 873 kilometres (542 miles) of flood plain had been mapped at a cost of approximately 3,125 dollars per kilometre (5,028 dollars per mile). Emphasis has been placed on determining the areal extent of the 200 year flood. Such maps are now available for 41 areas, including parts of the Columbia, Coquihalla, Elk, Fraser, Squamish and Thompson rivers.

When major floods occur both the federal and provincial governments have agencies ready to respond. Emergency Planning Canada is responsible for the co-ordination of the emergency planning of federal departments, agencies and Crown corporations. It also links their efforts with those of the British Columbian provincial government. Emergency Planning Canada also provides funds for approved emergency planning projects and sponsors courses and conferences at its study centre in Arnprior. One of the organization's most important functions is the administration of the federal Disaster Assistance Program. When a disaster occurs of sufficient scale to place an undue burden on any provincial economy assistance may be sought from the federal government. This implies that disaster relief costs are likely to exceed 2,500,000 dollars. The British Columbian government, for example, applied for aid under the Disaster Assistance Program as a result of both the 1980 Terrace and the 1980-81 Squamish floods.

Although Emergency Planning Canada has a significant role to play during flooding, most of the rescue work is co-ordinated by the Provincial

Emergency Program of the Ministry of Environment. The Provincial Emergency Program (P.E.P.), for example, is assisting in setting up a province-wide emergency communications network, and in the training of personnel capable of responding to emergencies. Emphasis is also placed on encouraging municipalities to prepare disaster plans for their own jurisdictions. The value of the Provincial Emergency Program was illustrated during the period October to December 1981 when major storms created debris-torrents in the Howe Sound area. Several threatened communities were evacuated by P.E.P. personnel, reducing life-loss.

Despite all federal and provincial efforts to reduce flood damage, losses continue to mount. As a result, both governments have begun to demonstrate a new willingness to widen the types of incentives available to those involved in the flood safety delivery system. Indeed, the recent acceptance by federal and provincial governments that the best way to prevent flood damage is often to stop development in high-risk areas represents a major change in philosophy. It is also becoming increasingly likely that the payment of relief to victims will be tied to more stringent conditions, hopefully precluding reconstruction on high risk sites.

AVALANCHE SAFETY NETWORK

In this century, avalanches have caused more deaths in British Columbia than any other natural hazard.[13] Between 1900 and 1979, 267 fatalities were recorded. The worst snow avalanche of this period took place on March 4, 1910, when a work train was derailed and buried in Roger's Pass. Sixty-six deaths resulted. In 1915 the Britannia Mine was struck, killing some 30 men. Similarly, in 1965, the Granduc Mine, north of Stewart, was hit by an avalanche which killed 26 miners. In the same year, an avalanche also destroyed housing in Ocean Falls. Seven lives were lost. However, only after the North Route Cafe, near Terrace, was destroyed on January 22, 1974, killing a further seven people, did the provincial government take the first steps in setting up a comprehensive safety delivery system for this hazard. This involved establishing a Department of Highways Avalanche Task Force. One of the major achievements of this group was the development of an avalanche hazard index. This was designed to illustrate the probability of a vehicle being caught in a snow slide and was based in part on traffic volume, as well as the magnitude and frequency of avalanches that a particular stretch of road is subjected to.[14] Such an index is necessary since, during a typical winter, three vehicles are completely buried, thirteen partially buried and eighteen trapped by avalanches on British Columbia's road system.[15] Snow avalanches are also a major problem in parks and other recreational areas. Skiers, particularly those who fly into remote areas by helicopter, hikers and climbers are frequently killed. Despite this, relatively little British Columbian legislation can be traced directly to major avalanches. The Land Registry Act of 1906 included certain clauses that were eventually used to reduce risk, including that from avalanches, in new subdivisions. This Act stated that in the case of land which was outside a municipality the approving officer could examine the site and have studies conducted to determine whether or not the proposed development was "against the public interest". If this were found to be the case, permission to subdivide could be refused. In 1979 the Land Titles Act was passed which gave the approving officer a clearer mandate to refuse development "if he considers that the land is subject, or reasonably be expected to be subject, to flooding, erosion, landslip, or *avalanche* . . .".

Naturally, since such legislation exists, the province has an obligation to determine and manage avalanche risk. For example, any subdivision proposal for an area that may be susceptible to snow avalanches is normally referred, by the approving officer, to the Snow Avalanche Section of the Ministry of Transportation and Highways. This section evaluates the site and makes recommendations based on the degree of risk involved. Two other programs are also administered by the Snow Avalanche Section. All British Columbia's highways are being risk mapped for avalanches, and ski area developments are being reviewed for the same hazard. In an effort to mitigate the avalanche hazard 13 atlases have been produced to identify high risk sites along all major highways. Highway managers can use these to reduce losses in their jurisdictions. The Section is heavily involved in increasing awareness about the snow avalanche threat, and has developed an avalanche management program for the province's road system and organized seminars for approving officers, land use planners and forest management personnel. The Section has constructed 94 weather stations to monitor snow conditions, and supplies rescue equipment and volunteers in emergencies.

There is also some federal government involvement in avalanche management in British Columbia. Parks Canada, for example, administers risk management programs within the national parks. However, the provincial Snow Avalanche Section is responsible for the very high hazard Rogers Pass section of the Trans-Canada Highway. Numerous research projects have been carried out in this area by members of the National Research Council of Canada, Division of Building Research, often in co-operation with the provincial government.[16] In certain very high risk areas, such as Rogers Pass, active hazard reduction takes place. Extensive avalanche defences have been built to modify the magnitude and intensity of impact of snow on the Trans-Canada Highway. In addition, artillery and explosives are used to increase the frequency and, hence, decrease the severity, of avalanches.

British Columbia is fortunate that, despite the frequency of avalanches it experiences, there is relatively little development in high risk zones. As the population continues to grow and winter sports increase in popularity, this situation may change. It is necessary, therefore, that avalanche tracks be identified and their history determined so that this information can be incorporated into land use planning. The whole area of risk-taking during recreation, especially the benefits and costs of helicopter skiing, also require more research.

SEISMIC SAFETY DELIVERY SYSTEM

The earliest known seismic event occurred in British Columbia in 1841. Since that date the west coast of the province has experienced one great earthquake (8.0 Richter magnitude) and five major earthquakes (7.0 -7.9). Earthquakes of lesser magnitude are commonplace. For example, between 1969 and 1978, 347 tremors with magnitude greater than 4.0 on the Richter scale were experienced. As a result, the seismic zoning map for Canada, based on a computer analysis of all earthquakes recorded in the country since 1899, demonstrates that much of western British Columbia lies within the zone in which maximum peak horizontal ground acceleration can be anticipated.[17] Although potential damage from large earthquakes is high, actual losses have been small. Wuorinen[18], for example, studied the damage caused by the 7.3 earthquake that affected Vancouver Island on June 23, 1946. Only one fatality occurred, a drowning caused by a slump-generated wave. In Victoria, breakages were largely confined to chimneys and cracked walls. The greatest damage occurred at the British Columbia Packers installation at Kildonan on the Island's west coast.[19] There pilings settled and had to be repaired later at a cost of 100,000 dollars. Simulations of damage in the Victoria area, however, indicate that large potential losses are likely, if and when the city experiences an earthquake capable of generating Modified Mercilli intensities of VIII or higher on 'normal ground'.[20] Older, often heritage structures, which are commonly constructed of brick, are particularly vulnerable, since these were built prior to the enforcement of any rigorous codes.

Tsunamis, on the other hand, have already caused serious damage in British Columbia. These are trains of seismically triggered sea waves, most frequently generated by submarine dip-slip faulting, which are capable of causing extensive damage in low-lying coastal areas. Although occurring in other oceans, tsunamis have been recorded most frequently within the Pacific basin where, between 1900 and 1970, 138 were reported, 34 of which were locally destructive, while nine also caused damage at great distances from their sources. The effects of such waves in coastal areas can be very destructive. In 1964, for example, 122 died and 200 were injured by seismic sea waves generated as the result of earth movements centred in Prince William Sound, Alaska.[21] This tsunami was the greatest of the 10 seismic sea wave trains recorded on the west coast of British Columbia during the 65 years of operation of the Tofino tide gauge. It is, therefore, not surprising that it caused widespread damage to many Vancouver Island coastal settlements. The greatest impact occurred at Port Alberni where the first wave arrived without any official warning. Including the damage to the MacMillan Bloedel and Powell River Company plants in Port Alberni, estimates of the total losses reached as high as 10 million dollars.[22] Houses were displaced up to 300 metres (1,146 feet) and logs moving at speeds in excess of 32 kilometres (20 miles) per hour were driven into buildings. As a result, 58 properties were completely destroyed, and 320 dwellings suffered damage.

Little of the hazard-related legislation in British Columbia is specifically designed to reduce earthquake risk. However, this hazard and its consequences could be addressed under the Environment Management Act (1982) that applies to all hazards, or through the Municipal Act (1957). The latter legislation provides for controls at the local level. Councils, for example, may pass zoning bylaws and regulate the size, shape and siting of buildings within the resulting zones. They also have the power to prohibit particular uses in specific zones. However, there has been little, if any, attempt to use this authority to avoid construction in high risk seismic locations, but some progress has been made in building design. The National Research Council, Division of Building Research, develops, assesses and amends building codes for the construction industry at the federal level. These have been consolidated into a National Building Code which was adopted in British Columbia in 1973 through an amendment to the Municipal Act. Administration and enforcement occurs at the local level. Since earthquake loading specifications are included, newer buildings should now be able to withstand the design earthquake. However, a problem still remains with thousands of older buildings which are very expensive to modify in an attempt to make them less damage prone. Federal government inspired restoration of four Vancouver heritage office buildings is currently under way. This program includes eight million dollars for earthquake protection. Additional steel columns and beams, wall strengthening and new elevator shafts are expected to improve their capacity to absorb seismic shocks.[22]

In contrast to earthquakes, the risk of tsunami destruction along the western Canadian coast has led to the development of a complex safety delivery system which includes organizations and individuals drawn from almost the entire Pacific Rim. They are involved because the tsunami warning network currently operational in British Columbia is an integral part of a larger international system covering the Pacific Ocean.[24] The headquarters of the Seismic Sea Wave Warning System is in Honolulu and is supplied with data by at least 30 seismological and some 50 tide stations. These include the Dominion Astrophysical Observatory near Victoria, and Tofino and Langara Island tide stations on Canada's Pacific coast. When an earthquake, registering 6.5 or greater on the Richter scale, is observed at any participating seismological station this information is immediately relayed to Honolulu. There the location of the epicentre is determined by analyzing data from several stations. If it is suspected that a tsunami may have been generated tide stations in the area are requested to monitor their gauges for evidence of anomalous readings. If the existence of a tsunami is verified warnings are issued to all participating countries, including Canada. Once the generation of a tsunami had been confirmed, the arrival time at various Pacific coastal points is estimated from travel time charts prepared for each tide station in the system. These travel times are considered accurate within approximately 2.3 percent. Except when close to the epicentre, threatened areas can then be evacuated and mobile infrastructures and fittings removed.

Several types of bulletins may be issued by the International Tsunami Information Centre. Watch bulletins report the occurrence of tsunami generating size earthquakes. Warning bulletins are issued when the existence of a tsunami has been confirmed, and this bulletin includes the estimated time of arrival at the tide stations. Tsunami warning supplements, giving information on reported wave heights, are issued at least hourly thereafter. A tsunami watch is cancelled when it is determined that no tsunami has been generated, but the "all clear" decision is left to the discretion of local dissemination agencies. The British Columbia Provincial Emergency Programme (PEP) receives these bulletins at its headquarters in Victoria.

If, as a result of the information received from Honolulu, the decision is made by the PEP Coordinator and the Provincial Secretary to alert coastal communities further dissemination of the information occurs. RCMP detachment commanders are, for example, responsible for alerting the local PEP Coordinator in each community and designated municipal officials. Bulletins are also passed to the Broadcast News Service which has a teletype service to coastal radio stations. The Raven Network, an independent Indian-owned and operated radio-telephone system centred at Bowser, south of Courtenay on Vancouver Island's east coast, is also requested to pass information to its subscribers.

As with any other system, this safety delivery network is only as strong as its weakest link. The British Columbia tsunami warning system is unlikely to function effectively because local governments and the public at risk are unprepared to respond to the information the network generates.[25] There are very few tsunami risk maps available which delineate those areas, in specific communities, which are likely to be flooded. Only one Vancouver Island municipality, Port Alberni, has detailed a local tsunami warning plan. Its most comprehensive fan-out system is used when the tsunami warning has not been preceded by a watch bulletin. Few, if any, other coastal communities in British Columbia have developed an official tsunami response plan, despite the fact that several other settlements have previously been inundated.

MASS MOVEMENT AND SUBSIDENCE SAFETY DELIVERY SYSTEMS

Instability is a recurrent feature of the slopes of British Columbia. Heavily jointed and faulted sedimentary and igneous rocks, commonly oversteepened by glacial erosion and mantled by tills or glacio-fluvial sediments, are often subjected to high precipitation and seismic disturbance. Riverine and marine undercutting also occurs frequently, as does disturbance by ubiquitous logging operations. As a consequence, slope failure often takes place. Major rock slides, for example, occurred on the Fraser River near Hell's Gate in 1913 and 1914, greatly impeding salmon runs up the river. In consequence, spawning was reduced and the 1917 and 1921 returns were substantially down, leading to the loss of many millions of dollars to the fishing industry.[26]

During the past 20 years, 34 people have been killed in the province by mass movement. Associated direct damage is estimated in the tens of millions of dollars.[27] The largest recent example was the Hope landslide which took place on January 9, 1965. This buried a provincial highway to depths of up to 100 metres (328 feet) with rock and killed four motorists. Other examples can be drawn from elsewhere. Port Alice, on the northwest coast of Vancouver Island, appears to have been built on a debris fan, and has been subjected to several severe mudslides. On December 15, 1973 damage totalling some 800,000 dollars was caused by a mud flow, originating at the 700 metre (2,300 foot) level, on steep slopes above the town. A further slide occurred in 1975.[28] The greatest life loss caused by mass movement recently occurred in the winter of 1981-82 in the area between Squamish and Vancouver. During the period October, 1981 to February, 1982, 10 people were killed by debris torrents and one by a rock fall. Mass movement is also a major threat to several dam sites in British Columbia. Hydro-electric projects are normally associated with large valley reservoirs, which may initiate or reactivate instability. If major mass movement occurs the dam may fail or be overtopped. This happened, for example, during the October 9, 1963 disaster at the Vaiont Dam in Italy, causing some 2,600 fatalities.[29] The problems associated with such phenomena in British Columbia are illustrated by the Downie Slide in the Revelstoke Dam pondage and the slope creep failures taking place upstream from the Seven Mile Dam in the Pend'Oreille Valley. Mass movement is also a major problem along the coast, where marine undercutting frequently reduces sheer strength. Areas of particular risk include the University of British Columbia cliff area, the east side of the Saanich Peninsula and parts of the Queen Charlotte Islands. Such erosion is particularly hazardous in areas where Pleistocene sediments are exposed to repeated storm waves which have developed over a long length of fetch. Major slumps normally occur during high tides and after periods of heavy rainfall.[30]

Subsidence, usually associated with glacio-lacustrine clays and silts, or sands and gravels that are underlain by them, occurs in certain interior drainage systems. Extensive damage has been caused by caving and collapse features in the southern Columbia, South Thompson and Okanagan river valleys. In 1976, for example, a relatively new public swimming pool in Kamloops was severely cracked by subsidence. Damage in excess of 280,000 dollars occurred and the facility had to be filled and paved over. It now serves as a tennis court. Buchanan recorded that residents of the South Thompson Valley cited numerous examples of driveway damage, foundation cracking and excessive settlement of buildings. Clearly, subsidence causes sustained, yet generally unrecorded, economic loss in certain interior settlements.

Loss due to mass movement and subsidence have been addressed by several pieces of British Columbian legislation. The Local Services Act of 1960, for example, was passed to assist in assuring the safe, healthful, efficient, economical and attractive subdivision of land. Natural hazards are not mentioned in the 1960 text, but provisions were made for the design of subdivision regulations, which became the 1970 British Columbia Regulation 262/70: Subdivision Regulations. These are applicable only in areas outside the Lower Mainland, which are not in a municipality or regulated by a regional district. However, where they apply, such regulations permit the refusal of subdivision applications where the land is subject to erosion, or may be affected by mass movement. These regulations also allow the government to request that any prospective developer provide a professional engineering report on the land's stability. Municipalities also have the right to preclude development on such high risk sites through their zoning powers under the Municipal Act (1957). Further legislative power against the mass movement hazard was provided in 1974 when the Land Registry Act of 1906 was amended to introduce hazard specific clauses to aid in defining "against the public interest". This gave the approving officer the right to examine, or have examined, land outside municipalities and refuse development rights where the land was subject to, or could reasonably be expected to be affected by, either erosion, floods or landslides. This Act was repealed in 1979 and replaced by the Land Titles Act which also included similar provisions.

Less than 10 percent of all subdivision applications submitted to approving officers in British Columbia are, in fact, referred to the Geotechnical and Materials Branch of the Ministry of Transport and Highways for hazard evaluation. This small percentage, however, involves some 200-300 applications annually, and these are reviewed for a diversity of hazards, including landslides, coastal erosion and subsidence. The Geotechnical and Materials Branch also undertakes large scale hazard evaluations in suspect areas, such as the northwest of Vancouver Island and the Lower Fraser Valley. The Terrestrial Studies Branch of the Ministry of Environment is also undertaking a hazard mapping program, through its Surficial Geology Unit. By 1981, approximately 35 percent of British Columbia had been covered by this program. The resulting maps were specifically designed for use in hazard assessment, especially mass movement and subsidence. Unfortunately, there is very little federal input into reducing losses from coastal erosion or mass movement phenomena. Within the National Building Code, only vague reference is made to the difficulties associated with construction on ground susceptible to landslides. The Code only requires that "existing stresses within the soil or rock shall be considered in design". However, because of the latitude this wording permits, building inspectors rarely enforce this section.

As can be seen from this review, the legislative and administrative thrust in British Columbia against mass movement and subsidence generally involves attempts to prevent development on high risk sites. The review process is not comprehensive, however, and errors have occurred. Where development already exists on hazardous sites engineering alternatives are normally applied. At Port Alice, on Vancouver Island, for example, the repeated damage from mud flows clearly required remedial action. A scale model of the town and adjacent slopes was, therefore, constructed. Mud flows were reproduced, using bentonite mud with a similar viscosity to the coarse, bouldery gravel actually involved. On the basis of this experiment, a dyking system was constructed at a cost of 250,000 dollars. This was designed to protect the town from flows of up to 2.5 times the volume of those occurring in 1973 and 1975. Any further mud flows will be diverted into unsettled areas.[31] One major exception to this engineering approach recently occurred in the Mount Garibaldi area, some 30 kilometres (19 miles) northeast of Squamish. There the area of highest risk lies on the east side of the Cheakamus River Valley, extending up Rubble Creek to a geological formation called the Barrier. The area has a complex history of major rock slides. In May, 1971, Cleveland Holdings initiated an application to create a 288 lot residential subdivision on the Rubble Creek rock fan. This was refused, after review by the Department of Highways, a decision upheld by the Supreme Court of British Columbia. A detailed risk analysis of the area was subsequently undertaken by a special advisory panel. This eventually led to the Ministry of Environment invoking powers provided by the Emergency Program Act. The area at risk from catastrophic rock slides was declared a civil defence zone, and land and buildings within it to be compulsorily purchased by the Minister of Environment. This order in council set aside a sum not exceeding seven million dollars for compensation, and directly affected 102 property owners. Against enormous local opposition, land was acquired and buildings threatened by future landsliding demolished.[32]

CONCLUSION

It can be seen from this brief review of geological and geomorphological hazard management in British Columbia that safety delivery system improvements are possible. Losses are increasing and there is still the potential for major earthquake or flood generated disasters. Legislation has been *ad hoc*, often crisis stimulated, and has tended to deal with hazards in a fragmented manner. Perhaps new comprehensive legislation is needed. This might seek to address all major hazards, promote total risk mapping and provide for greater integration of existing safety delivery systems. Insurance might also be used to ensure that those taking the highest risks do not continue to do so at public expense.

REFERENCES

1. HOLLAND, S.S., *Landforms of British Columbia: A Physiographic Outline*. Victoria: British Columbia Department of Mines and Petroleum Resources, Bulletin No. 48, 1964.

2. PRESS, F. and SIEVER, R., *Earth*. San Francisco: W.H. Freeman, 1982, pp. 441-463.

3. MILNE, W.G., *et. al.*, "Seismicity of Western Canada", *Canadian Journal of Earth Sciences*, 15, No. 7, July 1978, pp. 1170-1191.

4. HOLLAND, *op. cit.* ,p. 22.

5. FLINT, R.F., *Glacial and Pleistocene Geology*. New York: John Wiley, 1957, pp. 304-306.

6. FOSTER, H.D., "Reducing Disaster Losses: The Management of Environmental Hazards" in MITCHELL, B. and SEWELL, W.R.D., *Canadian Resource Policies: Problems and Prospects*. Toronto: Methuen, 1981, pp. 209-232.

7. BUCHANAN, R.G., *An Assessment of Natural Hazards Management in British Columbia*. Victoria: University of Victoria, Department of Geography, unpublished M.A. thesis, 1983.

8. CANADA, "Federal Flood Damage Reduction", *Emergency Planning Digest*. Ottawa: Environment Canada, 1975, pp. 2-7.

9. SEWELL, W.R.D., *Water Management and Floods in the Fraser River Basin*. Chicago: University of Chicago, Department of Geography, Research Paper No. 100, 1965.

10. CANADA-BRITISH COLUMBIA, FRASER RIVER JOINT ADVISORY BOARD, *Fraser River Upstream Storage Review Report*. Victoria: Queen's Printer, 1976.

11. See BUCHANAN, *op. cit.*, pp. 48-66, for a detailed review of British Columbia's hazard related legislation.

12. BRITISH COLUMBIA, *Annual Report, 1980-81*. Victoria: Ministry of Environment, 1982.

13. FREER, G.L., "Investigating Snow Avalanches" in Ministry of Transportation and Highways, *Geological Hazard Evaluation for Highway Personnel*. Victoria: Queen's Printer, 1980, pp. 59-66.

14. BRITISH COLUMBIA, *Avalanche Task Force: Report on Findings and Recommendations*. Victoria: Department of Highways, 1974.

15. BUCHANAN, *op. cit.*, p. 26.

16. See, for example, SCHAERER, P.A., "Terrain and Vegetation of Snow Avalanche Sites at Rogers Pass, British Columbia" in SLAYMAKER, O. and MCPHERSON, H.J. (eds.), *Mountain Geomorphology*. Vancouver: Tantalus, 1972, pp. 215-222.

17. WHITHAM, K., MILNE, W.G. and SMITH, W.E.T., "The New Seismic Zoning Map for Canada, 1970 Edition", *Geophysics*, June 1970, pp. 1-8.

18. WUORINEN, V., "Seismic Microzonation of Victoria: A Social Response to Risk", in FOSTER, H.D. (ed.), *Victoria: Physical Environment and Development*. Victoria: University of Victoria, Department of Geography, Western Geographical Series, Vol. 12, 1976, pp. 185-219.

19. ROGERS, G.C., "A Documentation of Soil Failure During the British Columbia Earthquake of 23 June, 1946", *Canadian Geotechnical Journal*, 17, No. 1, 1980, pp. 122-127.

20. FOSTER, H.D. and CAREY, R.F., "The Simulation of Earthquake Damage", in FOSTER, H.D. (ed.), *Victoria: Physical Environment and Development*. Victoria: University of Victoria, Department of Geography, Western Geographical Series, Vol. 12, 1976, pp. 221-240.

21. HANSEN, W.R. and ECKEL, E.B., "The Alaska Earthquake, March 27, 1964: Field Investigations and Reconstruction Effort", in TANK, R.W. (ed.), *Focus on Environmental Geology*. London: Oxford University Press, 1966, pp. 46-65.

22. WHITE, W.R.H., "The Alaska Earthquake: Its Effects in Canada", *Canadian Geographical Journal*, 72, 1966, pp. 210-219.

23. SCHREINER, J., "Earthquake Engineering: Canadians Give the Issue a Shake", *Financial Post*, Special Report, July 2, 1983, p. 12.

24. MURPHY, L.M. and EPPLEY, R.A., "Development Plans and the Pacific Tsunami Warning System" in ADAMS, W.M. (ed.), *Tsunamis in the Pacific Ocean*. Honolulu: International Union of Geodesy and Geophysics and East-West Centre, 1970, pp. 261-269.

25. FOSTER, H.D. and WUORINEN, V., "British Columbia's Tsunami Warning System: An Evaluation", *Syesis*, 9, 1976, pp. 113-122.

26. PETHICK, D., *British Columbia Disasters*. Langley, B.C.: Stagecoach Publishing, 1978.

27. BUCHANAN, *op. cit.*, p. 21, based on information from the files of the Ministry of Transportation and Highways.

28. ANDERSEN, J., IZARD, T., STEVENS, J. and TURNER, D., *The Port Alice Mudslide of 1973: A Case Study*. Victoria: University of Victoria, Department of Geography, unpublished paper prepared for Geography 377, 1974.

29. KIERSCH, G.A., "The Vaiont Reservoir Disaster", in TANK, R.W., *Focus on Environmental Geology*. New York: Oxford University Press, 1973, pp. 153-164.

30. FOSTER, H.D., "Relief, Drainage and Natural Hazards" in FORWARD, C.N. (ed.), *Vancouver Island: Land of Contrasts*. Victoria: University of Victoria, Department of Geography, Western Geographical Series, Vol. 17, 1979, pp. 60-64.

31. Details of this study and model were kindly provided by H.W. Nasmith who was personally involved.

32. BUCHANAN, *op. cit.*, pp. 113-150.

PLATE 5 Local storm activity in the Interior. *B.C. Government Photo* ▶

4 CLIMATE

Stanton E. Tuller
University of Victoria

The purpose of this chapter is to present a general summary of the patterns of mean climate in British Columbia. The province possesses a great variety of climates. The effects of land and sea distribution, latitude, elevation, relief, exposure and local surface type create a plethora of individual, local climates. Any summary discussion focussing on the gross patterns must make a number of simplifications and generalizations. The approach used is to concentrate on the lowland and valley bottom climates of areas where observational data are readily available. While this neglects much of the total area of the province, it does concentrate attention on the areas where most of the population resides. Climatic differences will be illustrated using maps with dot symbols representing a value at a particular station. This eliminates the problem of drawing isolines across areas of mountainous relief. Emphasis will also be placed on the mean climate as represented by the most recently available climatic normals (usually 1951-1980). Temporal variation is beyond the scope of this discussion, although it may be an important factor in many aspects of applied climatology. The controlling factors which operate to produce the climates of British Columbia are also quite complex. This chapter will highlight only the most important controls.

A series of other summaries of the climate of British Columbia have appeared in the literature over the years.[1] The present discussion should be viewed as an introduction to the diversity of climate in British Columbia that complements these other studies.

SOLAR AND NET RADIATION

The initial source of energy for all other climatic elements is solar radiation. Spatial variation in solar radiation is the driving force that produces the observed patterns of many other climatic processes and elements. Mean global solar radiation data for nine stations, covering periods of record from 7 to 21 years, are published in the climatic normals.[2] Data for six of these stations, representing different regions of the province, are reproduced in Figure 1,4.

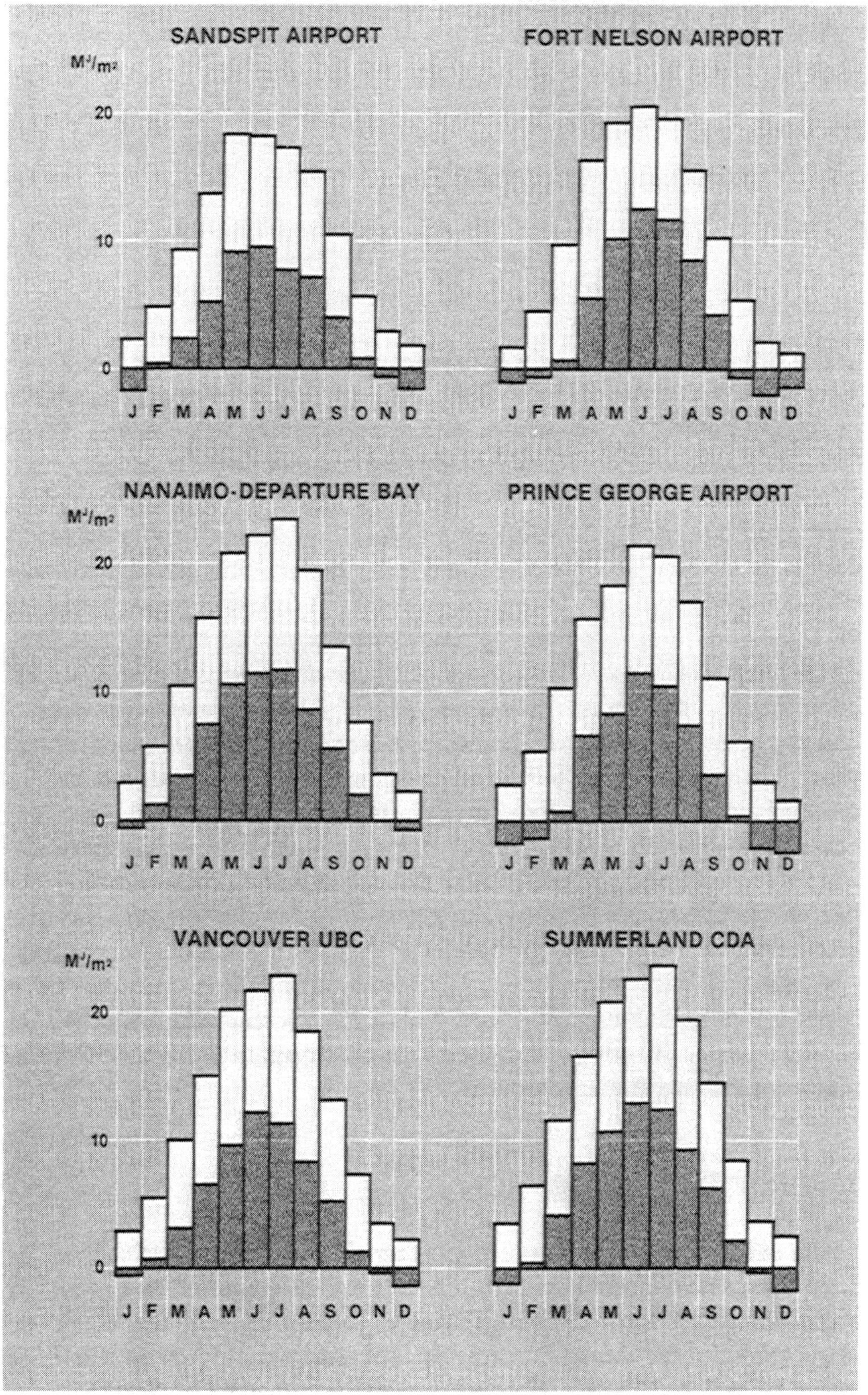

FIGURE 1,4 Solar radiation regimes at selected stations

Although there are many factors which control the quantity of solar radiation, its distribution in British Columbia can be summarized in terms of three basic controls: elevation, latitude and cloud cover.

The southern Interior benefits from low latitude, relatively high elevation and low amount of cloud cover compared with other regions of the province. It generally receives the most solar radiation throughout the year. Summerland CDA exemplifies the solar radiation regime in this region (Figure 1,4). The southern Interior is followed by the south coast region whose stations are somewhat in the lee of the mountains and are not subjected to excessive cloud cover (Nanaimo Departure Bay and Vancouver UBC). This region is also at relatively low latitude, but has a lower elevation than the southern Interior. Solar radiation is generally less in central and northern British Columbia than it is in the south. Fort Nelson A has the lowest total among the long term stations during winter when the latitudinal gradient of solar radiation is at its annual maximum. Stations on the outer coast, for example, Sandspit A, have lower solar radiation than stations in the central interior, for example, Prince George A, in all seasons. Sandspit's total is even less than Fort Nelson's in summer when the latitudinal gradient is reduced and the effects of lower elevation and greater cloud cover[3] become dominant.

British Columbia generally receives less global solar radiation than other regions of southern Canada in winter. One factor is that British Columbia is at a higher latitude than southern Ontario and Quebec, and the Atlantic Provinces. Another is that many British Columbia stations, especially those on the coast, are at lower elevations and have more cloud cover than stations on the Prairies. During the summer, however, when the effects of latitude are reduced, British Columbia stations receive more solar radiation than those in the Atlantic Provinces and Quebec, and slightly more than many southern Ontario stations.

The minimum mean monthly solar radiation occurs in December at all stations. This is the month of the winter solstice and of extensive cloud cover in all regions of the province. The maximum occurs in June at northern stations and in July at southern stations, where less cloud in July offsets the effects of lower solar altitude and shorter days.

Net all wave radiation (net radiation) is the net amount of radiant energy (both solar and long wave) available on the earth's surface. It was estimated for the six stations in Figure 1,4. Mean monthly surface albedoes representative of the regions surrounding the stations presented by Hay[4] were used, except for Prince George A where they were estimated by the author. Summer albedo ranged from .14 at Fort Nelson to .25 at Nanaimo, Sandspit and Vancouver. January albedo ranged from .26 (Vancouver) to .55 (Fort Nelson).

A number of equations that estimate the amount of downward clear sky atmospheric long wave radiation were tested against data from Canadian stations that measure global and reflected solar radiation and net radiation. In general, the equation of Idso and Jackson gave the best results during months when the air temperature was below freezing and that of Satterlund was best during the rest of the year.[5] These two equations were used during the appropriate months. Clear sky values were corrected for the effect of cloud cover using the method and cloud type coefficients attributed to Bolz.[6] Cloud cover data were taken from published sources[7] and cloud type distribution was taken as that expected at each station.

Upward terrestrial (long wave) radiation emitted by the earth's surface was computed using a ground surface temperature that varied from 2°K above the air temperature during mid-summer, to 2°K below air temperature in mid-winter.

The mean monthly net radiation at all stations is negative during the November-January period that surrounds the winter solstice (Figure 1,4). It is also negative at Prince George and Fort Nelson in February when surface snow cover produces a high albedo. Fort Nelson receives relatively low amounts of global solar and downward atmospheric long wave radiation and has a slightly negative net radiation in October. The variations in net solar and net long wave radiation are about equal between stations and usually counter each other in winter. Therefore, the difference in net radiation between stations is small. In general, however, it is slightly higher at the two south coast stations.

Summer differences in net radiation are produced primarily by the differences in solar radiation. The absolute range among stations in net solar radiation is about 3.5 times that in net long wave radiation. The stations with the highest summer net radiation are those with either a high global solar radiation, for example, Summerland CDA, or a low albedo, for example, Fort Nelson.

AIR TEMPERATURE

Air temperature is controlled primarily by the surface radiation and energy budgets of an area, by elevation, and by exposure to air advected from areas with other radiation and energy budget regimes. The pattern of air temperature in British Columbia can be viewed as a response to latitude, which influences the surface radiation budget, elevation, which primarily controls the density of the air and its efficiency in absorbing solar and terrestrial radiation, and the contrasting surface energy budgets of land and ocean surfaces.

The summer air temperature pattern is represented by the mean daily maximum, the temperature that influences most outdoor activities. The July mean daily maximum air temperature is highest in the interior valleys of south central British Columbia and decreases to the west (cool ocean influence), north (higher latitude) and east (higher elevation) (Figure 2,4).[8] The dominant controlling factor is the land-sea influence. Abundant ground heat flux density and evaporation on the ocean, along with upwelling of cool sub-surface water along the immediate coast, produce a low surface temperature compared to land surfaces. Mean July sea surface temperatures at stations on the outer coast range from about 11°C (52°F) to 13°C (55°F).[9] The resulting lower rate of heat transfer to the air through long wave radiation and sensible heat flux density produces a lower air temperature. The lowest July daily maximum air temperatures, therefore, are found on the outer coast.

The difference in summer warmth between inland and coastal stations is somewhat overestimated by the July temperature. July is usually the warmest month at inland stations, but August is often somewhat warmer at outer coast stations. The longer delay in time of annual temperature peak is a result of greater heat storage in the ocean and is a common feature that differentiates maritime and continental temperature regimes.

The range of summer temperature found in British Columbia is emphasized by comparisons with the rest of Canada. July mean daily maximum air temperatures in lower elevation, southern interior valleys are among the highest in Canada. Those on the outer coast, however, are among the lowest in southern Canada. Only coastal Newfoundland has stations whose values rival those found here. The ocean-land influence on air temperature is reversed in winter when the ocean surface is warmed by heat stored during summer or brought in by currents from warmer regions. The January mean sea surface temperature at stations on the outer coast ranges between about 6°C (43°F) and 8°C (46°F),[10] much higher than land surface temperatures at this time of year. The resulting heat transfer to the air gives outer coastal stations in the south the highest January mean daily air temperatures (Figure 3,4). The coastal mountain ranges interact with the ocean-land differences to accentuate the winter air temperature difference between coastal and interior stations. Coastal stations are better exposed to relatively warm southwesterly air flows. The Coast Mountains limit the inland penetration of mild oceanic air at low levels and, at the same time, are a barrier against the movement of cold continental air to the coast. Continental air that does reach the coast must descend during its migration and would thus undergo some adiabatic warming.

The January mean daily air temperature is below freezing throughout the province, except on the immediate coast. The lowest winter temperatures

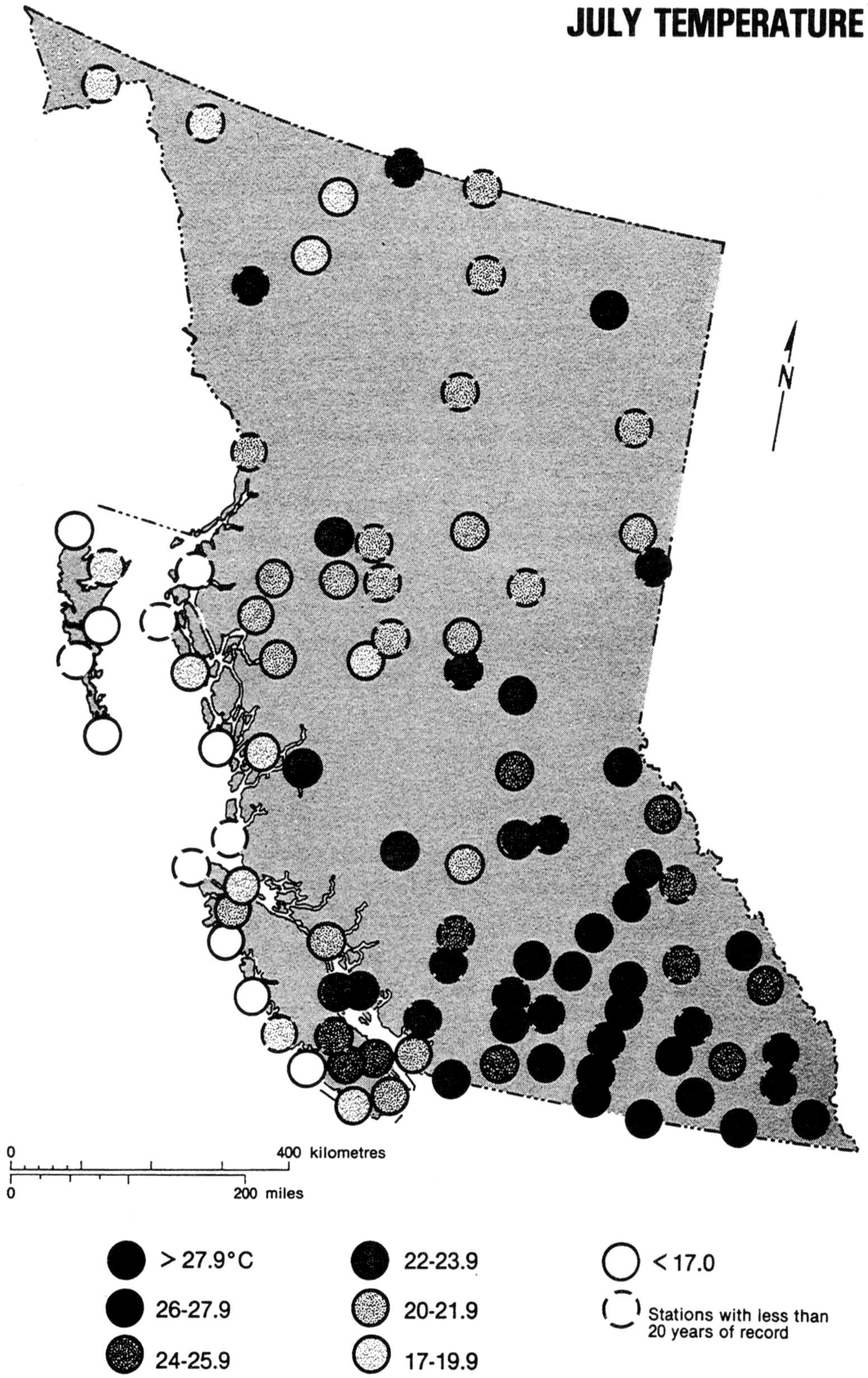

FIGURE 2,4 The summer air temperature pattern as indicated by the July mean daily maximum air temperatures

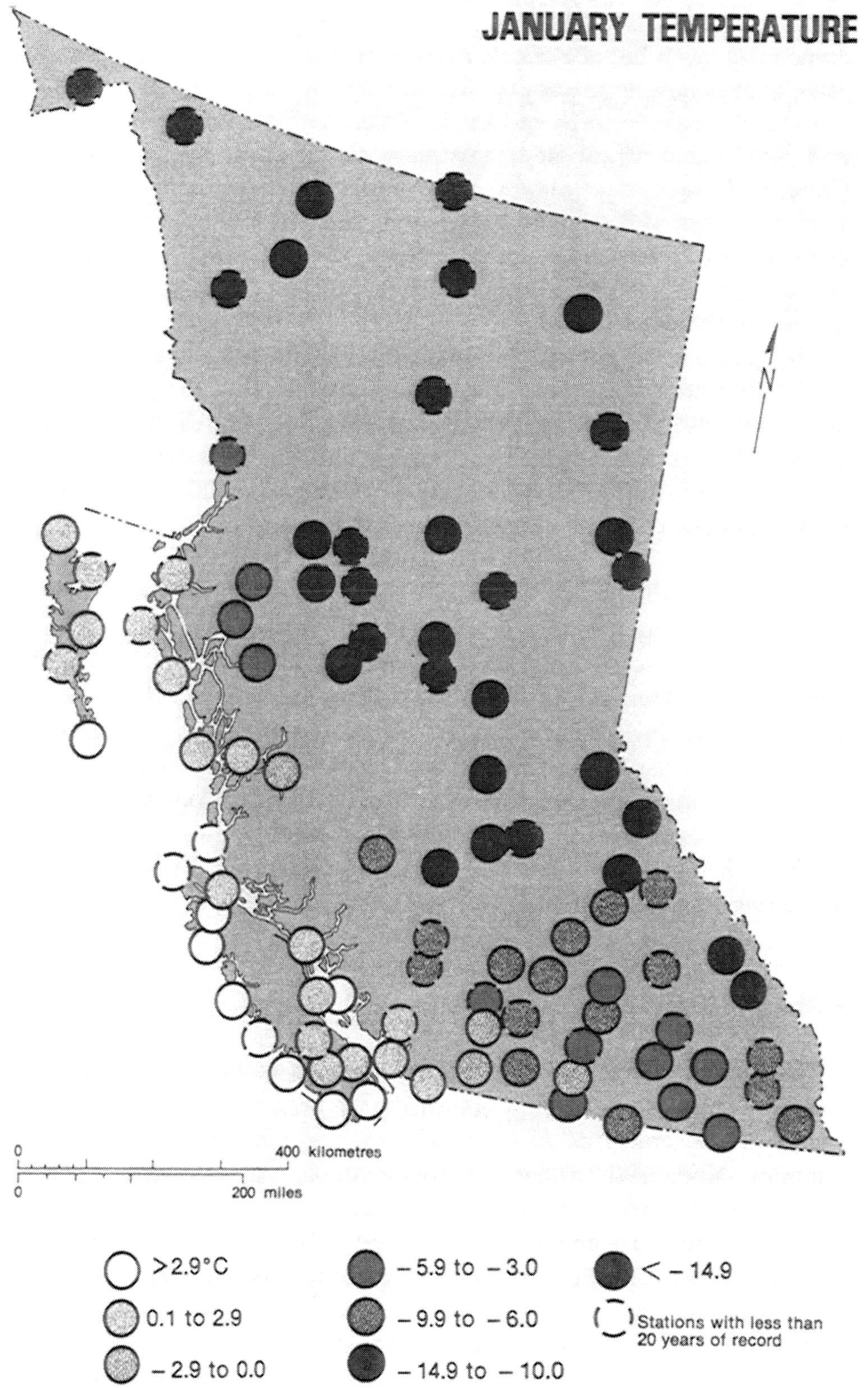

FIGURE 3,4 The winter air temperature pattern as indicated by the January mean daily air temperatures are found in the north. Stations in this part of the province more frequently come under the influence of cold Arctic air and receive much less solar radiation than stations in the south. The variation in air temperature throughout the province is much greater in winter than in summer (Figures 2,4 and 3,4). Mean winter air temperatures on the coast are the highest in Canada. Temperatures at stations in the southern valleys of the Interior are similar to those of southernmost Ontario. Stations in the far north are as cold as those found on the Prairies. Thus, British Columbia in winter, as well as summer, has temperature regimes that span the range found in the whole of southern Canada.

The effects of land and sea distribution, latitude and elevation are also seen in the variation of frost-free periods throughout the province. In addition, local relief has an important effect. Valley bottoms collect cold air that drains from adjacent slopes during the night and thus have shorter frost-free periods. The frost-free period is generally near or over 200 days at stations on the immediate coast, about 150 days in the southern Interior and less than 120 days in the north and northeast (Table 1,4).[11] The frost-free period is about 40 to 50 days shorter in the Rocky Mountain Trench (elevation 800 - 900 metres [2,625 - 2,450 feet]) than in the Okanagan and Thompson river valleys (elevation 350 - 450 metres [1,150 - 1,175 feet]). Stations with extremely long frost-free periods, for example, Victoria Gonzales Heights, 287 days, and Cape St. James, 272 days, combine the marine influence with free air drainage. Coastal stations have by far the longest frost-free periods in Canada. The frost-free periods at stations in the southern Interior are similar to those in southernmost Ontario or at coastal stations in the Maritimes. Those in northeastern British Columbia are similar to ones in the southern Prairies.

AIR PRESSURE AND WINDS

Mean upper air winds over British Columbia come from the western quadrant throughout the year, paralleling the pressure gradient created by the latitudinal gradients of solar radiation and temperature in the mid-latitudes. Mean 500 millibar geostrophic winds over the southern part of the province are from the west in both summer and winter. Those over the north are generally from the west-northwest in January and west-southwest in July.[12] The mean speed of the 500 millibar geostrophic wind is greater in winter than summer in response to the increased solar radiation, air temperature and, thus, upper air pressure gradients in this season. The speed is greater over the southern part of the province than over the north. The approximate mean January speeds are about 61 kilometres per hour (38 miles per hour) in the south and 36 kilometres per hour (22 miles per hour)

TABLE 1,4 FROST-FREE PERIOD AT SELECTED STATIONS

Station	*Frost-Free Period in Days*
Cape St. James	272
Estevan Point	229
Victoria Gonzales Heights	287
Vancouver International Airport	216
Chilliwack	216
Lytton	186
Kamloops Airport	149
Penticton Airport	148
Golden	115
Prince George Airport	85
Smithers Airport	91
Fort St. John Airport	115
Fort Nelson Airport	106

SOURCE: Canada, *Canadian Climate Normals, Volume 6, Frost, 1951-1980.* Ottawa: Environment Canada, Atmospheric Environment Service, 1982.

in the north. July values vary from 36 kilometres per hour (22 miles per hour) in the southeast to 11 kilometres per hour (7 miles per hour) in the northwest. The upper air westerlies are important in imparting the west-east direction of travel to weather systems that pass over British Columbia.

The ocean-land temperature difference is a major control of the mean sea level air pressure pattern over British Columbia. Cold, dense continental air in winter is associated with high sea level air pressure, whereas the warmer, less dense, oceanic air is associated with lower pressure. The frequent cyclonic storms in winter help to accentuate the low pressure over the Pacific. The highest mean winter sea level air pressure in the province is found in the northeast and the Rocky Mountain Trench area. The lowest is found on the outer north coast.[13] The general trend of the isobars varies between northwest-southeast and north-south over the province.

The pressure gradient reverses in summer. The warm land now has a lower sea level air pressure than the ocean. The subtropical high pressure cell moves northward and extends its influence along the British Columbia coast.[14] The highest mean sea level air pressure is found along the outer coast and the lowest in the northeast and in the warmest valleys of the southern Interior.[15] The northeast is well removed from the influence of the subtropical high pressure cell and far enough north to be located near the principal summer storm track.[16] The low pressure in the interior valleys is primarily a result of the high air temperature. The July isobars run in a general northwest-southeast trend over the province. The air pressure gradient in summer is only about one-half that of winter.

Surface winds at valley bottom stations are very strongly influenced by local relief. The prevailing wind directions usually parallel the trends of the valleys. In general terms, the mean winter pressure pattern would create surface winds with a southerly or easterly component. Winter winds at most stations follow this pattern, but the actual distribution varies from station to station. Northerly winds are common in the northern part of the province and Victoria Gonzales Heights has prevailing north and northeast winds as air flows past the station on its way to the Strait of Juan de Fuca and the open sea. Calms are relatively frequent at valley bottom stations during winter when cold, stable air can accumulate and decouple the surface air from the winds at higher elevation. These situations with stable, stagnant air can produce a high air pollution potential.

Surface winds following the generalized mean pressure gradient in summer would have a northerly or westerly component. Northwesterlies are common at stations on the outer coast and winds from the northern or western quadrants are also common at inland stations. Calms are generally less frequent during summer than during winter at interior stations where greater instability reduces the chance of calms. Most coastal stations, however, have more calms in summer, the season with fewer cyclonic storms and a smaller mean pressure gradient.

The seasonal pattern of wind speed varies widely from station to station. In general, however, coastal stations are the windiest in winter and have the lowest mean monthly wind speed in late summer or early fall.[17] Spring is the windiest season at most stations in northern British Columbia and the Rocky Mountain Trench and winter the least windy. The late summer-early fall minimum is found at many other interior stations, but the month with the maximum mean monthly wind speed is quite variable. Well-exposed coastal stations are as windy as any place in Canada. Cape St. James, a very well-exposed station on the southern tip of the Queen Charlotte Islands, has a mean annual wind speed of 33.7 kilometres per hour (20.9 miles per hour) and a December mean of 41.7 (25.9).[18]

It has recorded a maximum speed of over 120 kilometres per hour (75 miles per hour) in every month in the September-April period and its maximum recorded wind speed through 1980 was 177 (110). Mean wind speeds at most interior stations are lower than those on the southern Prairies or in the Atlantic Provinces and are about equal to, or somewhat lower than, those at most stations in southern Ontario and Quebec.

HUMIDITY

The term "humidity" is used here to describe the actual quantity of water vapor in the atmosphere. The vapor pressure, the partial pressure of water vapor, is the measure of humidity that is used.

The major source of water vapor for all of British Columbia is the Pacific Ocean. Stations near the coast are closest to this source and have the highest humidity throughout the year (Table 2,4).[19] Inland stations that receive more frequent incursions of low level Pacific air have higher humidity than those that are more sheltered.

Two additional controls of humidity are elevation and air temperature. Humidity decreases rapidly with elevation. The Okanagan (Penticton Airport) and Thompson River Valley (Kamloops Airport) have higher humidity than does the Rocky Mountain Trench (Kimberley Airport). The amount of water vapor the air can hold is limited by the air temperature. The low air temperature at northern stations puts an upper limit on winter humidity and combines with distance from an effective source of moisture to produce a low winter humidity.

Coastal stations in British Columbia have the highest winter humidity in Canada. The rest of the country is predominantly under the influence of dry continental air in this season. Winter humidity in the southern Interior is similar to that of the Atlantic Provinces and southernmost Ontario. The far north has a humidity regime like that of the Prairies. Summer humidity in British Columbia is relatively low, however. Evaporation from the eastern Pacific is somewhat limited by its low surface temperature. Even coastal stations have lower summer humidity than do those in Ontario, the Maritimes and southern Manitoba, which come under the influence of warmer water surfaces.

Although humidity near the surface is much higher in summer than in winter, the zonal transfer of tropospheric moisture from the Pacific Ocean is greater in winter, a result of the higher westerly wind speed. The maximum inflow of moisture during winter occurs over southern and central British Columbia, whereas the maximum in summer occurs over the northern part

TABLE 2.4 MEAN VAPOR PRESSURE AT SELECTED STATIONS DURING MID-SEASON MONTHS IN KILOPASCALS (1 kilopascal = 10 millibars)

Station	*Elevation (metres)*	*January*	*April*	*July*	*October*
Cape St. James	89	.72	.78	1.29	1.06
Prince Rupert	52	.64	.75	1.33	.97
Estevan Point	20	.77	.86	1.38	1.12
Victoria International Airport	19	.70	.83	1.34	1.03
Vancouver International Airport	3	.67	.85	1.47	1.07
Penticton Airport	344	.45	.63	1.22	.79
Kamloops Airport	346	.38	.59	1.16	.77
Kimberley Airport	914	.33	.54	1.08	.65
Prince George Airport	676	.27	.49	1.14	.66
Smithers Airport	523	.28	.50	1.13	.68
Fort St. John Airport	695	.17	.45	1.16	.55
Fort Nelson Airport	382	.09	.42	1.25	.49
Dease Lake	816	.15	.42	.99	.53

SOURCE: Canada, *Station, Sea Level and Vapor Pressure Normals Based on the Period 1953-1972.* Downsview, Ontario: Atmospheric Environment Service, 1976.

of the province.[20] The transport of moisture and the total amount of water vapor in the air are more important factors in precipitation than is the surface humidity.

PRECIPITATION

Precipitation measured at ground level (neglecting dew) is a result of the available water vapor in the air and dynamic lifting mechanisms that produce the cooling necessary to convert water vapor into water droplets or snowflakes large enough to fall to the surface. In most cases the presence or absence of the lifting mechanisms is the most important control of the distribution of precipitation over an area the size of British Columbia.

The most important lifting mechanism for precipitation in British Columbia is the mid-latitude cyclonic storm. Cyclonic storms affecting the province originate primarily in the central and western Pacific.[21] Those formed in the western Pacific reach the coast in their old age or occluded stage and usually bring light to moderate, but often long duration, precipitation. Those from the central Pacific are younger and more vigorous. An occasional intense storm migrates from south to north along the west coast of North America, reaching southwestern British Columbia. Many Pacific cyclones are followed by unstable marine air that brings unsettled weather and showers to the coast.[22]

The most obvious feature of the distribution of annual precipitation is the high amounts received on the windward sides of the major north-south mountain ranges (Figure 4,4).[23] Mountains on the outer coast that are perpendicular to the westerly and southwesterly flow that accompanies much of the precipitation in the province have particularly strong effects.[24] These well exposed stations on the outer coast have the highest annual precipitation in Canada. Stations in the rainshadow of the Coast Mountains have only about one-tenth the annual precipitation recorded at well exposed stations on the windward side of the range. Low precipitation is especially evident in the narrow valleys. The interior mountain systems also show the orographic influence. The Pacific air that reaches these, however, generally is drier air from higher elevations that passes over the Coast Mountains and, thus, orographic precipitation is not as extreme as that on the coast. It must be remembered that the lowland and valley bottom patterns depicted in Figure 4,4 do not reveal the much higher precipitation that occurs on the higher slopes. Orographic effects in the interior, for example, are especially evident in the Selkirk Mountains.[25] Revelstoke, on the windward side of the range, has an annual total precipitation of 1,064 millimetres (41.9 inches), but Brisco, on the leeward side, receives only 427 millimetres (16.8 inches). Two stations on the upper slopes of the Selkirks, Glacier NP Rogers Pass and Glacier NP Mt. Fidelity, receive 1,606 and 2,169 millimetres (63.2 and 85.4 inches), respectively.

Precipitation in northern British Columbia is limited by the dry Arctic air that is common, especially in winter and spring, and by the north's isolated location with respect to the southwesterly flow of moist Pacific air that accompanies much of the precipitation in the southern part of the province.

Precipitation follows a marked seasonal cycle at stations on the coast. Southern stations have a distinct maximum in winter. This coincides with the greatest frequency of mid-latitude cyclonic storms. The monthly maximum usually occurs in December. December is more humid than January, although storm frequency may be somewhat greater in the latter month. The maximum precipitation shifts to the fall on the far northern

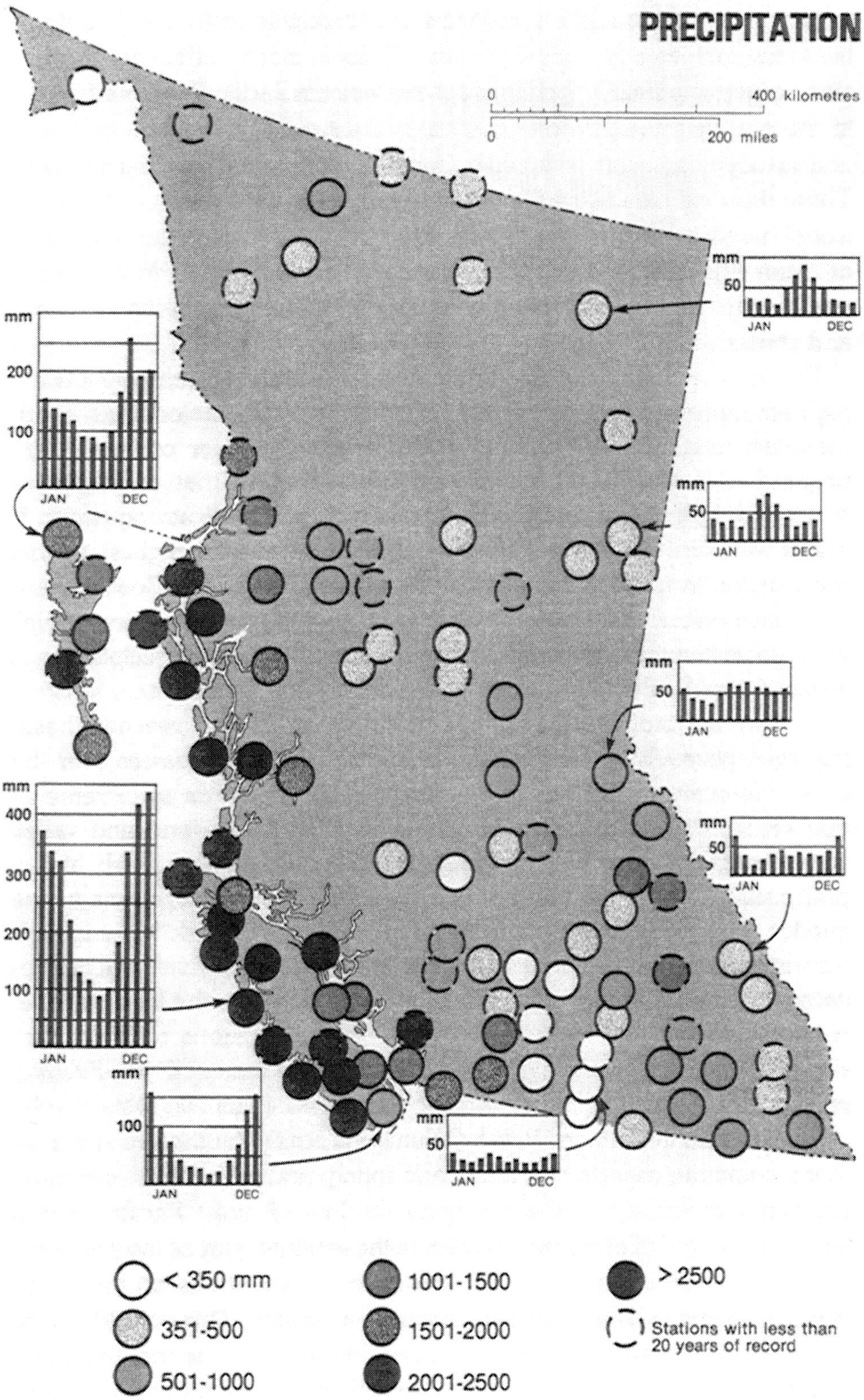

FIGURE 4.4 The distribution of mean annual precipitation

coast. Cyclonic storms are relatively numerous during that season and precipitable water vapor is much higher than it is in mid-winter.[26] Summer is the dry season all along the coast, with July being the driest month at most stations. Cyclonic storm frequency is low during the summer and the subtropical high pressure cell, with its subsiding air and subsidence inversion, frequently influences the south coast.

The northern interior displays the opposite seasonal pattern. The greatest precipitation occurs in summer when more moisture is available and the air is more unstable than in other seasons. July is the wettest month at most stations (Figure 4,4). The driest season is spring, April being the driest month at many stations. Spring is the time between the cyclonic storm regime of winter and the instability of summer. Dry Arctic air dominates northern British Columbia in that season.[27]

The most even distribution of precipitation throughout the year is found in the southern Interior. Most stations have a primary maximum in winter (cyclonic storm regime), a secondary maximum in summer (instability and convective precipitation) and a relatively dry spring and fall. The bulk of annual precipitation throughout the province is associated with frontal systems,[28] and even in the Interior they act as important trigger factors for summer convectional showers.

High intensity convectional rainfall is not common at lowland and valley bottom stations in British Columbia. Over most of the province the 15-minute rainfall totals with a 10-year return period are only about one-half of those on the southern Prairies or in southern Ontario.[29] The Peace River region has the highest values, followed by the Rocky Mountain Trench and outer coast.

One-day rainfall intensities can be relatively high on the outer coast, however, when orographic effects combine with a vigorous cyclonic storm to produce heavy, sustained rain. The majority of well-exposed stations have recorded over 150 millimetres (6 inches) of precipitation in one observing day and a few, where orographic effects are pronounced, have received over 200 millimetres (8 inches).[30] Even higher values can be expected on the upper slopes. The outer coast has the highest one-day rainfall intensity of any region in Canada. Nova Scotia, which can occasionally receive heavy rainfall from tropical cyclones, is the only area that comes close to matching the outer coast of British Columbia. The maximum at most of the Nova Scotia stations with very high values was recorded in August, compared with fall or winter at most British Columbia coastal stations. This reflects the seasonal distribution of the major storm systems responsible for the greatest recorded 24 hour precipitation.

The large variation in winter temperature and precipitation throughout the province gives rise to a large variation in snowfall regimes. Snowfall is irregular in occurrence and generally melts quickly along the south coast. Mean annual snowfall varies between about 40 and 75 centimetres (16 and 30 inches) and falls on 5 to 15 days per year during the October-March period. The annual snowfall on the south coast is the lowest in southern Canada. Annual snowfall on the north coast can range up to about 100 centimetres (40 inches) at lowland stations well exposed to ocean influences and up to 200 centimetres (80 inches) on cooler sites at higher elevations or in protected inlets. These values are similar to those on the Prairies, but a continuous snow cover throughout the winter is not expected. Snow in the Interior, however, can be expected to persist, on average, although in light precipitation areas a great deal of drifting can occur, leaving some areas bare of snow. The annual total reflects the distribution of annual precipitation. The dry valleys of the southern Interior receive between 50 and 100 centimetres (20 and 40 inches) of snow per year. The Rocky Mountain Trench receives between 100 and 200 centimetres (40 and 80 inches), but Revelstoke averages over 400 centimetres (160 inches). Annual snowfall varies from about 130 to near 400 centimetres (50 to 160 inches) at northern Interior stations. The Peace River and Fort Nelson regions receive about 200 centimetres (80 inches) of snow a year, falling on between 45 and 75 days.

MAY-SEPTEMBER CLIMATIC WATER BALANCE

Summer is relatively dry throughout most of British Columbia. Potential evapotranspiration exceeds precipitation, leading to a climatic moisture deficit[31] that must be alleviated by soil moisture, ground water, irrigation, or some other source of water to ensure adequate growth of shallow-rooted plants. In order to illustrate the regional variation in the growing season (May-September) climatic moisture balance, the potential evapotranspiration (PE) was computed using the Priestley and Taylor method[32] (α = 1.25) for the six stations for which net radiation (the major independent variable in the Priestley and Taylor formula) was determined (Figure 1,4). Ground heat flux density was taken to decline from about five percent of net radiation at the beginning of the period to zero at the end. Station to station differences in May-September PE reflect the differences in net radiation. Values range from a high of 508 millimetres (20.0 inches) at Summerland CDA to a low of 327 millimetres (12.9 inches) at Sandspit Airport. Other values for individual stations are: Nanaimo Airport, 450 (17.7); Vancouver UBC, 435 (17.1); Fort Nelson Airport, 424 (16.7); and Prince George Airport, 363 (14.3).

Through extrapolation of the individual station results to larger regions, one would expect the greatest PE to occur in the southern interior valleys, followed by the lower Fraser Valley and southeastern Vancouver Island. Northeastern British Columbia would have a relatively high growing season PE. The minimum would occur on the west coast, especially in those areas with frequent summer cloud or fog.

The regional pattern of growing season climatic moisture deficit follows that of PE. Deficit values range from 40 millimetres (1.6 inches) at Sandspit Airport to over 380 millimetres (15.0 inches) at Summerland CDA. July is the month of maximum deficit at all stations, except for Fort Nelson Airport where it is June. Most of the province has a growing season climatic water deficit. The major exception is the exposed west coast which has relatively low PE and sufficient summer precipitation.

TEMPERATURE-PRECIPITATION REGIONS

Published summaries of the climate of British Columbia have utilized a number of different regionalization schemes. Kendrew and Kerr, for example, discussed the climate of British Columbia in terms of the broad subjective regions: the littoral, southern interior and northern interior.[33] Schaefer and Chilton employed physiographic regions, a concept that has some validity in an area where relief is a major control of climate.[34] Chapman used a more climatic approach. He primarily employed the Koppen classification.[35]

In recent years a number of studies have appeared which use factor analysis or principal components analysis as the basis for a largely objective classification scheme based on observed climatic data.[36] These methods reduce a large number of climatic variables to a smaller number of uncorrelated variables or factors. The factor scores for each station are then used as input into a cluster analysis which produces groupings of roughly similar stations. Except for the selection of the actual grouping algorithm to be used (different algorithms can give different results) and the number of discrete groups, the procedure is largely objective and based solely on the observed climate.

A simple factor analysis procedure was used to determine a preliminary regionalization of coastal and valley bottom stations in British Columbia. Monthly mean daily air temperature and total precipitation were used as input variables in order to maximize the number of available stations and simplify the regional pattern. The final analysis utilizes 61 stations located throughout the province. A cautionary note is that the use of different input variables or cluster analysis algorithms could give rise to somewhat different regional groupings.

Six major temperature-precipitation regions were identified, four of which were divided into smaller sub-regions. The coastal area is composed of two temperature-precipitation regions (Figure 5,4). Region 1 is the most distinct of all and consists of those stations which have a very pronounced maritime temperature regime and very high precipitation. Tasu Sound on the west coast of the Queen Charlotte Islands is the wettest of the stations used and represents a sub-region of even heavier precipitation.

Other coastal stations are grouped in Region 2. These stations still have a largely maritime temperature regime, but less orographic effect and precipitation than stations in Region 1. Three sub-groups with somewhat different air temperature regimes and rain shadow effects comprise this region. Sub-region 2A includes the cooler, wetter stations on the northern or western portions of Vancouver Island and the Queen Charlottes which have only minor rain shadow effects. Sub-region 2B is much more in a rain shadow area, is at a lower latitude and has distinctly lower precipitation and warmer summers than Sub-region 2A. The two Victoria stations, Gonzales Heights and International Airport, are in the lee of the Olympic Mountains and are far enough south and west to be less affected by cold air during the winter. They are drier, have cooler summers, but warmer winters, than stations in 2B, and form sub-region 2C.

The factor and cluster analyses indicated four temperature-precipitation regions in the Interior. Stations in the narrow, lower elevation valleys of the southern Interior comprised Region 3. This group has the warmest summers in the province, cool winters and low total precipitation. Hope is transitional between this region and Region 2B. It was included in Region 3 under the "distance between groups" criterion used, although it might be considered a separate sub-region under a somewhat different criterion.

Region 4 includes stations in the central Interior plus higher elevation stations in the south. Stations in Sub-region 4A are characterized by cold winters, short warm summers and relatively low precipitation. Stations in central British Columbia which are transitional between the coastal and interior climatic regimes are grouped in Sub-region 4B. These stations are somewhat in the lee of the Coast Mountains, but are in valleys or fiords which provide relatively easy access for marine air. They generally have warmer winters, slightly cooler summers and more precipitation than stations in Region 4A.

Region 5 includes most stations in northwestern British Columbia, along with some higher elevation stations to the south. The climate of this region is characterized by cool summers, cold winters and low precipitation.

The northeastern portion of the province is in Region 6. This area is in the lee of the Rocky Mountains and physiographically and climatically is

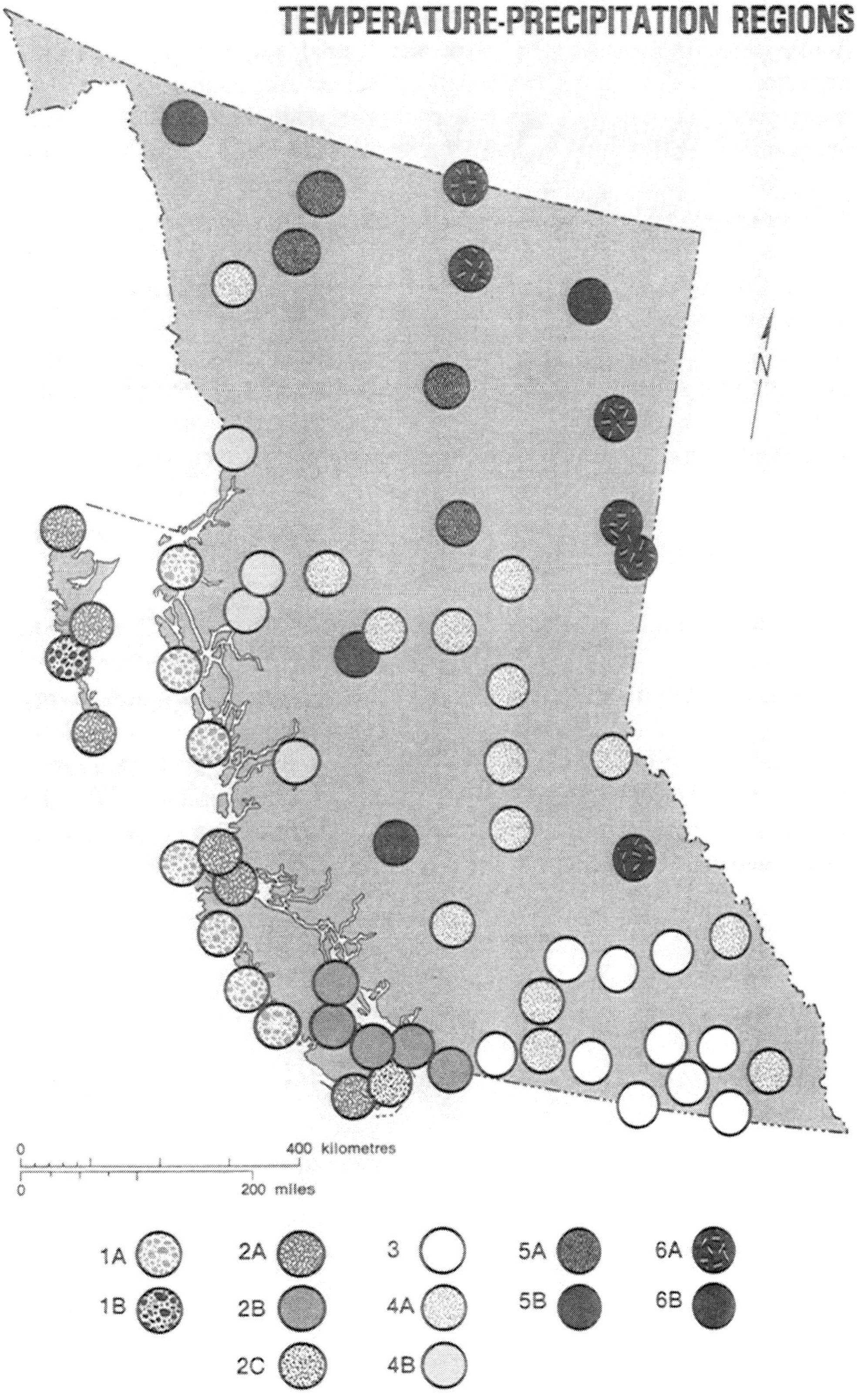

FIGURE 5.4 Temperature-precipitation regions and sub-regions

part of the Prairie regime. The temperature-precipitation pattern is distinctly continental with cold winters, short, mild summers and low precipitation, with a summer maximum. Fort Nelson Airport has warmer summers, greater annual air temperature and precipitation ranges, and slightly more annual precipitation than other stations analyzed. It was, therefore, put in a separate sub-region.

In summary, the effects of latitude, elevation, land and sea distribution and relief produce a number of distinct temperature-precipitation regions in British Columbia. The results of the factor analysis approach used are not surprising and agree closely with those of other schemes used in the past. The major differences between the regions presented here and the simple physiographic regions utilized by Schaefer and Chilton are the finer differentiations achieved by more detailed analysis and the placing of higher elevation southern stations into more northerly climatic regions.

CONCLUSION

British Columbia is a land of great topographic diversity and moderate latitudinal extent. The effects of this on radiation budgets, advection of air masses of different characteristics, and surface-air moisture and energy exchanges, create a wide variety of climatic conditions, even among lowland and valley bottom stations. The diversity of climates presents a range of opportunities and challenges for successful human occupance. With the exception of tropical and polar climates, there should be something for nearly everyone somewhere in the climate of British Columbia.

REFERENCES

1. Examples include: KERR, D., "Climate of British Columbia", *Canadian Geographical Journal*, 45, 1952, pp. 143-158; CHAPMAN, J.D., *The Climate of British Columbia*. Paper presented to the Fifth British Columbia Natural Resources Conference, 1952, 47 pp.; KENDREW, W.G. and KERR, D., *The Climate of British Columbia and the Yukon Territory*. Ottawa: Queen's Printer, 1955; SCHAEFER, D.G., "Climate", in VALENTINE, K.W.G., *et. al.* (eds.), *The Soil Landscape of British Columbia*. Victoria: Ministry of Environment, 1978; CHILTON, R.R.H., *A Summary of Climatic Regimes of British Columbia*. Victoria: Ministry of Environment, 1981.

2. CANADA, *Canadian Climate Normals, Volume 1, Solar Radiation, 1951-1980*. Ottawa: Environment Canada, Atmospheric Environment Service, 1982.

3. CANADA, *Climatic Normals, Volume 3, Sunshine, Cloud, Pressure and Thunderstorms*. Toronto: Department of Transport, Meteorological Branch, 1968.

4. HAY, J.E., *An Analysis of Solar Radiation Data for British Columbia*. Victoria: Ministry of Environment, 1979.

5. IDSO, S.B. and JACKSON, R.D., "Thermal Radiation From the Atmosphere", *Journal of Geophysical Research*, 74, 1969, pp. 5397-5403; SATTERLUND, D.R., "An Improved Equation for Estimating Long-wave Radiation From the Atmosphere", *Water Resources Research*, 15, 1979, pp. 1649-1650.

6. See GEIGER, R., *The Climate Near the Ground*. Cambridge, Mass.: Harvard University Press, 1966; MORGAN, D.L., PRUITT, W.O. and LOURENCE, F.J., "Estimation of Atmospheric Radiation", *Journal of Applied Meteorology*, 10, 1971, pp. 463-468; ARNFIELD, A.J., "Evaluation of Empirical Expressions for the Estimation of Hourly Conditions", *Quarterly Journal of the Royal Meteorological Society*, 105, 1979, pp. 1041-1052.

7. CANADA, *op. cit.*, reference 2; HAY, J.E., *op. cit.*

8. CANADA, *Canadian Climate Normals, 1951-1980, Temperature and Precipitation, British Columbia.* Ottawa: Environment Canada, Atmospheric Environment Service, 1982.

9. HOLLISTER, H.J. and SANDNES, A.M., *Sea Surface Temperatures and Salinities at Shore Stations on the British Columbia Coast, 1914-1970.* Victoria: Environment Canada, Marine Sciences Directorate, 1972.

10. *Ibid.*

11. CANADA, *Canadian Climate Normals, Volume 6, Frost, 1951-1980.* Ottawa: Environment Canada, Atmospheric Environment Service, 1982.

12. LAHEY, J.F., BRYSON, R.A., WAHL, E.W., HORH, L.D. and HENDERSON, V.D., *Atlas of 500 mb Wind Characteristics for the Northern Hemisphere.* Madison: University of Wisconsin Press, 1958.

13. CANADA, *Atlas of Climatic Maps.* Ottawa: Department of Transport, Meteorological Branch, 1970, Series 3; CANADA, *Station, Sea Level and Vapor Pressure Normals Based on the Period 1953-1972.* Downsview, Ontario: Atmospheric Environment Service, 1976.

14. NEIBURGER, M., JOHNSON, D.S. and CHIEN, C-W., *Studies of the Structure of Atmosphere Over the Eastern Pacific Ocean in Summer, I. The Inversion Over the Eastern North Pacific Ocean.* Berkeley: University of California Press, 1961.

15. CANADA, *op. cit.*, reference 13.

16. KLEIN, W.H., *Principal Tracks and Mean Frequencies of Cyclones and Anticyclones in the Northern Hemisphere.* Washington: U.S. Government Printing Office, 1957; REED, R.J., "Principal Frontal Zones of the Northern Hemisphere in Winter and Summer", *Bulletin of the American Meteorological Society*, 41, 1960, pp. 591-598; REITAN, C.H., "Frequencies of Cyclones and Cyclogenesis for North America, 1951-1970", *Monthly Weather Review*, 102, 1974, pp. 861-868.

17. CANADA, *Canadian Climate Normals, Volume 5, Wind, 1951-1980*. Ottawa: Environment Canada, Atmospheric Environment Service, 1982.

18. *Ibid.*

19. CANADA, 1976, *op. cit.*, reference 13.

20. HARE, F.K. and HAY, J.E., "The Climate of Canada and Alaska", in BRYSON, R.A. and HARE, F.K. (eds.), *Climates of North America*. New York: Elsevier, 1974, pp. 84-87.

21. KLEIN, W.H., *op. cit.*; REITAN, C.H., *op. cit.*, reference 16.

22. HARE, F.K. and HAY, J.E., *op. cit.*, p. 89.

23. CANADA, *op. cit.*, reference 8.

24. WALKER, E.R., *A Synoptic Climatology of Parts of the Western Cordillera*. Montreal: McGill University, Arctic Meteorology Research Group, Publication in Meteorology No. 35, 1961.

25. KENDREW, W.G. and KERR, D., *op. cit.*, reference 1.

26. HAY, J.E., "Precipitable Water Over Canada: II. Distribution", *Atmosphere*, 9, 1971, pp. 101-111; TULLER, S.E., "Mean Monthly and Annual Precipitable Water Vapour in Canada", *Weather*, 27, 1972, pp. 278-289.

27. BRYSON, R.A. and HARE, F.K., "The Climates of North America" in BRYSON, R.A. and HARE, F.K. (eds.), *Climates of North America*. New York: Elsevier, 1974, p. 23.

28. WALKER, E.R., *op. cit.*

29. NATIONAL RESEARCH COUNCIL, *Climatic Information for Building Design in Canada*. Ottawa: National Research Council, 1975.

30. CANADA, *op. cit.*, reference 8.

31. WALLIS, C.H., *Calculating Irrigation Requirements for Eastern Vancouver Island*. Victoria: Ministry of Environment, 1982.

32. PRIESTLEY, C.H.B. and TAYLOR, R.J., "On the Assessment of Surface Heat Flux and Evaporation Using Large Scale Parameters", *Monthly Weather Review*, 100, 1972, pp. 81-92.

33. KENDREW, W.G. and KERR, D., *op. cit.*, reference 1.

34. SCHAEFER, D.G., *op. cit.*, reference 1; CHILTON, R.R.H., *op. cit.*, reference 1.

35. CHAPMAN, J.D., *op. cit.*, reference 1.

36. For example: McBOYLE, G.R., "Climate Classification of Australia by Computer", *Australian Geographical Studies*, 9, 1971, pp. 1-14; McBOYLE, G.R., "Factor Analytic Approach to a Climatic Classification of Europe", *Climatological Bulletin*, 12, 1972, pp. 3-11; POWELL, J.M. and MacIVER, D.C., *A Summer Climate Classification for the Forested Area of the Prairie Provinces using Factor Analysis*. Edmonton: Northern Forest Research Centre, 1977; WILLMOTT, C.J., "A Component Analytic Approach to Precipitation Regionalization in California", *Archiv fur Meteorologie, Geophysik und Bioklimatologie*, 24B, 1977, pp. 269-281; WILLMOTT, C.J., "P-mode Principal Components Analysis, Grouping and Precipitation Regions in California", *Archiv fur Meteorologie, Geophysik und Bioklimatologie*, 26B, 1978, pp. 277-295.

PLATE 6 Garry oaks on southern Vancouver Island. *Geography Dept. Photo* ▶

PLATE 7 Coastal rain forest. *B.C. Government Photo* ▶

5 VEGETATION

Michael C.R. Edgell
University of Victoria

An enduring image of British Columbia is that of a mountainous province covered with an endless sea of coniferous forest, broken only by widely scattered alpine areas and open sagebrush grasslands in the southern Interior. But this image belies the floristic diversity and vegetational complexity of the province. In addition to conifers, over 2,000 species of flowering plants are found within its boundaries. Admittedly, forests do cover about 55 percent of the provincial area and over 90 percent of those forests are dominated by evergreen conifers. But deciduous trees are a significant component of many forest types which, encompassing such wide latitudinal and longitudinal extent and topographic diversity, are anything but uniform in floristics and structure.

INTRODUCTION

British Columbia is the meeting place of three great North American forest realms. The forests of the Pacific slopes are part of a continuous 3,600 kilometre (2,300 mile) coastal belt of tall coniferous forests extending from northern California to Alaska, variously dominated throughout their range by western hemlock, western red cedar, sitka spruce, Douglas fir and coast redwood. Extending across the southern two-thirds of the province from the lee slopes of the Coast Mountains to the Rockies are diverse mixed coniferous forests that stretch southward through the Cordillera to Mexico. Finally, across the northern part of the province is a belt of forests structurally and floristically akin to the vast boreal forests extending eastward to Labrador. To this forested vegetation mosaic must be added considerable areas of treeless alpine tundra and, in the central Fraser, lower Thompson, Similkameen, Nicola and Okanagan valleys, the dry sagebrush grasslands and open savannas which are northern extensions of the Palouse Prairie and Sonovan desert of the United States.

Each of these major realms contains many different vegetation types, reflecting complex, long-evolving and selective interactions between flora and environment during the past 12,000 to 15,000 years. The major influences governing these interactions are climatic and edaphic diversity which, in turn, are dependent on the topographic complexity and latitudinal-longitudinal extent of the province. Even a cursory glance at a vegetation map shows the obvious northwest-southeast zonation of vegetation types aligned parallel to climatic and topographic contrasts. At a more detailed level, differences in local climate, soil type and drainage impart finer-grained textures and differences to the vegetation cover.

These plant-environment systems are not merely static reflections of present climate, topography and soils. They are dynamic, responding both to internal successional processes inherent in the vegetation itself, and to externally imposed environmental changes that have influenced the availability and distribution of suitable habitats or migration routes.

The broad outlines of this extraordinarily complex vegetation cover can be examined from two viewpoints — the floristic and the ecological. The first of these provides a picture of the geographical distribution of taxonomic groups, particularly species. In contrast, the ecological viewpoint focuses on the discrete vegetation assemblages or communities in the landscape that have evolved from selective interactions between the flora and the environment.

FLORISTIC PATTERNS

Generally, the climatically mild southern areas of the province are floristically more diverse (having more species per unit area) than northern areas. Sixty percent of all tree species in British Columbia can be found in the coast-interior ecotone of the southwest corner of the province, and the south Kootenays are almost as rich.[1] At least four geographical "floristic elements" or groupings of species can be recognized within British Columbia. These elements reflect both the influence of species' origins with respect to more general floristic regions of North America, and the organization of them into distinctive spatial patterns within the province according to their general ecological requirements, availability of suitable habitats, migration and establishment, and competition with other species. Selected examples of these four elements are shown in Figure 1,5.

The boreal-transcontinental element (Figure 1a,5) consists of species that are widespread across North America in the boreal subarctic or temperate zones. In British Columbia they have a decidedly northern or eastern distribution. Some species, such as tamarack (*Larix laricina*), jack pine (*Pinus banksiana*) and many shrub willows, are restricted to the far northern or northeastern sections of the province. Others, such as white and black spruce (*Picea glauca, P. mariana*), extend farther south into the Interior Plateau.

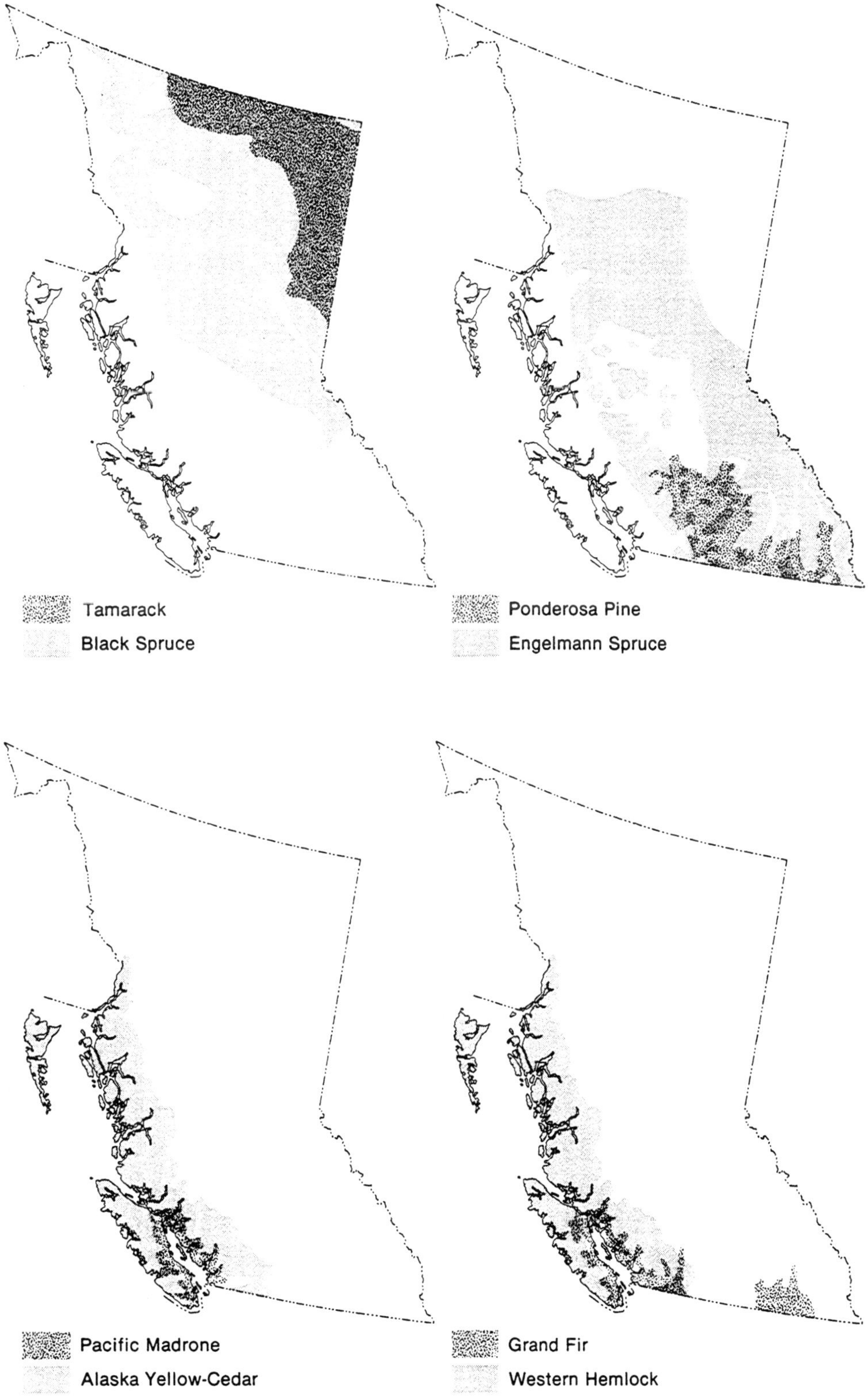

FIGURE 1,5 Selected examples of the floristic elements

Species centred in the mountain and valley complexes east of the coastal divide form a cordilleran element (Figure 1b,5). The distribution of some species in this group, such as Engelmann spruce (*Picea engelmanii*) and whitebark pine (*Pinus albicaulis*), overlap the southern edges of the boreal element. Others, such as ponderosa pine (*P. ponderosa*), western larch and subalpine larch (*L. occidentalis, L. lyallii*), are southern or south-eastern in distribution, and are widespread farther south in the cordilleran systems of the United States. Distribution patterns in this element are complex, due to the topographic diversity of the interior mountain and valley systems. Many shrub species, such as sumac (*Rhus glabra*) and poison ivy (*R. radicans*), sagebrush and some wormwoods (*Artemesia* spp.), which are limited to or centred in the drier interior regions, are also included in the cordilleran element.

Species restricted to lowland and mountain areas west of the coastal divide constitute the Pacific floristic element (Figure 1c,5), which contains two main sub-groups. First are those species, such as yellow cedar (*Chamaecyparis nootkatensis*), amabilis fir (*Abies amabilis*), sitka spruce (*P. sitchensis*) and red alder (*Alnus rubra*), which are widespread throughout the high precipitation zones of the coastal region. Second is a distinctly southern, or Gulf Islands element, including flowering dogwood (*Cornus nuttallii*), garry oak (*Quercus garryana*), arbutus (*Arbutus menziesii*), hairy manzanita (*Arctostaphylos columbiana*) and a number of flowering herbs, such as Easter Lily (*C. Erithronium oreganum*), which are restricted to the drier rain shadow areas of the southwest coast.

An important group of trees and shrubs shows a dual or disjunct distribution pattern, occurring in both the Pacific region and the wetter Columbia and western Rocky Mountain areas of the Cordillera. Characteristic of this Pacific-cordilleran element, in addition to those shown in Figure 1d,5, are western red cedar (*Thuja plicata*), Pacific yew (*Taxus brevifolia*), cascara (*Rhamnus purshiana*) and the appropriately named shrub, devil's club (*Oplopanax horridum*). Some of the flowering plants of the Gulf Islands coastal zone also show a disjunct distribution, reappearing in the drier areas of the southern Interior.

In contrast to these floristic groups are species which are more ubiquitous in their distribution. These include different races of lodgepole pine (*P. contorta*) which is the most widely distributed conifer in the province; subalpine fir (*A. lasiocarpa*) which occurs in boreal, as well as sub-alpine, forests; Douglas fir (*Pseudostuga menziesii*) which in its coastal and interior varieties extends from west to east across the southern third of the province; and trembling aspen (*Populus tremuloides*) and paper birch (*Betula papyrifera*) which, although of limited occurrence on the coast, are widespread throughout the rest of the province.

The actual range of a species is often less than its potential range, due to competition from other species or barriers to its migration. Thus, no true firs (*Abies*) have colonized the Queen Charlottes, although the islands are climatically suited to both amabilis fir and probably subalpine fir. Members of the genus *Populus* are also absent from the Queen Charlottes, apart from an isolated stand of black cottonwood (*Populus trichocarpa*). Conversely, some species may occur in small isolated areas, or refuges, outside their main ranges, such as madrone on the west coast of Vancouver Island at Kyuquot Sound. Isolated patches of Rocky Mountain juniper (*Juniperus scopulorum*) in northwestern British Columbia, and tamarack in the Prince George area, are hundreds of kilometres from their respective major ranges.

In many parts of the province, especially the more densely settled areas that have been occupied for many years, the native flora has been augmented by successfully established exotic species. Approximately one-third of the species on the Saanich Peninsula have been introduced, mainly from Europe; 36 percent of the grasses (family Gramineae), 60 percent of Cruciferae, 50 percent of Fabaceae and 44 percent of Asteraceae are of foreign origin.[2] The latter two families, in particular, include persistent and invasive weed species. Scotch broom (*Cytisus scoparius*) has run rampant on the coast, following its introduction in the 1850s, and is also established in the Kootenays. Because of its aggressive growth habits, efficient seed dispersal and nitrogen fixing ability, it is able to supplant native species in disturbed or dry sites, and is difficult to eradicate. In the southern Interior, two composites of the introduced genus *Centaurea*, or knapweeds (*C. nigra*, *C. repens*), have invaded thousands of hectares of grazing land, seriously reducing range land carrying capacity.

VEGETATION PATTERNS

Floristic maps, while giving an indication of what species may be found where, give no impression of the actual plant communities in the landscape or the qualitative nature of the vegetation cover of the province. Species are not distributed evenly within their ranges shown in Figure 1,5. Rather, within its range, each species is organized quantitatively in the landscape, according to its ecological requirements. This organization, in response to the environmental complex of topography, climate, soil, drainage, fire and competition, is multidirectional and multidimensional. It is not, however, random. Species with similar ecological requirements and the ability to co-exist through at least part of their life spans will tend to occupy similar habitats. Other groups of species will occupy different habitats. These species groups, which have evolved as fairly discrete units and which reappear in similar habitats, are the plant communities which collectively constitute the vegetation of a given area. Thus, alpine tundra vegetation is composed of a mosaic of communities variously dominated by species groups of lichens, mosses, grasses, heaths or broad leaved herbs. The forest vegetation of the Gulf Islands zone is characterized collectively by relatively dry and open woodland and savanna communities in which there are varying proportions of Douglas fir, lodgepole pine, arbutus, garry oak and colorful spring flowering forbs.

The pattern of communities in the landscape, and their collective expression as the vegetation, is, therefore, much more complex than that of individual species. The use of an hierarchical biogeoclimatic classification system, progressing from generalized categories to more detailed, is one way of understanding this complexity.[3]

At the most generalized level, the vegetation of the province can be divided into four formations (Table 1,5). Mesothermal (moderate temperature) forests grow in mild and sub-humid to very humid areas of the coast; microthermal (cold temperature) forests occupy a wide range of cold, humid, snowy winter climates, although summers may be warm and dry; treeless alpine tundra occupies high elevation areas where extreme climatic harshness precludes normal tree growth; and semi-arid, cold steppe occurs in cold winter, warm summer and low precipitation areas in which tree growth is limited to the most favorable sites.

Each of these formations can be further subdivided into regions, which more specifically define climatic types resulting from the longitudinal transition from maritime to continental, the latitudinal change from temperate to sub-boreal or boreal, and altitudinal influences (Table 1,5). The altitude of the tree line varies considerably across the province. It is about 2,250 metres (7,950 feet) in the continental southeast, 1,600 to 1,800 (5,650 to 6,355) in the maritime southwest, 1,400 (4,950) in the northeast and 1,000 (3,530) on the northern coast (Figure 2,5). Because of this variation the terms "sub-montane", "montane" and "sub-alpine" used to denote regions and forest vegetation types below the tree line encompass different altitudinal ranges in each of the regions.

Within each region are one or more biogeoclimatic zones (Table 1,5) which are distinguished by a combination of relatively uniform climate, soils, and vegetation. The communities which collectively make up the vegetation in any zone are dominated by a limited number of characteristic major species. Climax communities, dominated by self-perpetuating species and occupying "mesic" habitats intermediate in moisture and nutrient status, are assumed to reflect the overriding influence of climate, and characterize each zone.

Table 1,5 SIMPLIFIED CLASSIFICATION AND SELECTED CLIMATIC CHARACTERISTICS OF BIOGEOCLIMATIC UNITS

BIOGEOCLIMATIC UNITS			*CLIMATIC CHARACTERISTICS*		*EXTENT*
Formation	*Region*	*Zone*	*Accumulated Degree Days Above* 6°C	*Mean Annual Precipitation in Centimetres*	*Square Kilometres*
Mesothermal Forest	PACIFIC COAST TEMPERATE sub-montane/ montane: a. wetter b. drier	a. Coastal western hemlock	850-2150	73-665	56,000
		b. Coastal Douglas fir	1400-1900	65-174	8,500
Microthermal Forest	SOUTHERN CORDILLERAN TEMPERATE montane: a. wetter b. drier	a. Interior western hemlock	1030-1920	57-145	47,000
		b. Interior Douglas fir	890-1700	35-57	48,000
	COAST MOUNTAINS BOREAL sub-alpine	Mountain hemlock	250-900	178-440	42,000
	CORDILLERAN BOREAL sub-alpine: a. southern-central b. northern	a. Engelmann spruce/ sub-alpine fir	250-1000	40-220	131,500
		b. Spruce/willow/ birch	200-500	63-90	65,000
	HIGH LATITUDE BOREAL sub-montane/ montane: a. northern b. central	a. Boreal white and black spruce	550-1080	28-127	205,000
		b. Sub-boreal spruce	650-1040	45-62	170,000
Alpine	ALL AREAS ABOVE TREE LINE	Alpine tundra	0-250	70-350	150,000
Semi-arid Cold Steppe	SOUTHERN CORDILLERA SEMI-ARID/ TEMPERATE sub-montane	Ponderosa pine/ bunch grass	1250-2200	20-45	11,500

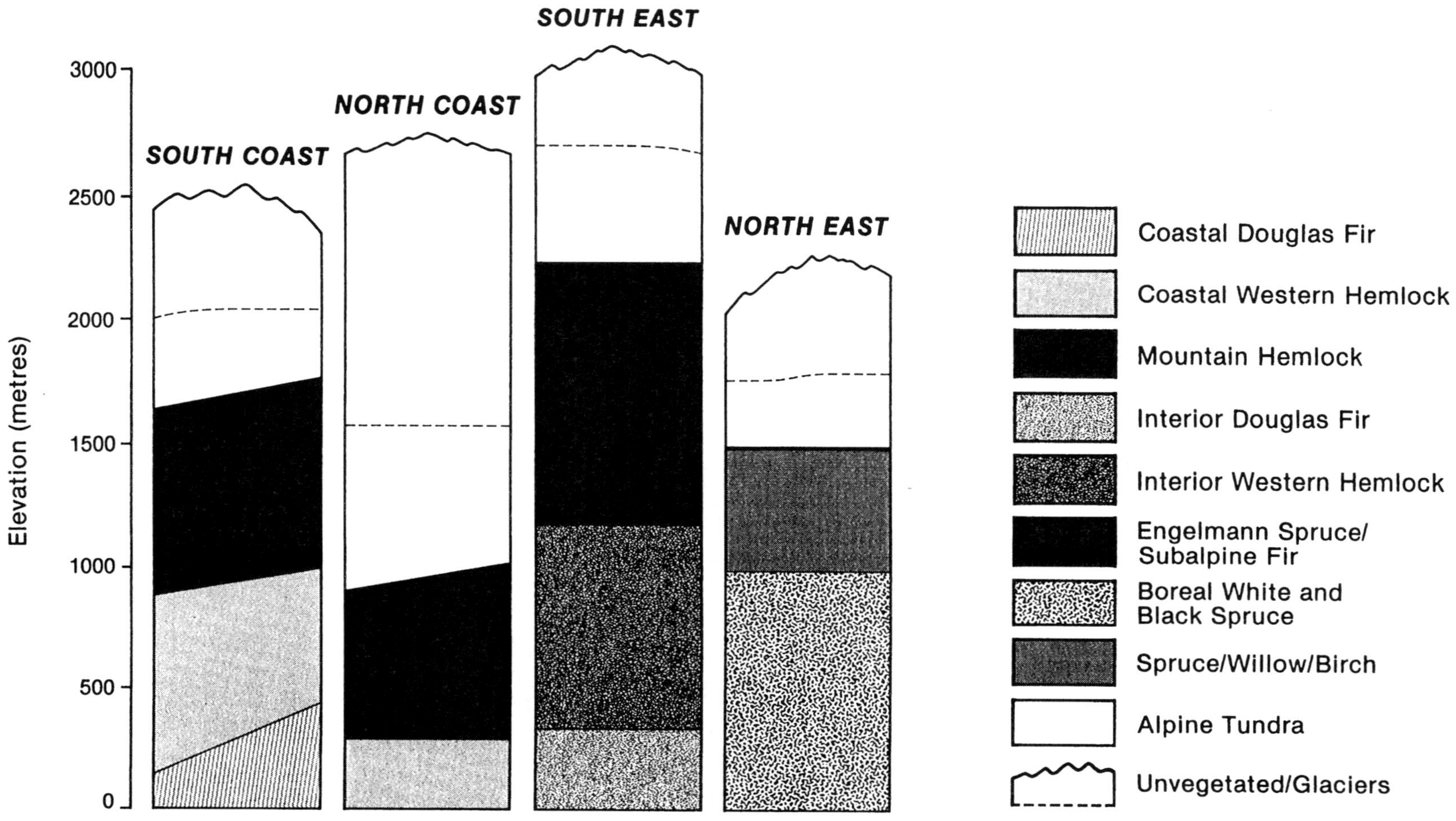

FIGURE 2,5 Variations in treelines and altitudinal forest zones

Biogeoclimatic zones provide the framework for the following overview of vegetation in the province (Figure 3,5). At the level of dominant tree species, the concept of the biogeoclimatic zone is a useful summary, but it does have some disadvantages. Communities of very wet (hygric) or dry (xeric) habitats are often as abundant as those of mesic ones, so that it is necessary to further subdivide each zone into smaller units to more truly reflect what is seen in the landscape. Perhaps more important in many areas of the province is the fact that climax communities are the exception rather than the rule. They have been replaced by "seral" or regrowth communities following the destruction of the original communities by agricultural clearing, logging, grazing, drainage and other human activities. These seral communities in, for instance, the coastal Douglas fir and ponderosa pine-bunch grass zones are more important landscape elements than the few remaining patches of climax vegetation.

Mesothermal Forests

These forests, the most luxuriant of which are the coastal rain forests, occupy sub-montane to montane elevations [0 to 1,000 metres (3,530 feet) in the south, 0 to 300 (1,060) in the north] on the Pacific slopes of the Coast and Insular mountains. Winters are mild and wet, and summers generally cool, although in the rain shadow of the Olympic and Insular mountains on the southwest coast summers are warm and dry. They contain two biogeoclimatic zones.

COASTAL WESTERN HEMLOCK ZONE

Occupying the wettest outer and more northerly parts of the coast, forests dominated by western hemlock also extend inland along the major valleys of the Coast Mountains. Although western hemlock is the characteristic species throughout the zone, western red cedar is often co-dominant. Douglas fir occurs as a seral species in the south, while amabilis fir becomes important farther north. A narrow coastal fringe on the west coast of Vancouver Island is characterized by sitka spruce, and this species becomes more important on the north coast and in the Queen Charlotte Islands. On extreme exposed coasts of northern Vancouver Island, the northern mainland and the western Queen Charlottes, stunted forests or woodlands of western hemlock, lodgepole pine and yellow cedar form complex mosaics with blanket bogs. Apart from the characteristic tree species, other features of these forests are a very dense shrub layer, consisting mainly of salal (*Gaultheria shalon*) and *Vaccinium* species, an abundance of bryophytes and a paucity of herbs. The zone contains the

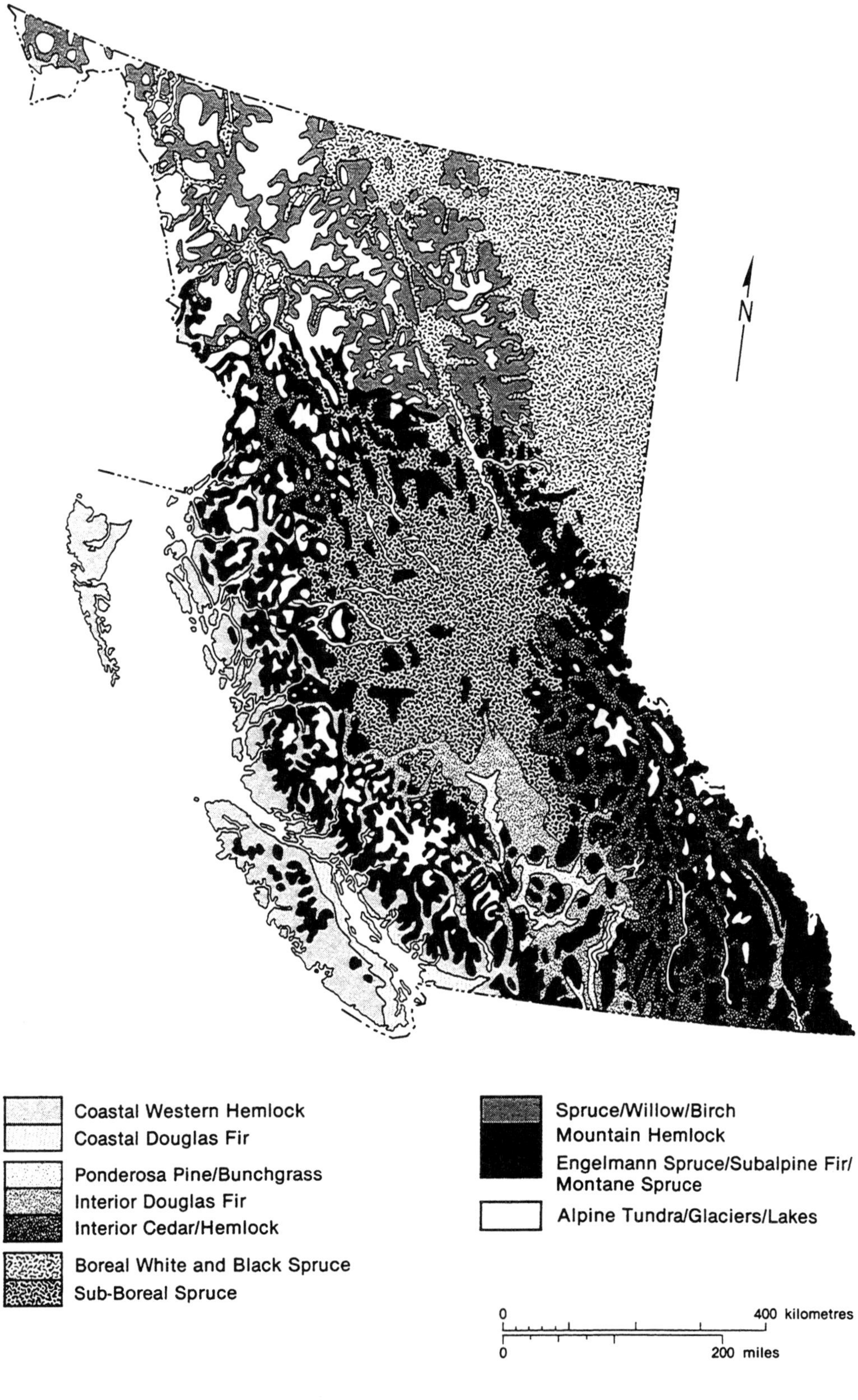

FIGURE 3,5 The generalized pattern of biogeoclimatic zones most productive forest land in Canada, and has been extensively logged in the last thirty years, especially on the south and central coast, so that seral (regrowth) stands, often dominated by red alder, are widespread.

COASTAL DOUGLAS FIR ZONE

Forests in which Douglas fir is, or was, a dominant constituent are limited on the coast to the rain shadow area of Georgia Strait, at elevations up to 450 metres (1,590 feet) on both eastern Vancouver Island and the Lower Mainland. Climatically, the zone is characterized as modified Mediterranean, with a well marked summer dry period of varying intensity. Although Douglas fir is the climax species, almost all the original forests have disappeared, as they, along with some drier forests of the western hemlock zone, were almost the only forests utilized by the forest industry up to the 1940s. A mosaic of regrowth communities occupies those areas that have not been permanently cleared for agriculture or urban development. Western red cedar, grand fir (*Abies grandis*), lodgepole pine, madrone, flowering dogwood and bigleaf maple (*Acer macrophyllum*) are variously mixed with Douglas fir, according to differences in soil type and successional stage. In the drier sub-zone centred on the Gulf Islands and southeast Vancouver Island, arbutus becomes more abundant and garry oak occurs in open woodlands or grassy savanna, with a rich and attractive flora of spring flowering forbes unique to British Columbia. These mixed, floristically rich, open forests and woodlands are among the most aesthetically pleasing and diverse in the province.

Microthermal Forests

Collectively, these forests are the most widespread, occupying the boreal latitudes, and montane to sub-alpine elevations of the Cordillera and Coast Mountains. All areas have cold, snowy winters and short, though sometimes dry and hot, summers. Because of the wide latitudinal and longitudinal range, and associated climatic, topographic and floristic diversity, microthermal forests are extremely varied. With the exception of the wetter types of the Columbia Mountains, they are structurally and floristically distinct from the coastal mesothermal forests.

MONTANE FORESTS

There are two zones of montane forests, both in the southern third of the province, at relatively low elevations of 300 to 1200 metres (1,060 to 4,240 feet). The two are climatically contrasted by differences in available moisture.

Interior western hemlock zone

Occupying the lower slopes of the Columbia Mountains and the lower, western slopes of the Rocky Mountains in the southeast of the province, these forests are sometimes called the interior wet belt. Winters are cool and wet, due to orographic effects on eastward moving air masses which still retain enough moisture to produce heavy precipitation. Although precipitation is lower than that on the coast, vegetation structure and floristics are very similar. Because quality and volume of timber in these forests are second only to those of the coastal stands, logging has removed many of the original climax communities of western hemlock and western red cedar. Spruces, subalpine fir, western larch, Douglas fir and western white pine (*P. monticola*) are widespread, both in the remaining climax and in many seral stands. The forests of that part of the zone centred in West Kootenay are among the most floristically diverse in the province, exceeded only by those of southeast Vancouver Island and the montane areas of the Lower Mainland.

Interior Douglas fir zone

Apart from an isolated area in the southern Rocky Mountain Trench, this zone essentially extends westward from the interior hemlock zone across the montane elevations of the southern Interior Plateau. It, therefore, lies in the rain shadow of the coastal mountains, and although temperatures suggest a long growing season, the dry summers often impose severe water stress. Like the interior hemlock zone, interior Douglas fir forests extend southward into Washington, Oregon, Idaho and Montana. Although the zone is named after the interior variety (var. *glauca*) of Douglas fir, which forms open to closed forests over large areas, forest types are complex, due to both evironmental and human influences. Trembling aspen and spruce are common, and ponderosa pine, lodgepole pine and western larch are frequent xeric or seral species in the southeast. Grand fir, hemlock and western white pine occur in the wetter areas. Almost pure seral stands of ponderosa and lodgepole pine frequently replace Douglas fir forests. In marked rain shadow areas, dry grasslands are common, and have been extended by fire and persistent grazing, especially in the southern Chilcotin.

SUB-ALPINE FORESTS

These forests occur between montane forests and the altitudinal tree line throughout British Columbia, both in the Pacific and cordilleran regions. Elevations range from 400 to 1,000 metres (1,400 to 3500 feet) in the northwest coastal mountains to 1,250 to 2,250 (4,400 to 7,950) in the dry, continental sub-alpine regions of the southeast.

Mountain hemlock zone

This is the mildest sub-alpine zone, restricted to the coastal mountains above 900 metres (3,175 feet) in the south and 400 (1,400) in the north. It is, however, characterized by very heavy snowfalls which last into spring and prevent the soil from freezing, but also cause a short growing season. The forests are dominated by mixtures of mountain hemlock, amabilis fir and yellow cedar, the latter especially on wetter soils. Subalpine fir and whitebark pine are present near timber line, and Douglas fir and western white pine intrude in southern regions or at lower elevations. Tree growth is retarded at the highest elevations, and closed forests give way to a parkland of trees in favored habitats interspersed with more open areas of alpine heath or meadow. Like the lower elevation coastal western hemlock zone, these forests are rich in mosses, but the shrub layer of salal is replaced by one dominated mainly by *Vacciniums*, *Rhododendrum albiflorum* and the mountain heathers, *Phyllodoce* and *Cassiope* spp. These shrubs become more abundant in the parkland zone transitional to alpine tundra where floristically rich and colorful herb meadows develop in moist and sheltered habitats. Although tree growth rates are low at the upper elevations, more accessible stands have been cleared, and some logging has extended to tree line.

Engelmann spruce-subalpine fir zone

This zone occupies an inverted U-shaped crescent in the southern two-thirds of the Cordillera, from the Rockies west to the eastern part of the Coast Mountains and north to the southern part of the Skeena and Omineca mountains. Altitudinally it lies between 1,250 and 2,250 metres (4,400 and 7,950 feet) in the southwest and 1,200 and 2,100 (4,250 and 7,400) in the north. The continental climate results in less protective snow cover than in the mountain hemlock zone, and soils may become frozen. In such a wide latitudinal range there are many tree species, although mixtures of Engelmann spruce and subalpine fir are characteristic. To the south the zone grades into the two montane forest zones and is characterized also by whitebark pine and alpine larch. In the north, Engelmann spruce is replaced by hybrid Engelmann-white spruce in a transition to boreal and sub-boreal forests. Throughout the zone, as in so many other forests, lodgepole pine is a dominant seral species. Parklands are important at the upper limits of tree growth, varying from grasslands on dry slopes, through heaths, to lush herb meadows in the more favored areas.

Spruce-willow-birch zone

Restricted to sub-alpine elevations of 1,000 to 1,700 metres (3,550 to 6,000 feet) north of 57 degrees, this zone is centred on the Cassiar, Skeena and Omineca mountains, extending west to the St. Elias Mountains and east to the northern Rockies. Due to a combination of altitude and latitude, it is climatically the most extreme of the forest zones. Over large upland areas, it is characterized by a scrub of willows and birch, with only scattered subalpine fir. At lower elevations, forests of white spruce and subalpine fir contain occasional black spruce, lodgepole pine and aspen. Some areas show the interesting phenomenon of an inverted tree line. Large valleys, into which cold air drains and accumulates, sometimes causing permafrost, are too harsh for tree growth and are vegetated only by grasslands, heaths or low shrubs. The lower slopes of the valleys, away from cold air drainage, are occupied by trees which, in turn, at the more exposed higher elevations, are replaced by low shrubs.

BOREAL FORESTS

Forests typical of the great trans-Canadian boreal forest formation are restricted in British Columbia to sub-montane and montane elevations [200 to 950 metres (700 to 3,350 feet)] north of 54 degrees. Two zones are recognized, one a true boreal zone, and the other transitional to more southern microthermal forests.

Boreal white and black spruce zone

This zone is centred on the Alberta Plateau region east of the Rocky Mountains. It occurs also in valleys to the west of the Rockies, but there it does not extend south of 57 degrees. Dominance by Arctic air masses produces a very short growing season, harsh winters, and generally poor soil development which is often associated with permafrost and/or poor drainage. Forests are dominated by white and black spruce, subalpine fir, lodgepole pine, tamarack, trembling aspen, balsam poplar and birch. Lower, poorly drained areas contain muskeg, tamarack (only northeast of the Rockies) and black spruce. True forests, in which white spruce and trembling aspen are the most important species, are limited to the better drained upper or alluvial habitats, where the species balance is largely influenced by fires, an integral part of the boreal environment. The generally poor growth and limited access to these forests means that large areas are still unlogged: aspen, however, represents an important future resource.

Sub-boreal spruce zone

Lying to the south of the true boreal forests, this zone extends from 52 to 57 degrees in the central Interior, and dominates much of the Nechako and Fraser plateaus. Climatically less severe than the white and black spruce zone, the area is transitional to the more southerly forest types and sub-alpine forests. White spruce, hybrid Engelmann-white spruce and subalpine fir are dominant. Seral stands of lodgepole pine and aspen are widespread, and lower, wetter areas are characterized by paper birch and cottonwood. Douglas fir becomes important in the south. In the northern Chilcotin district, climatic harshness and fire have produced extensive lodgepole pine forests. Currently, the forests in this zone are a crucial resource for the Interior forest industry, and are among the most heavily and destructively logged in the province.

Alpine Tundra

At elevations ranging from 1,000 metres (3,500 feet) on the north coast to 2,250 (7,950) in the southeastern mountains, increased climatic severity ultimately prevents tree growth, and the open parklands of the upper sub-alpine zone grade into alpine tundra. Stunted or "krummholz" trees sometimes extend into the lower elevations of the alpine tundra zone, but true tundra communities are dominated by low growing shrubs, herbs, grasses, mosses and lichens. Often they form complex mosaics in response to minor differences in topography, exposure, bedrock geology and length of snow cover. Only lichens or xeric mosses will grow in the most extreme habitats, and thin grasslands of *Festuca* and *Carex* species are frequent on dry, exposed slopes. Denser plant cover includes shrub communities dominated by dwarf willows, *Vaccinium*, *Cassiope*, *Phyllodoce* and *Dryas*, and in sheltered or moist areas, luxuriant herb meadows. Adaptions to extreme dessication, cold and physical abrasion include the prostrate growth of many dwarf shrubs and the well known cushion form of a wide range of different plants. Large areas are bare of plant cover and alpine ecosystems in general are fragile and prone to damage from the heavy recreational use that they often attract.

Semi-Arid Cold Steppe

This formation is limited to elevations to 250 to 750 metres (885 to 2,650 feet) in the valleys of the southern Interior. There, the combination of continentality and rain shadow effects produces the warmest and driest climate of the province, characterized by the ponderosa pine-bunch grass zone. Tree growth is marginal over much of the area, and consists of drought and fire resistant ponderosa pine, supplemented by Douglas fir and aspen on soils with better water supply. Lower and drier elevations were originally occupied by grassland or shrub steppe, in which bunch grass species of *Agropyron* and *Poa*, along with sagebrush (*Artemesia tridentata*), were dominant. Overgrazing has significantly altered the original communities, which are now often dominated by "increaser" or invading species, such as rabbit bush (*Chrysothamnus nauseosus*), pussytoes (*Antennaria* spp.), cactus (*Opuntia fragilis*) and *Centaurea* spp. At higher elevations, ponderosa pine forms open forest or woodland. Few areas remain in their natural state.

REFERENCES

1. KRAJINA, V.J., KLINKA, K., and WORRALL, J., *Distribution and Ecological Characteristics of Trees and Shrubs of British Columbia*. Vancouver: University of British Columbia, Faculty of Forestry, 1982.

2. SZCZAWINSKI, A.F., and HARRISON, A.S., *Flora of the Saanich Peninsula*. Victoria: Provincial Museum, Occasional Papers No. 16, 1972.

3. KRAJINA, V.J., "Ecology of Forest Trees in British Columbia", *Ecology of Western North America*, 2, 1969, 1-146.

PLATE 8 Log harvesting in the Interior. *B.C. Government Photo* ▶

6 FORESTRY

Michael C.R. Edgell
University of Victoria

INTRODUCTION

Forests occupy 520,000 square kilometres (200,720 square miles), or 55 percent of British Columbia's total area (Figure 1,6). Their role in the geography of the province extends far beyond mere landscape dominance, for they are integral to the province's social and environmental health and have been a pivotal factor in its economic development. During much of this century, but especially since the 1950s, British Columbia's growth has been attained largely by the continued liquidation of a standing stock of mature timber and its subsequent processing and export to continental and overseas markets in the form of various wood, pulp and paper products.

Although by 1945 the forest industry had grown to a significant size following its early coastal beginnings in the latter part of the nineteenth century, its arrival at its present status really dates from the 1950s, which saw the implementation of "sustained yield management" and the establishment of tenure and licence systems that opened up vast new forest areas to harvesting.[1] From 1950 to the mid-1970s, the industry operated in an optimistic and expansionist milieu, supported by the buoyant world economy and apparent assumptions that timber resources were virtually inexhaustible and expanding markets would always be available. During this period, the provincial government, viewing the forest resource as a ready source of revenue and a basis for economic stability and growth, encouraged expansion of the industry to the limits of the available timber supply.

The economic results and benefits of this expansionist period were enormous and the industry has long been credited with generating 50 percent of all economic activity in the province.[2] From 1964 to 1979, the combined forest industry (logging, wood processing, and pulp and paper production) directly accounted for an average of 10.8 percent of the provincial gross domestic product (GDP) in constant 1971 dollars, and 10 percent of the total work force.[3] Additionally, as a basic industry, forestry indirectly supported another 15 percent of the provincial work force in service industries.

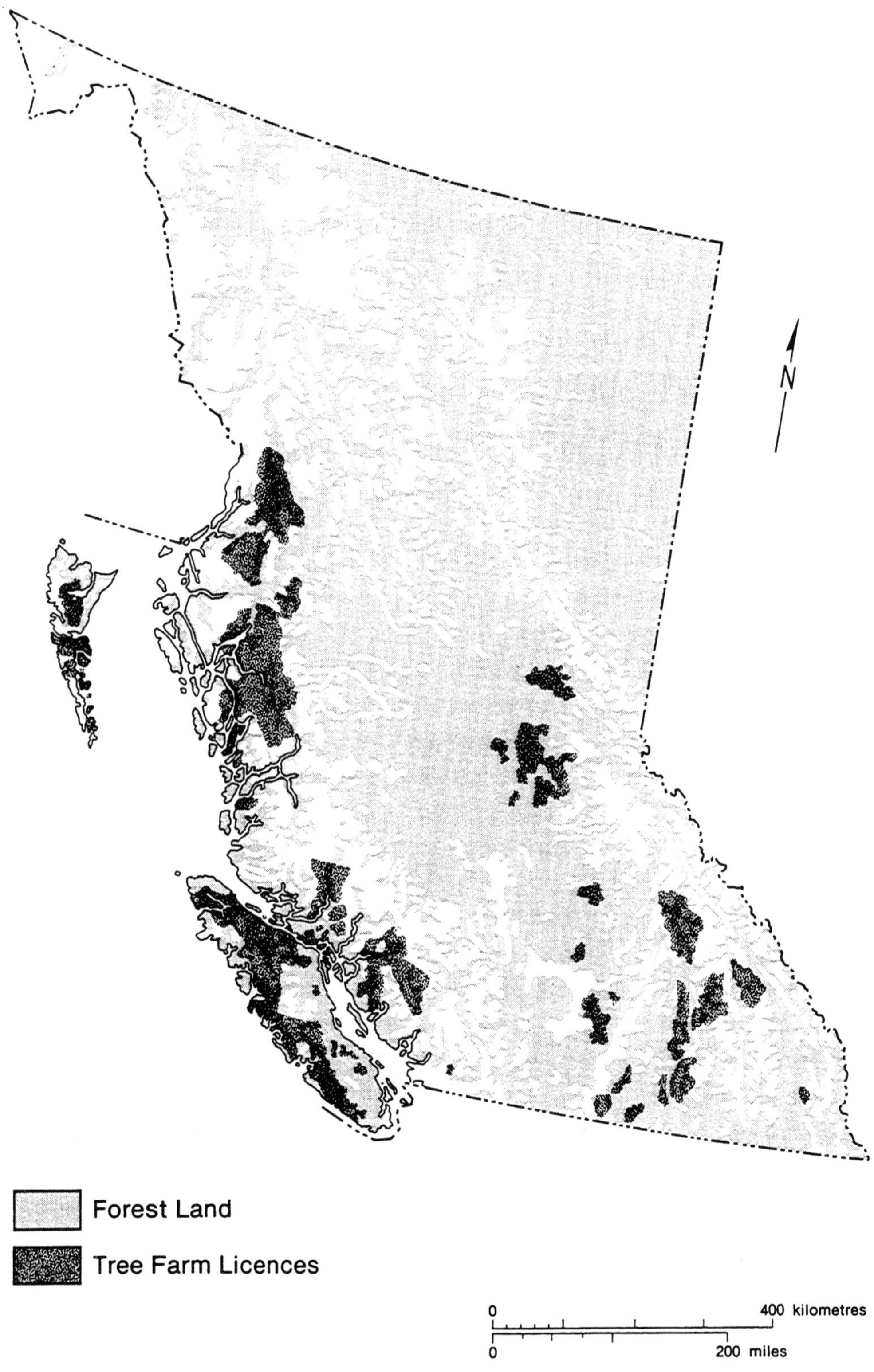

FIGURE 1,6 Forest land and tree farm licences

Thus, 25 percent of employment in British Columbia was generated by the forest industry. Logging accounted for an average of 34 percent of the GDP in the primary sector (agriculture, logging, fishing, and mining) and 40 percent of its employment. Wood processing and pulp and paper production have supported 45 percent of the manufacturing work force, and their share of the provincial manufacturing GDP increased from 39 to 46 percent between 1964 and 1979.[4] These impressive performances were made possible by the continuing and growing export of wood, paper and pulp products which accounted for approximately 60 percent of total provincial exports in the 1960s and 1970s — a crucial contribution to the resource-based and market-oriented economy.

This period of growth was temporarily reversed by the global economic crisis of 1975, compounded by provincial labor problems. But by 1979 the industry was experiencing record harvests, manufacturing activity and earnings. However, following hard on the heels of that record year, the two foundation stones of the industry — availability of markets and assured supplies of timber — were shaken to the core.

The global 1981-82 recession, high interest rates, and reduced demand for primary resources in traditional markets hit suddenly and hard. As the provincial Ministry of Finance aptly expressed the situation, "The buoyant economic conditions of the 1970s disguised fundamental changes that were taking place, and few were prepared for what was to follow".[5] With drastically reduced overseas markets, resource exports dwindled and activity in the forest industry, especially the logging and wood industry sectors, was sharply curtailed (Table 1,6). The contribution of the forest industry to the provincial GDP fell from 10.4 percent in 1979 to 8.9 in 1982; and its share of provincial employment dropped from 9.8 percent to 7.0.[6]

The repercussions that spread through the provincial economy were catastrophic, and as mills closed the foundations of many forestry-based communities evaporated. The post-recession performance of the industry has differed significantly from recoveries following previous periodic downturns. Characteristically, recovery has been led by the sawmill industry, responding to increased demand, mainly from the United States market. A resurgence in sawmill activity did take place in 1983 when provincial lumber production increased by 29 percent over the previous year. But 1984 production, although an all time record, was only 0.3 percent higher than that of 1983, due largely to a six percent decline in output from coastal mills which continued to be depressed during the gradually improving conditions of 1983-84. Subsequent 1985 production fluctuated widely from month to month.[7] Significantly, prices obtained for the 1984 record output were 13 percent lower than in 1983, a reflection of the continuing decline in all forest product export prices through the post-1982 recovery.

Table 1,6 VALUE OF SHIPMENTS AND EMPLOYMENT IN THE FOREST INDUSTRY, 1979 AND 1982

	1979		*1982*	
	Value of Shipments in Dollars	*Employment*	*Value of Shipments in Dollars*	*Employment*
Logging	3,175,170	24,474	1,933,445	16,429
Wood Industries	4,840,706	51,369	3,505,826	40,230
Paper and Allied Industries	2,532,492	20,998	2,702,269	18,875
TOTAL	10,553,368	96,841	8,141,540	75,534

SOURCE: *Canadian Forest Statistics*. Ottawa: Statistics Canada, 1979 and 1982.

There is a generally emerging consensus that this trend "may not be . . . an aberration from a longer-term historical perspective . . .",[8] and that the decline from the giddy heights of the 1970s is permanent. Increased protectionist sentiments directed against imports of cheap "subsidized" British Columbia lumber into the United States add another dimension to the uncertainty of that traditional market.

With an overwhelming emphasis on harvesting and processing, and a disproportionately small effort directed toward forest management and basic reforestation, the Ministry of Forests and the forest industry appear to have operated almost on blind faith that the resource would always be there. The long established policy of harvesting only the best and most accessible oldgrowth timber made economic sense to the industry, and was seen by the Forest Service as an essential step in creating a "normal" productive regrowth forest mosaic from an unproductive mature or overmature natural forest. By the early 1970s, however, concern was being expressed about future timber supplies, especially on the coast where many remaining oldgrowth stands were of lower quality than those that had been previously harvested, and were becoming geographically and economically difficult of access. Reforestation was failing to keep pace with extraction and few of the established regrowth stands were seen as a viable immediate resource. Where would the timber come from to support current levels of industrial activity once the oldgrowth had gone? Although about 55 percent of available timber in the early 1980s was still unlogged and mature, the Ministry of Forests in two major reports argued that, under current harvesting, utilization and management standards, future timber supplies in remaining oldgrowth and lower volume regrowth stands throughout much of the province would be inadequate to support established levels of industrial activity.[9]

By 1983, descriptions of forestry as a "sunset industry" and financial liability were surfacing at cabinet level, perhaps not surprisingly in light of the previous revenues and unquestioned support of the provincial economy to which politicians had become accustomed. Premature and inappropriate to the tasks at hand, such perceptions do reflect the seriousness of real changes in market availability and apparent changes in wood supply. In whatever manner the forest industry and government forest managers adjust to those changes, the transition will undoubtedly alter the structure of the industry itself and the role that it plays in the provincial economy.

RESOURCE CONTROL AND TENURE

Three percent of British Columbia's forest is privately owned, one percent is federal Crown land, and the remaining 96 percent is provincial Crown land. The small area of private land — mainly Crown grants in the Georgia Strait region — provided the security of ownership and high quality timber necessary for the early establishment and development of the forest industry. Supplemented by timber from various harvesting rights on Crown land, especially early timber and pulp leases and licences now collectively called Timber Licences and later Timber Sale Licences (TSLs), private timber was the mainstay of the coastal forest industry until the 1940s. But the subsequent emergence of the modern forest industry, and its extension from the coast to the Interior, was made possible only by access to previously unallocated and untouched "public" timber on large areas of provincial Crown land.

That access was effected by the implementation, in the 1947 amendments to the 1912 Forest Act, of recommendations put forward by the Sloan Royal Commission of 1945.[10] The government established two major tenure and licence mechanisms — Tree Farm Licences (TFLs) and Public Sustained Yield Units (PSYUs) — in order to: (1) ensure industrial access to guaranteed long-term timber supplies; (2) stimulate capital investment in processing plants and therefore assure economic stability and development and; (3) bring forests under "sustained yield management". The combined area encompassed by TFLs and PSYUs now totals approximately 889,000 square kilometres (343,150 square miles), or about 93 percent of the province's area. Only about 500,000 square kilometres (193,000 square miles) are actually covered by forests, and of this area about 85 percent is occupied by "productive forest" capable of growing a commercial crop in a reasonable length of time.

TFLs were granted mainly to large companies with existing private holdings or old temporary tenures (now Timber Licences) on Crown land. To these "Schedule A" lands were added large areas of Crown land (Schedule B) over which companies were granted exclusive harvesting rights in 21 year renewable licences. Companies are directly responsible for management, restoration and protection of the forest resource in TFLs, and for developing inventories, annual allowable cuts (AACs) and harvesting schedules under sustained yield guidelines monitored and approved by the Forest Service. Forty-one TFLs had been granted by 1966, mostly on the coast where the industry was well established and timber values were highest. A few were also established in the better timber areas of the interior wet forests in the southeastern part of the province (Figure 1,6). The original number was consolidated to 32 by the 1980s, occupying about 10 percent of the province's productive forest land, but containing approximately 17 percent of the mature timber volume.

PSYUs were established on the remaining 91 percent of the forest land. Also guided by sustained yield principles, they were, however, managed directly by the Forest Service, which had the responsibility for inventory and the setting of AACs and harvest levels. Timber in each PSYU was committed to a number of companies in a complex of licence and harvest arrangements, the most important of which was the Timber Sale Harvest Licence (TSHL). Many companies with TFLs were successful in acquiring TSHLs. Fifty-eight PSYUs had been established by 1957. Along with Pulpwood Harvesting Areas and later additional PSYUs, they were the basis of interior logging and sawmill expansion after 1950, and of the establishment of the interior pulp and paper industry during the 1960s.

These tenure and licence arrangements were both consolidated and given new direction by the passage of the 1979 Forest Act and Ministry of Forests Act, and by subsequent amendments. TFLs were retained and their renewal period extended to 25 years. The 85 PSYUs were rationalized into 33 Timber Supply Areas (TSAs),[11] and the new Forest Licence (FL) was introduced to phase out the previous TSLs and TSHLs. Soon after the passing of the 1979 acts the industry slid into the 1981 depression and, with the fall in resource revenues between 1980 and 1984, Ministry of Forests expenditures fell by 23 percent in constant dollars. Both the size and activities of the Forest Service were affected and in the same period two further developments of great significance occurred.

First, under the provisions of a 1982 amendment to the Forest Act, holders of FLs, TSLs and TSHLs could apply for them to be converted to TFLs. By 1984, two applications had been approved and about 40 were being considered. In the next few years, therefore, it is likely that large areas of TSAs will be converted to Tree Farm Licences and that the already strong role of industry in direct management and control of the public forest resource will increase significantly. Second, and of equal importance, was the establishment of a "Forest Management Partnership Agreement" (now called the "Subsidiary Agreement" which had the stated aim of reducing the size of the Ministry of Forests and increasing the role of the private sector in order to ". . . enhance and support the economic recovery of the forest sector to the greatest possible degree."[12]

The government, as forest landowner, is gradually relinquishing much of its direct managerial and economic responsibility for the public forests in existing and possible new TFLs, and in a move toward "privatization" is allocating much of that responsibility to its industrial tenants. This move has been welcomed by the industry, for it is assured an increased control of what happens to the forest resources on public land. But it has not yet been clarified how, or if, with the decreased responsibilities and abilities of the Forest Service, the performance of industry in meeting acceptable social, environmental and silvicultural objectives will be monitored.

RESOURCE BASE

Extent and Types of Forest

Only 432,646 square kilometres (167,000 square miles), or 48 percent of the total TSA and TFL area, are potentially capable of producing commercial timber crops. However, not all of this "gross forest land base" is actually suitable or available for the production of commercial timber. The Ministry excludes approximately 147,139 square kilometres (56,790 square miles) which are deemed inaccessible or occupied by non-commercial species; 35,603 (13,740) which are not stocked with usable timber because of fire and logging, the so called "not satisfactorily restocked" or NSR; and 36,799 (13,818) which it is estimated will be withdrawn for non-forest uses in the 20 years following 1984. The remaining 212,905 (82,180), or less than one-quarter of the province's area, is available and suitable for timber production and is classed as the "net forest land base".[13] It is assumed by the Ministry that this net forest land, together with the small area of privately owned land, will support the future forest industry. All predictions about available timber volumes, AACs and harvests are based on this net, rather than gross, forest land base.

Figure 2,6 indicates, within forest regions, the varied composition of mature forests on the gross land base. The distinction between coastal and interior forests should be emphasized, for not only is it based on contrasting

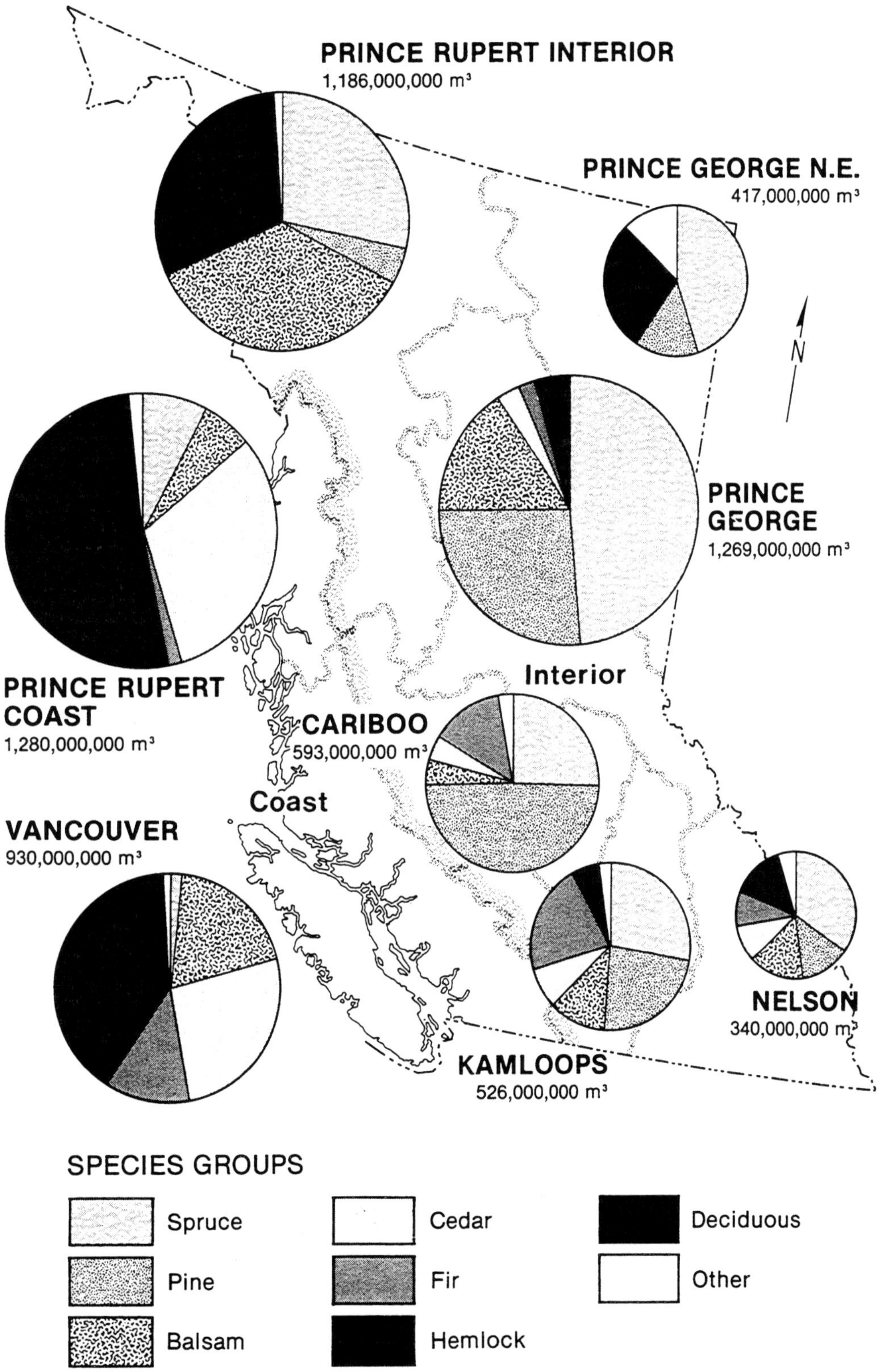

FIGURE 2,6 Forest regions and composition of mature forests characteristics of the forests themselves, but it also reflects differences in historical development and structure of the two major geographical components of the forest industry.

Although occupying only 15 percent of the gross forest land, coastal forests contain 33 percent of the province's mature timber volume. They are dominated mainly by western hemlock, western red cedar, true fir ("balsam") and, in the Queen Charlotte Islands, also by sitka spruce. Douglas fir, the species on which the early development of the industry was founded, has a relatively limited extent. Because coastal forests supported the industry well into the 1950s, the better quality and more accessible stands have been logged and replaced by secondary regrowth stands. Site qualities and annual growth rates in these regrowth stands, especially at lower elevations, can be high. Total forest biomass and size of trees can be impressive in some of the remaining mature and oldgrowth stands, but large areas contain low quality timber and occupy steep or inaccessible terrain.

Interior forests, occupying 85 percent of the gross forest land, account for only 67 percent of the mature timber volume. They contain a great variety of types, encompassing northern plateau boreal spruce-deciduous forests, a wide latitudinal range of mixed subalpine forests, semi-arid open pine-fir parkland, and moist Columbia or "interior wet belt" forests of the higher precipitation areas in the Selkirk, Monashee and Purcell mountains. For the interior forests as a whole, the major species are white, black and Engelmann spruce and lodgepole pine. Balsam, Douglas fir and yellow pine are of lesser importance, while western hemlock and western red cedar are limited to the wet Columbia forests. Two deciduous species, aspen and poplar, are widespread in many forest types, especially those of the Prince George northeast district. Site productivity and growth rates are lower than in coastal forests, and trees are smaller. Interior forests have been extensively logged only since the 1950s and supported most of the rapid growth of the forest industry in the 1960s and 1970s.

Timber Supplies

In 1984, the total volume of growing timber on the gross land base was estimated at 8.79 billion cubic metres (310.42 billion cubic feet), and on the net land base at 5.49 billion (193.88 billion).[14] Ninety-five percent of this volume is softwood, dominated by hemlock, spruce, balsam and lodgepole pine. Aspen is the most important species in the remaining five percent of hardwood volume. During the 100 years from 1983 to 2083, the volume of timber on the net land base is projected to decline by a provincial average of 44 percent (Figure 3,6), with declines slightly higher on the coast than

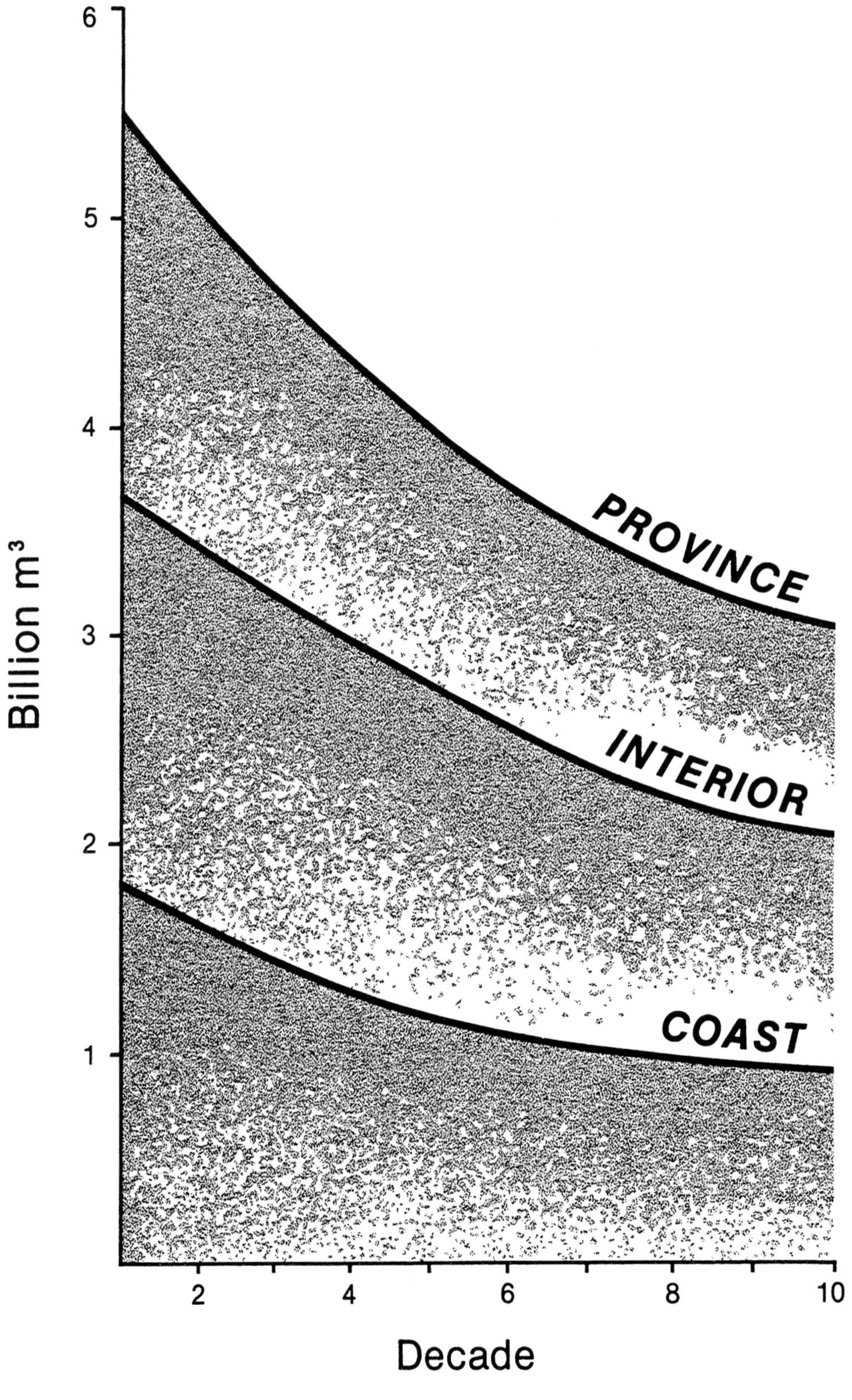

FIGURE 3,6 Volume of timber growing stock on the net forest land base, 1983-2083

those in the interior. These declines will result from the continuing priority given in harvesting schedules to the removal of oldgrowth forests. As the mature and oldgrowth stands disappear, they are replaced by lower volume regrowth stands or NSR.

By 2083, if current harvest and management practices continue, it is expected that only 16 percent of forests on the net forest land will be mature (over 120 years old), compared with 45 percent now.[15] The area of oldgrowth timber (more than 250 years old) will decline to about one-quarter of its 1983 extent, although large areas will remain on the "unavailable" land base outside of the net forest land. Declining timber volumes will be paralleled by declining availability of some of the more desirable commercial species. On the coast, volumes of western hemlock, red cedar and sitka spruce will decline rapidly in the next 50 years, although Douglas fir will increase slightly, due to the extensive area planted to fir in the last 40 years. Volumes of spruce will decline sharply in the Interior.

In response to these declines, the annual allowable cut and actual harvest rates will also fall, although not necessarily by the same magnitude. Preliminary yield analysis of TSAs has led the Forest Service to predict that over the next 100 years "future harvestable timber volumes will decline by at least 10 percent as the oldgrowth stock is removed."[16]

Harvest Levels and Annual Allowable Cut

Following the establishment of TFLs and PSYUs, the provincial harvest steadily increased from 22 million cubic metres (777 million cubic feet) in 1950 to 54.7 million (1.932 billion) in 1970. This upward trend in harvest levels was interrupted briefly in 1975 (Figure 4,6) and on a more prolonged basis by the recession from 1980 to 1983 when the provincial harvest fell by 20 percent and the coastal portion of the harvest suffered an even greater 32 percent drop. In 1983-84 the harvest rebounded to exceed the previous 1979 record, but this resurgence will not continue, due to constraints on timber availability. About 90 percent of the annual harvests through the 1970s came from regulated cutting in TSAs and Schedule A and B lands of TFLs, and the rest came from unregulated cutting on private land.

The coastal share of the harvest dropped from 63 percent in 1970 to 37 percent in 1983 and was exceeded by interior production in 1972 (Figure 4,6). Thirty-three percent of the total harvest, however, still comes from the Vancouver forest region. Coastal logging is mainly accomplished by clearcut high lead (cable) systems, with trees cut by chainsaws. The harvest is dominated by hemlock and balsam ("hembal mix") which are the major resource of the sawmills and pulp and paper mills, and by cedar. Sitka spruce is harvested almost exclusively from the Queen Charlotte Islands.

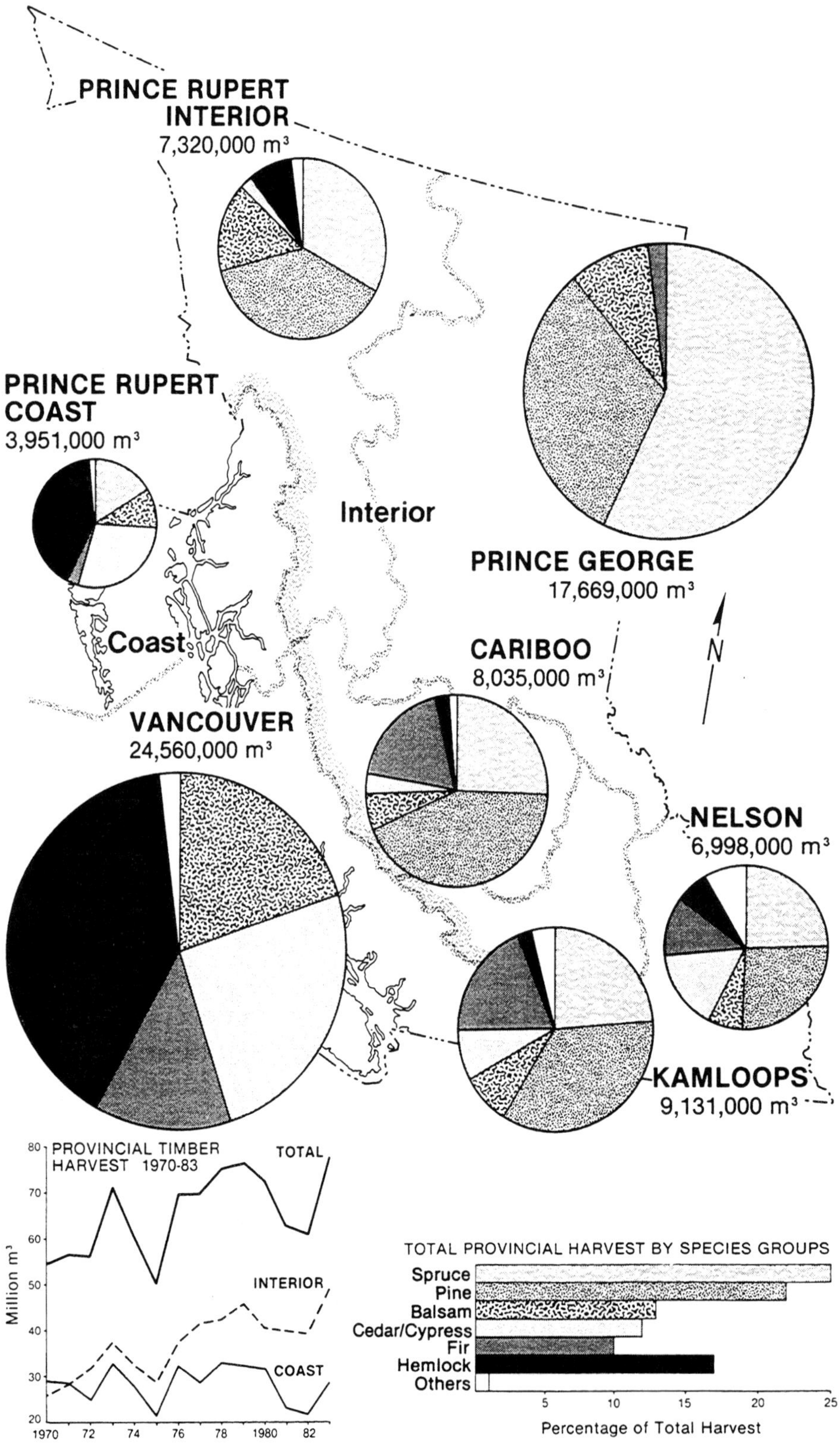

FIGURE 4,6 Total harvest by forest regions in 1983

Douglas fir now accounts for a small proportion of the harvest. Interior logging is also accomplished mainly by clearcutting, but is highly mechanized due to easier topographical conditions than those on the coast, and 50 percent of falling is done by mechanical shears. The harvest is dominated by spruces and lodgepole pine, while "coastal" species are locally important in the interior wet belt.

Annual and longer term average harvest rates on Crown land are regulated by the AAC. Established by the Ministry, the AAC is a volume of timber that can be harvested consistent with the "orderly liquidation" of mature timber and a sustained yield. It is initially established from estimates of standing mature timber volumes, length of the rotation period, and annual growth increments in immature stands. The volume thus obtained is then adjusted to allow for such factors as land alienated to non-forest use, regeneration delays, expected fire and disease losses, or losses during harvesting. Because of the high volume of timber in mature and oldgrowth stands the AAC has been set at high levels. In 1972 it was estimated at just over 99 million cubic metres (3.496 billion cubic feet) on Crown land and the small area of private forest land, although not all of this AAC was committed under tenure. This amount was considerably more than the 1972 harvest of 56.4 million cubic metres (1.992 billion cubic feet). During the 1970s, however, improved forest inventories resulted in lowered estimates of usable timber volumes. In conjunction with the failure of reforestation programs to keep pace with rate of extraction, reductions in the available forest land base, and greater than expected timber losses to fire and disease, these lowered estimates were instrumental in reducing the foreseeable amount of harvestable timber. By 1983, the AAC on Crown lands had been reduced to 68 million cubic metres (2.401 billion cubic feet), with an additional 3 million (106 million) estimated for regulated private land. In that year the harvest on Crown lands was 71.2 million (2.514 billion), and on regulated and unregulated private lands was 6.5 million (230 million).

The total regulated harvest, therefore, now equals the provincial AAC. But due to the declining volume of growing stock, the AAC will be further reduced if current management and harvesting policies continue. In 1980 the Ministry of Forests predicted a possible fall of the AAC to a long run sustainable yield of about 50 million cubic metres (1.766 billion cubic feet), or roughly two-thirds of its 1980 level.[17] Apart from any other impacts, such a decline would reduce direct employment in logging and wood processing by over 30,000, with an equal reduction in related service employment. More recent Ministry predictions, however, are for a decline of 13 percent in both AAC and harvest levels over the next 100 years (Table 2,6).

Projected declines are highest on the coast (26 percent for AAC and 24 percent for harvest) where the supply situation is also exacerbated by

Table 2.6 PROJECTED AAC AND HARVEST FROM REGULATED FOREST LAND*
(Volume in thousands of cubic metres)

		1983	*2003*	*2033*	*2083*
Interior:	AAC	41,292	41,288	40,068	39,433
	Harvest**	44,129	43,481	42,196	41,528
Coast:	AAC	26,754	24,957	21,274	19,605
	Harvest	27,057	26,283	22,404	20,446
Provincial Total:	AAC	68,046	66,245	61,342	59,038
	Harvest	71,180	69,764	64,600	62,174

*Excludes the harvest from federal land, private land and Indian reserves, (totalling about 6.5 million cubic metres or 230 million cubic feet), but includes the harvest from private lands in TFLs.

**Harvest amounts include unsalvaged timber losses, while AAC represents the net harvest.

SOURCE: British Columbia, *Forest and Range Resource Analysis, op. cit.*, reference 9, Table D-4 (revised).

the fact that, as logging moves into marginal stands, both quality and accessibility of timber decline. The already high costs of delivering the harvest to the mills escalate sharply as attempts are made to harvest the last 20 percent of the supply, and as much as 10 to 15 percent of "available" timber is inaccessible under present and foreseeable technological and economic conditions. In contrast, the interior harvest and AAC are projected to decline overall by only five percent, and in the Cariboo and Prince George NE regions will stay virtually constant. There are accessibility and quality problems in the Interior, however, and the cut from lower quality stands, especially of small diameter lodgepole pine, will increase from 20 to 33 percent of the harvest, while the percentage of spruce will decline from 25 to 11 percent.

A more pessimistic scenario is presented in a report prepared by the Forest and Policy Analysis Project.[18] This report generally agrees with Ministry predictions of an AAC of 67 to 75 million cubic metres (2.366 to 2.649 billion cubic feet) in the year 2000. But it suggests that if current trends in timber quality, accessibility, delivered wood costs and selling prices continue, one-third of the interior AAC will be too expensive to harvest. The economically viable provincial harvest may be as low as 40 to 44 million cubic metres (1.413 to 1.554 billion cubic feet), which is only 51 to 56 percent of the 1983 harvest.

From the often conflicting data and interpretations that are aired in the current debate on the future of forestry in British Columbia, all indications are that the physical limits to oldgrowth and mature timber supplies are now in sight and that the volume of timber available for harvest will decline. It is also clear that without an improved level of forest renewal and silvicultural management of regrowth stands, and in the absence of increased selling prices, there is no possibility of improving on future timber supply predictions. The changing nature of the resource base dictates that the past expansion and present level of activity in the forest industry cannot be maintained into the future.

TIMBER PROCESSING

The processing of timber is the single most important group of manufacturing activities in British Columbia. Even in the depressed year of 1982, processed timber exports were worth 4.2 billion dollars, exceeding the value of mineral exports by almost 1.9 billion dollars.[19] Timber processing is divided into the wood industries and the paper and allied industries, which are dominated, respectively, by sawmills and pulp and paper mills (Table 3,6).

Wood Industries

The wood industries generate a wide range of products for a variety of markets. Sawmills and planing mills have been the pivotal sector of the wood industries, and indeed the mainspring of the forest industry as a whole. Because they utilize about 80 percent of the log harvest trends in sawmill production and harvest levels are closely linked (Figures 4,6 and 5,6). They also supply 90 percent of pulp and paper raw material in the form of wood chips. In 1984 there were 182 sawmills and planing mills operating in British Columbia (Figure 5,6) with a total estimated annual capacity of 35 million cubic metres (1.236 billion cubic feet).[20] Half of this total was contained in the 45 mills with individual annual capacities of over 230,600 cubic metres (8,143,639 cubic feet). These large mills are concentrated in the Vancouver and Prince George forest regions. The trend throughout the province, especially since the late 1960s, has been toward fewer and larger mills.[21] For instance, in 1960 there were 401 mills in the Prince George region with an annual production of 1.45 million cubic metres (51.21 million cubic feet); by 1977 the number had dropped to 110, while production had increased to 6.11 million (215.77 million).[22]

Table 3,6 DISTRIBUTION OF VALUE ADDED, EMPLOYMENT AND NUMBER OF PLANTS IN THE FOREST PRODUCTS INDUSTRY, 1982

	Value Added		*Employment*		
	thousands of dollars	*percent*	*number*	*percent*	*Number of plants*
WOOD INDUSTRIES					
Sawmills and planing mills	871,987	38.3	28,943	48.9	320
Veneer and plywood mills	152,439	6.6	5,887	10.0	29
Shingle mills	33,631	1.5	1,114	1.9	86
Sash, door and millwork	102,109	4.5	3,122	5.3	116
Other wood industries	68,605	3.0	1,164	2.0	113
Subtotal	1,202,780	53.9	40,230	68.1	730
PAPER AND ALLIED INDUSTRIES					
Pulp and paper mills	966,209	42.5	16,779	28.4	23
Other paper and allied industries	84,902	3.6	2,096	3.5	58
Subtotal	1,069,249	46.1	18,875	31.9	71
TOTAL FOREST PRODUCTS INDUSTRY	2,272,029	100.0	59,105	100.0	801

SOURCE: Canada, *Manufacturing Industries of Canada: National and Provincial Areas*. Ottawa: Statistics Canada, 1982.

Forty-two percent of rated capacity is contained in the 75 mills located in the coastal division of the industry (Vancouver and Prince Rupert regions). However, in 1983, coastal sawmills actually produced only 29 percent of the provincial lumber total, a decline from 49 percent in 1970. The coastal division has long been oriented to overseas markets, which receive 40 percent of its output. It is characterized by a number of old "large log" mills that have been partly supplanted or supplemented with newer mills designed to handle the smaller logs that have become more common as the nature of the resource supply has changed. The coastal sawmill industry in general faces serious economic problems associated with the operation of obsolete or inefficient mills. In addition, the costs of delivering logs to the mills, composed of logging, transportation, stumpage and overheads, rose by 300 percent from 1971 to 1982 and are still rising.[23] These costs represent 52 percent of the total cost of lumber production.

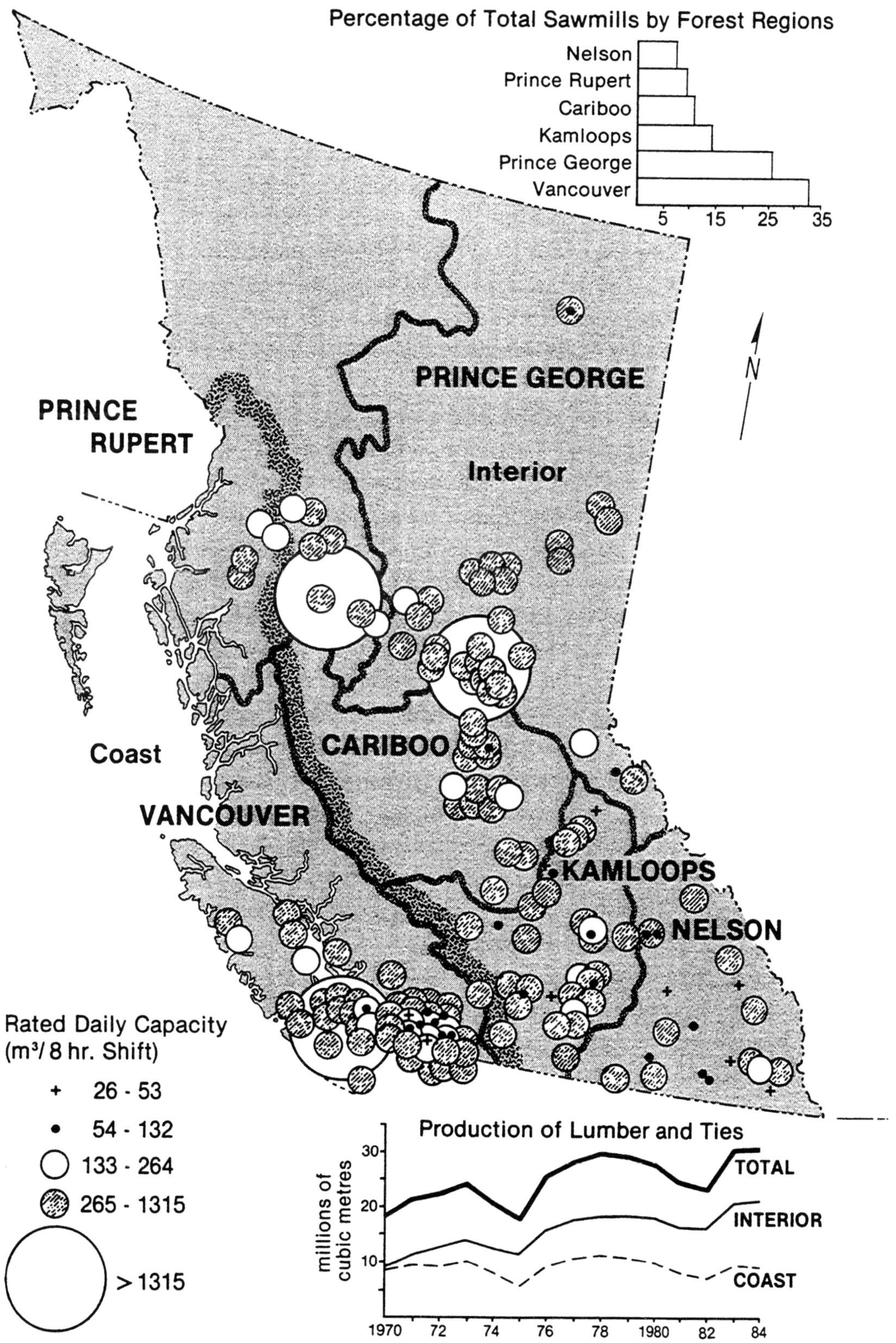

FIGURE 5,6 Location of sawmills by production volume

The interior sawmill industry developed mainly after 1950, subsequent to the establishment of PSYUs, and is dominated by the production of dimension lumber for the domestic and United States markets. Only five percent of the output is destined to overseas markets. Many of the sawmills were built partly to supply chips to the interior pulp mills established in Pulpwood Harvesting Areas and, generally, they are designed to utilize smaller logs than their coastal counterparts. They are also more mechanized and efficient in their use of labor and raw materials. Delivered wood costs rose by only 85 percent from 1970 to 1984, and the interior share of provincial lumber production increased from 51 to 71 percent during that period, although by 1984 interior mills accounted for only 58 percent of provincial rated capacity.

The production of plywood and veneer is a declining industry in British Columbia, and the coastal section, in particular, is considered to be in a terminal stage of decline. Plagued by rising costs, decreasing supplies of high quality peeler logs, and competition from alternative panelboard products, six mills representing about 15 percent of the total 1982 rated capacity closed between 1981 and 1984. Twenty-five mills were operating in 1984 with a total annual rated capacity of three million cubic metres (106 million cubic feet) (Figure 6,6), of which 1.84 million (64.98 million) was plywood capacity and the remainder veneer. Actual plywood production in 1984 was 1.76 million cubic metres (62.15 million cubic feet), slightly less than capacity. Unlike the situation in the lumber and pulp and paper sectors, 84 percent of the plywood market is domestic and there is also severe competition from other Canadian sources, especially from waferboard (oriented strand board), which is produced from aspen at about 70 percent of the cost of plywood. Only four percent of the British Columbia mill capacity is devoted to cheaper panelboards of this type, and this situation is unlikely to change significantly.

Shakes and shingles, although the least valuable of the basic wood products in terms of value added, now exceed plywood in export earnings. Mills are typically small scale operations — often on a "cottage" scale — and in 1984, only nine out of 80 companies had daily rated capacities in excess of 100 squares. Almost 90 percent of the annual production capacity of 5.8 million squares is concentrated in the Vancouver Forest District, chiefly on the Lower Mainland. The remaining capacity is located in the interior wet forests of the Kamloops and Nelson forest districts.

Paper and Allied Industries

There was a small pulp and paper industry in British Columbia before the 1940s, but the large scale industry developed on the coast from the

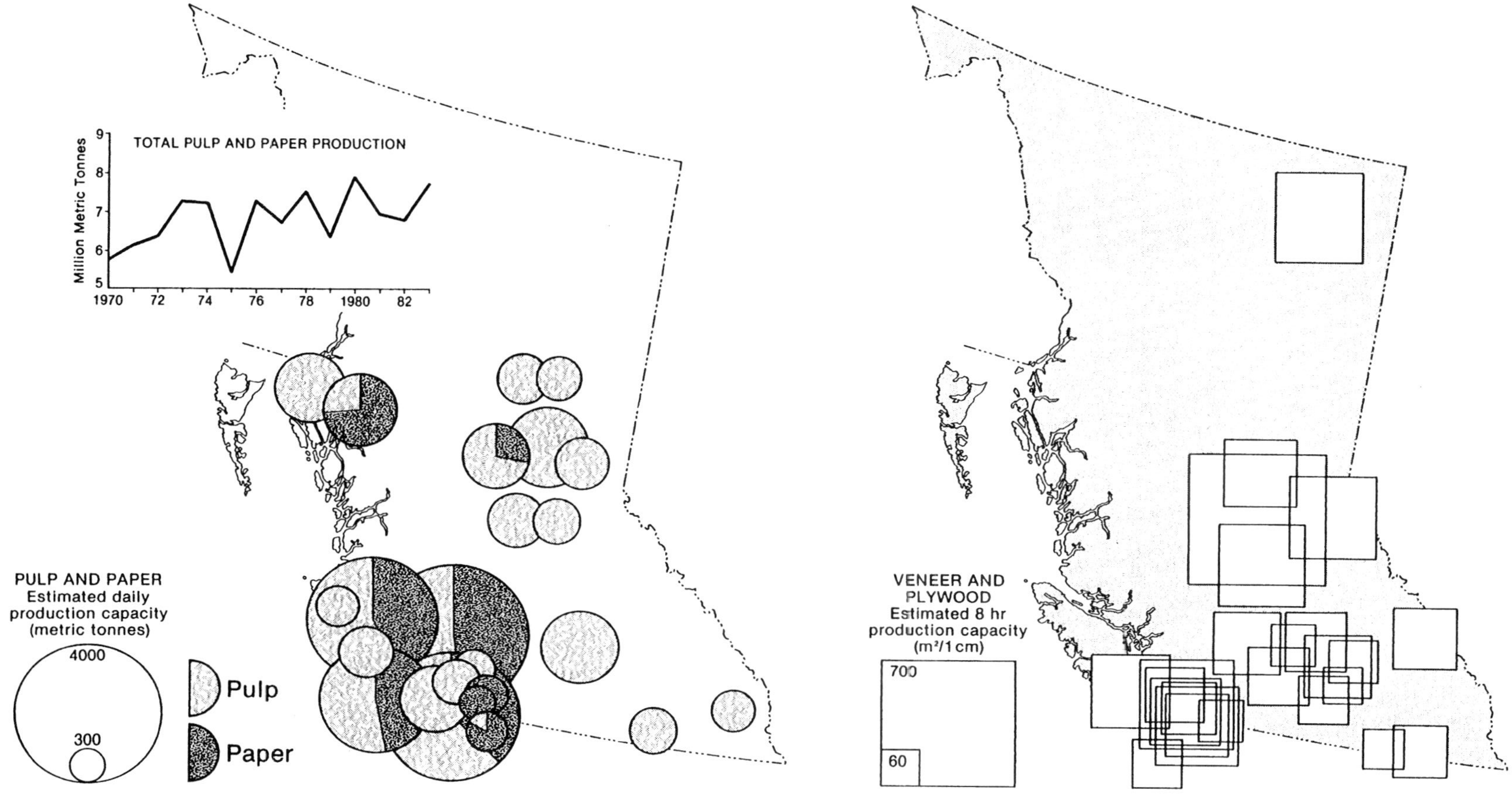

FIGURE 6,6 Location of pulp, paper and plywood mills by production volume

1950s onward, and in the interior from the 1960s. In both areas, access to assured timber under the new licence and tenure agreements and rising world demands for both chemical pulp and newsprint were key factors. Both coastal and interior mills are highly dependent on the sawmill sector for wood chip supplies; roundwood forms a small percentage of the resources intake.

Although roughly equal to sawmills in terms of value added, pulp and paper mills produce only about 60 percent of the value of shipments originating from the wood industries. Because it is capital intensive, the pulp and paper industry employs fewer than half as many people as the wood industries. Of the 24 mills in British Columbia, 14 are exclusively pulp producers, two are exclusively paper producers, and the remainder are integrated pulp and paper operations (Figure 6,6). Seventy-one percent of the rated annual capacity of 9.7 million tonnes (10.7 million tons) is devoted to pulp, which accounts for about 55 percent of the total value of shipments from the paper and allied industries. Production is dominated by bleached kraft (sulphate) pulp; mechanical pulp is secondary and sulphite pulp is produced in one mill only. The other 21 percent of rated mill capacity is devoted to paper and paperboard (dominantly newsprint) which account for roughly 35 percent of the value of shipments. The remaining 10 percent of shipments is composed of miscellaneous box, paper container and asphalt roofing products.

The older coastal mills contain 67 percent of total capacity, and virtually all of the paper capacity (Figure 6,6). Four of these (Crofton, Duncan Bay, Harmac, and Powell River) are integrated operations that produce chemical and mechanical pulp for their own newsprint production, as well as for export as market pulp. The generally newer interior mills, with only 33 percent of provincial capacity, are almost exclusively geared to chemical kraft pulp production, except for a minor paper output from the mill operated by Prince George Pulp and Paper.

Fluctuations in output and market performance have been less pronounced in the pulp and paper sector than in the sawmill sector (Figures 5,6 and 6,6), and the 1981 depression did not hit pulp and paper producers so early nor so hard. But the industry has long been characterized by an excess capacity: in 1983 only 8.7 million tonnes (9.6 million tons) of pulp and paper were produced from a rated capacity of 9.7 million (10.7 million) and only 75 percent of newsprint capacity was used. Much of the capacity, especially on the coast, is also in varying stages of obsolescence, in spite of large capital investments during the 1970s to upgrade performance, efficiency, and environmental protection standards. The coastal mills are already marginal, high cost producers relative to competing Swedish, eastern

Canadian and American mills, and relative production costs will rise further by the year 2000.[24] Dominated by chemical pulp and newsprint, which are relatively crude, low value added products, and facing continental and overseas competition from less expensive and more competitive pulp and paper grades, the industry increased its higher grade paper capacity by 30 percent from 1976 to 1984. But traditional low value added products will continue to be the main outputs, and it will be difficult for the industry to attract the capital investment needed to further modernize obsolete and inefficient capacity, especially on the coast.

It may be concluded, therefore, that all of the timber processing industries in British Columbia face an uncertain future. All predictions agree that the restructured forest industry at the end of the century will be smaller than it is now in response to changing resource supplies, changing global demand for wood products, and declining competitive position in continental and overseas markets. The recent study by Nillsen presented a number of possible forest industry performance levels in the year 2000 under alternate scenarios, ranging from a continuation of present economic, market and supply conditions to the most optimistic improvement in those conditions.[25] Two of those alternatives, the "base case" which continues the present trend, and a moderately "improved case" which assumes an overall 15 percent improvement in the industry's competitive position, are summarized in Table 4,6.

Even under the most optimistic scenario presented by Nillsen, in which there is complete utilization of the 67 million cubic metres (2.366 billion cubic feet) AAC, 20 percent of total processing capacity will still be obsolete in the year 2000; pulp production will be 5 to 10 percent below present levels, lumber production will be 5 percent lower and employment in the forest industry will be 7,000 lower.

Although world demand for forest products is expected to increase in the foreseeable future, the rates of increase now predicted are considerably lower than they were 10 or 20 years ago. The great challenge facing the industry is to maintain its relative position as a supplier in the increasingly competitive world market. Although the industry is now incapable of generating investment, infusions of capital are needed to improve technology and efficiency. For instance, British Columbia mills are 20 percent less efficient than Scandinavian mills in man-hour output; only 40 percent of log volume is utilized in sawmills compared with 55 percent in Sweden; and a British Columbian two by four contains 32 percent less wood than its Scandinavian counterpart. Higher value added output is seen as one route to improving the future competitive position of the industry, but it is likely that only 15 to 20 percent of capacity can be converted, and distance to markets offsets some of the benefits of higher value added output.

Table 4,6 POSSIBLE FOREST INDUSTRY PERFORMANCE IN THE YEAR 2000 UNDER TWO DIFFERENT SCENARIOS, EACH ASSUMING AN AAC OF 67 MILLION CUBIC METRES

	Base Case	*Improved Case*
HARVEST	Economical optimum harvest is 59 percent of the AAC or 40 million cubic metres - 51 percent of the 1983 harvest.	Economical optimum harvest is 85 percent of the AAC or 57 million cubic metres - 73 percent of the 1984 harvest.
LUMBER	1. Total production of 12.8 million cubic metres - 32 percent of 1983 production. 2. 40 percent of total capacity utilized.	1. Total production of 23.6 million cubic metres - 77 percent of 1984 production. 2. 80 percent of coastal and 65 percent of interior capacity utilized.
PULP	1. Total production of 3,210,000 tonnes - 57 percent of 1983 production. 2. 20 percent of chemical and 70 percent of mechanical pulp capacities utilized on coast. 3. 65 percent of chemical pulp capacity utilized in the interior.	1. Total production of 4,225,000 tonnes - 75 percent of 1983 and 130 percent of base case production. 2. 40 percent of chemical and 80 percent of mechanical pulp capacities utilized on coast. 3. 90 percent of chemical pulp capacity utilized in the interior.
PAPER	1. Total production of 1,786,000 tonnes - 81 percent of 1983 production.	1. Total production of 2,646,000 tonnes - 116 percent of 1984 production.
PLYWOOD	1. Only two mills remain, one coastal and one interior.	1. Same as base case.
	Overall poor economic performance.	Overall 110 percent increase in economic performance compared with the base case.

FOREST MANAGEMENT

Sustained yield emerged as the governing precept of the Ministry of Forests' management policy after the 1945 Royal Commission. Why, 40 years later, is the industry facing possible reductions or "falldowns" in the AAC, ranging from 13 to 30 percent? What has happened to "sustained yield"? One answer is that while sustained yield may be a governing precept of forest management, actual management has not been on a sustained yield basis. There are two main reasons for this.

First, sustained yield in the classical sense — a perpetual and equal volume of harvest based on an annual cut of mature timber that equals the annual growth of younger trees — is only possible in a "normal forest". The normal forest is composed of equal areas of all age classes in the rotation or harvest period, and as the mature age class is harvested it is replaced by an equal replanted area. The varied mature and oldgrowth forests of British Columbia do not constitute a normal forest. Therefore, the provincial "sustained yield" strategy, in effect, is simply the regulation of the removal of the oldgrowth, which will eventually be replaced by younger regrowth stands with an approximately equal age class distribution. Only then will sustained yield be possible. British Columbia has not yet attained sustained yield, and inherent in the progressive liquidation of the high volume oldgrowth is a gradual reduction in timber volumes to a lower "sustainable yield".

Second is the fact that this natural trend to lower volumes has been intensified by the failure of renewal (reforestation) programs to keep pace with the rate of harvest and replace the oldgrowth with a new crop. The Ministry, which has been preoccupied with utilizing the forests as a source of revenue, and the industry, which has concerned itself mainly with profit, have given insufficient attention to silvicultural management beyond minimal and often inefficient reforestation. The basic premise underlying sound forest management — that forest renewal should be the first charge against income obtained from the sale of forest products — has been ignored.[26] This failure to invest in the forests at a time when profits were high meant that areas of denuded and low productivity land have increased, and that intensive management of established regrowth stands has been virtually non-existent.

On average, 50 percent of the area denuded by harvesting and fire each year is replanted, and the remainder is left to regenerate naturally. Natural regeneration is very inefficient on some sites, with at least one-third of the land producing unsatisfactorily restocked forests. Brushing and weeding programs on the restocked areas are insufficient, so that the productivity levels of many regrowth stands are below potential. Basic silviculture — the prompt renewal and tending of currently denuded areas — has fallen short of its minimal objectives, but future timber supply projections assume basic silvicultural objectives. Intensive silviculture, which includes the rehabilitation and planting of backlog denuded areas and such regrowth management practices as conifer release, thinning, spacing, and fertilization, has accounted for a small percentage of the silvicultural budget.

In 1979 the Ministry of Forests instituted a five year management plan[27] aimed at alleviating the emerging falldown, and signed a joint cost-sharing agreement under the Canada-British Columbia General Development

Program.[28] The level of reforestation has doubled during the time of this agreement and nursery output has increased. But in 1984, the Ministry of Forests' allocation for silviculture in TSAs, although the largest single item in the budget, was 11 percent lower than in 1983, and the proportion of the budget allocated to intensive silviculture had fallen by 55 percent, although the proportion allocated to basic silviculture had risen by 8 percent. Silvicultural management on TFLs is largely influenced by Section 88 of the Forest Act, which gives companies credit against stumpage for silvicultural expenses that they incur. Because of decreased revenues, the amount allocated to Section 88 between 1982 and 1984 dropped from 90 million dollars to 59, a significant impact on basic silviculture. The most recent five year plan removes all intensive silviculture except for backlog planting.[29] The recently signed (1985) five year Federal-British Columbia Subsidiary Agreement on Intensive Forest Management has allocated 300 million dollars to forest management, but some foresters argue that the amount should be 50 to 60 percent higher if forest management is to be improved and industrial activity maintained.

The accumulated result of these trends is what has been described by a prominent forester as a "national and provincial disgrace"[30]: approximately 35,539 square kilometres (13,860 square miles), or 13 percent of the potentially productive forest land, is classed as not satisfactorily restocked (NSR), disturbed-stocking doubtful (DSD) or occupied by noncommercial brush (NC). Even when the NSR is adjusted to account for expected natural regeneration, there remain 16,345 square kilometres (6,375 square miles) of NSR, of which 7,380 (2,878) are on good and medium sites. If replanted, these sites would increase the annual growth rate by 2.2 million cubic metres (77.7 million cubic feet), or three percent of the 1983 AAC from regulated lands.[31] It has been estimated that with a full scale basic and intensive management program the AAC could be increased to 100 million cubic metres (3.531 billion cubic feet), but that if there is no planting and stand treatment the AAC will fall to 50 million (1.766 billion). An intermediate commitment to silvicultural management would result in a sustainable AAC of 75 million cubic metres (2.649 billion cubic feet), roughly equivalent to the current harvest: even this intermediate level is not being attained now.[32]

Because of limited research on the timber yield benefits that would result from increased intensive silvicultural management in British Columbia the advantages of such increased management cannot be stated precisely.

The economic benefits are also open to debate, although limited cost-benefit studies indicate a possible 5 to 10 percent return on investment.[33] However, the general consensus held by the Canadian Forest Service, provincial Ministry of Forests, professional foresters and the industry is that significantly increased levels of investment are necessary, both to rectify past mistakes and neglect, and to ensure a healthy future for the provincial forests and the forest industry.

REFERENCES

1. For overviews of the forest industry up to the late 1970s, see HARDWICK, W.G., *Geography of the Forest Industry of Coastal British Columbia*. Vancouver: Tantalus Research, Occasional Papers in Geography, No. 5, 1963; FARLEY, A.L., "The Forest Resource", in ROBINSON, J.L. (ed.), *British Columbia*. Toronto: University of Toronto Press, Studies in Canadian Geography, 1972, pp. 87-118; EDGELL, M.C.R., "Forest Industry", in FORWARD, C.N. (ed.), *Vancouver Island: Land of Contrasts*. Victoria: Department of Geography, University of Victoria, Western Geographical Series, Vol. 17, 1979, pp. 93-139.

2. The economic role of the forest industry varies from region to region in British Columbia, but complete information is difficult to obtain from published statistical data. See CANADA, *Manufacturing Industries of Canada: Sub-Provincial Areas*. Ottawa: Statistics Canada, Manufacturing and Primary Industries Division, annual; FARLEY, A.L., *Atlas of British Columbia*. Vancouver: University of British Columbia Press, 1979, Map 51, pp. 106-107.

3. BRITISH COLUMBIA, *British Columbia Economic Accounts*. Victoria: Ministry of Industry and Small Business Development, Central Statistics Bureau, annual.

4. *Ibid.*

5. BRITISH COLUMBIA, *The Economy in a Changing World*. Victoria: Ministry of Finance, 1985, p. 5.

6. BRITISH COLUMBIA, *British Columbia Economic Accounts, op. cit.*; BRITISH COLUMBIA, *Financial and Economic Review*. Victoria: Ministry of Finance, 1984.

7. CANADA, *Production, Shipments and Stocks on Hand of Sawmills in British Columbia*. Ottawa: Statistics Canada, Industry Division, monthly.

8. BRITISH COLUMBIA, *The Economy in a Changing World*, *op. cit.*, p. 12.

9. BRITISH COLUMBIA, *Forest and Range Resource Analysis Technical Report*. Victoria: Ministry of Forests, 1980; BRITISH COLUMBIA, *Forest and Range Resource Analysis*. Victoria: Ministry of Forests, Strategic Studies Branch, 1984.

10. SLOAN, G.M., *Report of the Commissioner Relating to the Forest Resources of British Columbia, 1945*. Victoria: King's Printer, 1945.

11. PSYUs were changed to TSAs in order that timber supplies could be rationalized with regard to the changes in processing plant location and function that had developed since the original establishment of the PSYUs.

12. BRITISH COLUMBIA, *Forest Management Partnership Proposal, Tree Farm Licenses*. Victoria: Ministry of Forests, Discussion Paper, September, 1983.

13. BRITISH COLUMBIA, *Forest and Range Resource Analysis*, *op. cit.*, reference 9, p. 84.

14. *Ibid.*, pp. B28-29.

15. *Ibid.*, p. B22.

16. *Ibid.*, p. B21.

17. BRITISH COLUMBIA, *Forest and Range Resource Analysis Technical Report*, *op. cit.* reference 9; see also YOUNG, W., Address to the Wood Forum, Vancouver, January 22, 1985 (mimeo).

18. NILLSEN, S., *An Analysis of the British Columbia Forest Sector Around the Year 2000, Vol. 1*. Vancouver: Forest Economics and Policy Analysis Project, Information Report 85-1, 1985.

19. A contentious issue is the export of raw logs, which in 1982 rose by 51 percent over the previous year, and increased by 800

percent from 1972 to 1982. While the Ministry and industry argue that the earnings from these exports are critical in poor economic times, the unions point to lost employment in the processing industry.

20. BRITISH COLUMBIA, *Major Primary Timber Processing Plants in B.C. - 1984*. Victoria: Ministry of Forests, Timber Management Branch, May, 1985.

21. This trend is indicated by comparison of Figure 5 with Map 31 in FARLEY, A.L., *op. cit.*, reference 1, pp. 66-67.

22. BRITISH COLUMBIA, *Forest and Range Resource Analysis Technical Report*, *op. cit.*, reference 9, p. 349.

23. BRITISH COLUMBIA, *Forest and Range Resource Analysis*, *op. cit.*, reference 9, Table H2.

24. NILLSEN, S., *op. cit.*, pp. 113-116.

25. *Ibid.*, p. 62.

26. REED, F.L.C., "Reshaping Forest Policy in British Columbia", Public lecture at Vancouver Institute, Saturday Evening Lecture Series, Feb. 16, 1985 (mimeo).

27. BRITISH COLUMBIA, *Five-Year Forest and Range Resource Management Program*. Victoria: Ministry of Forests, 1979.

28. CANADA AND BRITISH COLUMBIA, *Subsidiary Agreement on Intensive Forest Management*. Ottawa and Victoria: Department of Regional Economic Expansion and Ministry of Forests, 1979.

29. BRITISH COLUMBIA, *Five-Year Forest and Range Resource Management Program, 1984-1989*. Victoria: Ministry of Forests, 1984.

30. KIMMINS, J.P., "Letter to the Editor", *The Sun*, Vancouver, Feb. 18, 1985, p. A5.

31. CANADA, *The Need for Forest Renewal and Management in British Columbia*. Ottawa: Environment Canada, Canadian Forest Service, Background Paper for the Canada-British Columbia Agreement on Forest Renewal and Management, 1984.

32. BRITISH COLUMBIA, *Forest and Range Resource Analysis*, *op. cit.*, reference 9, p. I-17.

33. CANADA, *The Need for Forest Renewal and Management in British Columbia*, *op. cit.*, p. 32.

PLATE 9 Lettuce harvesting on the Lower Mainland. *B.C. Government Photo* ▶

7 AGRICULTURE

Colin J.B. Wood
University of Victoria

INTRODUCTION

In a province dominated by mountain ranges, valleys filled with swift flowing rivers and a fiorded coastline, good farmland is a scarce resource, yet, despite these constraints, agriculture plays an important part in the economy and the way of life in British Columbia. Although forestry, mining, fishing and tourism dominate the economy, agricultural output in 1982 produced cash receipts of over 900 million dollars, an increase of 4.7 percent over 1981.[1] At this rate of growth the industry should top the billion dollar sales level by mid-decade. The industry produces only 40 to 45 percent of the province's current food needs, a figure which suggests that more growth is possible.[2]

Despite the opportunities which exist for expansion in some sectors, producers face a wide range of problems, such as high input costs, price uncertainties, fluctuating production and intense competition nationally and internationally. In this chapter the resource base, and the evolution and operation of the industry will be outlined. The specialized character of sectoral and regional activities will be described and some of the more pressing issues facing agriculture in the province today will be discussed.[3]

RESOURCE BASE

The agricultural resource has been estimated by the Environment and Land Use Committee Secretariat as indicated in Table 1,7. Much of the best agricultural land is located in valleys, particularly in the lower Fraser area, the southern Interior and the southeastern coastal plain of Vancouver Island. An outlier of the Prairie growing environment exists in the Peace River region (Figure 1,7). Valley soils are generally deep and fertile and often originate from drained wetlands; elsewhere, soils are extremely variable, due mainly to glacial or glacio-fluvial derivation. The main problem with valley locations, however, is the occasional threat of flooding.

Table 1.7 CLASSIFICATION OF SOILS WITH CAPABILITY FOR AGRICULTURE

Land Capability Type	*Characteristics*	*Area in Hectares*	*Percent of Total Land Area*
1	No limits	69,948	.07
2	Moderate limits	397,688	.40
3	Moderately severe limits	999,778	1.00
4	Severe limits (e.g. stone and rocks)	2,131,867	2.30
5	Very severe limits –	6,138,294	6.60
6	Perennial forage only	5,357,900	5.70
7	No capability	14,900,513	16.00
Total Land Area		93,056,135	

SOURCE: British Columbia, *Agricultural Land Capability in British Columbia*. Victoria: Environment and Land Use Committee Secretariat, 1976.

Agricultural potential is not based solely on pedological attributes, for climate and location are key elements too. The length of growing season or number of degree days and the available moisture are key variables in productivity. Vancouver Island and the extreme southwest mainland generally have mild winters, which in some years will permit almost continuous grass growth, and sufficient precipitation for all the temperate zone crops, although irrigation is practised to increase crop yields. In the remainder of the province the length of growing season decreases quickly eastward and northward. However, the degree-days increase substantially in the southern interior valleys which, due to the rain shadow effect of the Coast Mountains, experience hot, dry summers. Therefore, apart from the coastal region, most of the province has cold winters and summer drought, so that winter growth does not occur and summer irrigation is necessary for crops and winter fodder. The Interior region, which is a mixture of upland forest, grassland and in some areas semi-desert vegetation, is suitable for rangeland, while intensive, irrigated agriculture is practised in some valleys. The Peace River region has extensive, gently undulating terrain, but the range of agricultural activity is restricted by a short growing season.

FIGURE 1,7 Agricultural land reserves and growing seasons

Sufficient precipitation usually occurs to permit grain growing, but drought is an occasional problem, as in 1982. In the northern part of this region the occurrence of cloudiness during harvesting requires farmers to have drying equipment.

Location is a significant feature in potential production. There is clearly a greater economic incentive to farm good land more intensively if it is close to the main urban centres rather than in less accessible areas. Unfortunately, because urban settlement, transportation routes, utility lines and dam sites are competing for the same valleys as agriculture, farmers have to operate in a highly competitive land market and conflicts occur frequently. The rate of urbanization increased dramatically during the last two decades, with more than 4,000 hectares (9,600 acres) of high quality land being lost annually to other uses.[4] In the 1970s this conversion rate became a political issue and resulted in the institution of an agricultural land reserve system. Even with this designation, which prevents subdivision and conversion, agricultural land prices still remain the highest in the country. Thus, in the Fraser Valley farmers intensify operations rather than expand through land purchase. The expanding suburbs have created problems of trespass and complaints of smell and noise. This has led to "green zone" committees to pressure municipalities to recognize farmers' rights. Consequently, with the limited resource base, good land is at a premium and competition for it is intense.[5] In the more remote parts of the province conflicts are usually between livestock and wildlife. This may take the form of wolf predation or conflict over irrigation needs and water levels for fish spawning.

EVOLUTION

The early settlement of British Columbia by Europeans was stimulated by the trade in furs and the rush for gold.[6] It was to provide food for the trading posts and later the gold and coal miners that farming initially evolved. Many prospecters stayed after the gold rush ended and settled down to ranching or subsistence farming in the interior valleys and along the coast. In the 1870s agricultural settlers from Britain and the United States began pioneering along the southeast coast of Vancouver Island, in the lower Fraser Valley and in small pockets of level land along the mainland coast. Much of this farming was subsistent in nature, with some produce being sent to the growing coastal towns and the coal mining communities of the Nanaimo region. The indigenous native people did not resist this settlement, mainly because they were small in number and poorly organized and their economy revolved around the fisheries.

The real surge in development came with the era of Canadian Pacific Railway construction during the 1880s which allowed contact with the East, stimulated settlement and permitted the easier movement of produce to the growing lumber towns and mining areas. As transportation improved, produce was sent eastward to the growing settlements of the Prairies, particularly during the winter months. A boom in greenhouse construction occurred in the early part of the present century, as growers on the coast took advantage of the mild climate and cheap fuel. In the lower Fraser Valley and on the Saanich Peninsula on Vancouver Island the introduction of electric railways provided a means of getting produce to market quickly — a transportation method which was soon to be superseded by trucks. The southern Interior region had initially developed as a ranching enterprise area, but fruit growing in the valleys quickly became established where sufficient water was available for irrigation. The development of irrigation systems and fruit farms frequently were organized by land companies which easily attracted immigrants drawn by the prospect of prosperity in a beautiful location. A small number of very large land holdings modelled on European estates which had ranching, farming and fruit activities in one unit were established at places such as Coldstream and Invermere. In the Chilcotin and Cariboo regions, a livestock economy gradually evolved around very large ranches, such as the Gang Ranch where stock were ranged on private and Crown leased land and at roundup time the operators drove their cattle to the railheads.

The early decades of the present century saw the completion of the Grand Trunk Pacific Railway to Prince George and Prince Rupert which opened up the Bulkley Valley for settlement. The Peace River region in the northeast remained isolated until the Northern Alberta Railway reached Dawson Creek in 1931; even so, the area was isolated from the rest of British Columbia until the British Columbia Railway (originally the Pacific Great Eastern) was extended to the region in 1958. The northeast region is in many ways still a pioneer fringe with strong ties to Alberta.

The pioneering phase in the south ended with the specialization that gradually developed in the first two decades of the present century, as improved transportation and modern marketing began to take shape. However, these improvements also meant increased competition in fruit and vegetable production from other regions, including California. The advent of refrigerated rail cars meant that produce from California, where the growing season was much longer and economies of scale made better use of evolving technology, could be shipped north and east across the continent to compete with British Columbia's produce. The outcome of these changes was the formation by the late 1930s of the first agricultural commodity boards. These government regulatory bodies were set up to stabilize production and sales. They had been pre-dated by co-operatives and grower associations, which had already attempted, albeit less successfully, to manage production and co-ordinate marketing in the face of strong external competition.[7] During World War II there was a boom in seed growing which declined as cheaper European seeds reappeared in the postwar era.

RECENT DEVELOPMENTS 1971-1981

The 1981 census revealed that, with the exception of dairy and tree fruit farms, all sectors of the industry experienced substantial growth during the previous decade. The total value of farm capital grew at a rate of 43 percent per annum which, even accounting for inflation, is a remarkable figure, while the rate of farm employment grew annually by five percent. Another significant change has been the increase in the number of farms (defined as operations with sales greater than 2,500 dollars) by 55 percent, which suggests that the industry is viewed as sufficiently profitable to attract new operators. The cultivated land area has increased by three percent per annum, even though substantial amounts of good land have been subdivided. The fertilized area has grown by 17 percent per annum, demonstrating the intensification of operations and increased reinvestment in the resource base.[8]

The census information of 1971 and 1981 are but indicators of more complex developments during this decade, both in agriculture and the economy of the province as a whole. Among the more significant recent issues have been loss of land to urbanization, declining prices, production controls, foreign competition and dumping, high interest rates, unsafe working conditions, consumer dissatisfaction with the retail food industry's structure and pricing, changes in consumer taste and, finally, environmental problems. That the provincial government, farmer associations and farm worker unions are acutely aware of them is shown in the reports of the provincial legislature's Standing Committee on Agriculture, the debates in the legislature, discussion in the media, the farming press and protest demonstrations. Some of the details behind these issues are described in the following pages.

AGRICULTURE IN BRITISH COLUMBIA IN A NATIONAL PERSPECTIVE

There are approximately one-third of a million farms in Canada of varying sizes and types of operation, from large grain producers to small chicken farms. British Columbia accounts for six percent of this total.[9] That greater capital investment and higher land prices are present in British Columbia are shown by the 1981 census figures: capital value is 6.5 percent and value of land and buildings is 7 percent of the national total. In capital value per farm the three western provinces are all above the national average, with British Columbia ranking third (Table 2,7).

However, the value of products sold per farm shows that the province, at 39,000 dollars, is well below the national average of 49,000. This figure has to be appreciated in the context of there being a large number of small, part-time and hobby farms in the province, particularly in the Greater Victoria and Lower Mainland regions. Thus, British Columbia has double the national average of farms with sales of less than 2,500 dollars and fully 70 percent of the farms have sales of less than 25,000 dollars, compared with the national figure of 50 percent. The cash sales figures do give an indication of the variation in output by province, however, many small farms in British Columbia rely on roadside sales and may not be reporting total income. Generally, the province is mid-way between the high production Prairie Provinces and the low production Maritimes. If productivity per hectare is used as a basis for differentiation, sales per farm as a ratio to average farmed area are high and illustrate the intensity of production. Further evidence to support this is the fact that the average total number of weeks of paid labor is high. In the national context, the agriculture of the province can be summarized as being both specialized and diverse, ranging from the growing of flowers to cereal crops and from seed potatoes to kiwi fruit.

REGIONAL VARIATION AND SPECIALTY SECTORS

The specific mixture of settlement history, economic development and political structure, when set against the rugged panorama of British Columbia's landscape, has resulted in the evolution of four types of specialties, each of which tends to have a regional association. They are:

1. dairying and medium intensity mixed farming (Lower Fraser Valley and southeast Vancouver Island);
2. extensive livestock ranching (Interior grasslands);
3. large-scale grain farming (Peace River region);
4. specialty agriculture — intensive berry and horticultural crops (Lower Fraser Valley, Saanich Peninsula of Vancouver Island), and grapes and tree fruits (southern Interior valleys, especially the Okanagan Valley).

Table 2,7 AGRICULTURAL CHARACTERISTICS BY PROVINCE

	British Columbia	*Alberta*	*Saskatchewan*	*Manitoba*	*Ontario*	*Quebec*	*Maritimes*
Total number of farms (thousands)	20	58	67	29	82	48	12.6
Total farm area (millions of hectares)	1.5	12	17	5	4	3	.9
Average farm size (hectares)	108	329	385	258	73	78	84
Area rented (millions of hectares)	.6	6.0	8.0	2.0	1.4	4	.16
Total produce (billions of dollars)	.8	3.2	3.0	1.4	4.6	2.0	.6
Value of produce sold per farm (thousands of dollars)	39	56	45	49	57	42	36
Average capital value (thousands of dollars)	426	634	466	355	381	197	183
Total weeks of paid labor per reporting farm	61	31	20	26	50	39	55

SOURCE: *Census of Canada*, 1981.

The number of operations per census division and the regional specialties are depicted in Figure 2,7 where it is apparent that, with the exception of wheat and small grain production in the Peace River region, most census divisions have representation in each category besides the regional specialty.[10]

Livestock

Animal products account for 60 percent of farm cash receipts in British Columbia (average 1982-1984). The total is composed of dairy products, 40 percent, cattle, 25, poultry, 15, eggs, 10 and pork, 10. As a proportion of the national total the livestock industry is relatively modest, accounting for 10 percent of poultry, 8 of cattle and 2.5 of pork.[11] Total cattle numbers increased gradually during the period 1961-1981, but are now entering a downturn phase, reflecting the general oversupply of beef internationally and the declining consumer preference for red meat. Again, changing diet and an aging population are reflected in the decreasing demand for dairy products. These general trends are apparent in Table 3,7.

Dairying

The livestock industry is comprised of both dairying and beef operations: some sheep and pork enterprises also exist, but are relatively small scale in nature. While most urban centres in the province have local dairy operations, the bulk of the industry is concentrated in the lower Fraser Valley and on southeast Vancouver Island, supplying the main urban centres of Vancouver, Victoria and Nanaimo with fresh milk and other dairy products.

Dairy farmers of the coastal region are able to capitalize on the mild winters, long growing season and rich pasture lands of the river valleys and deltas. Improvements in breeding and management methods, such as irrigation for hay production, have contributed to a steady increase in ouput per animal from 6,000 pounds (2,725 kilograms) in the 1920s to 15,000 (6,800) today.[12] Outputs in many herds are among the highest in Canada, but herein lies a problem. The very efficiency and skill exhibited by dairy farmers has produced periodic gluts on the market. The difficulties of over-production and the producers' relationships with creameries and retailers led to significant conflict as early as the 1920s between dairy farmers, independent creamery operators and the co-operatives, ultimately requiring provincial government inquiries (1927) and legislation (1929).[13] Further commissions of inquiry and legislation (1955) and studies (1966) showed

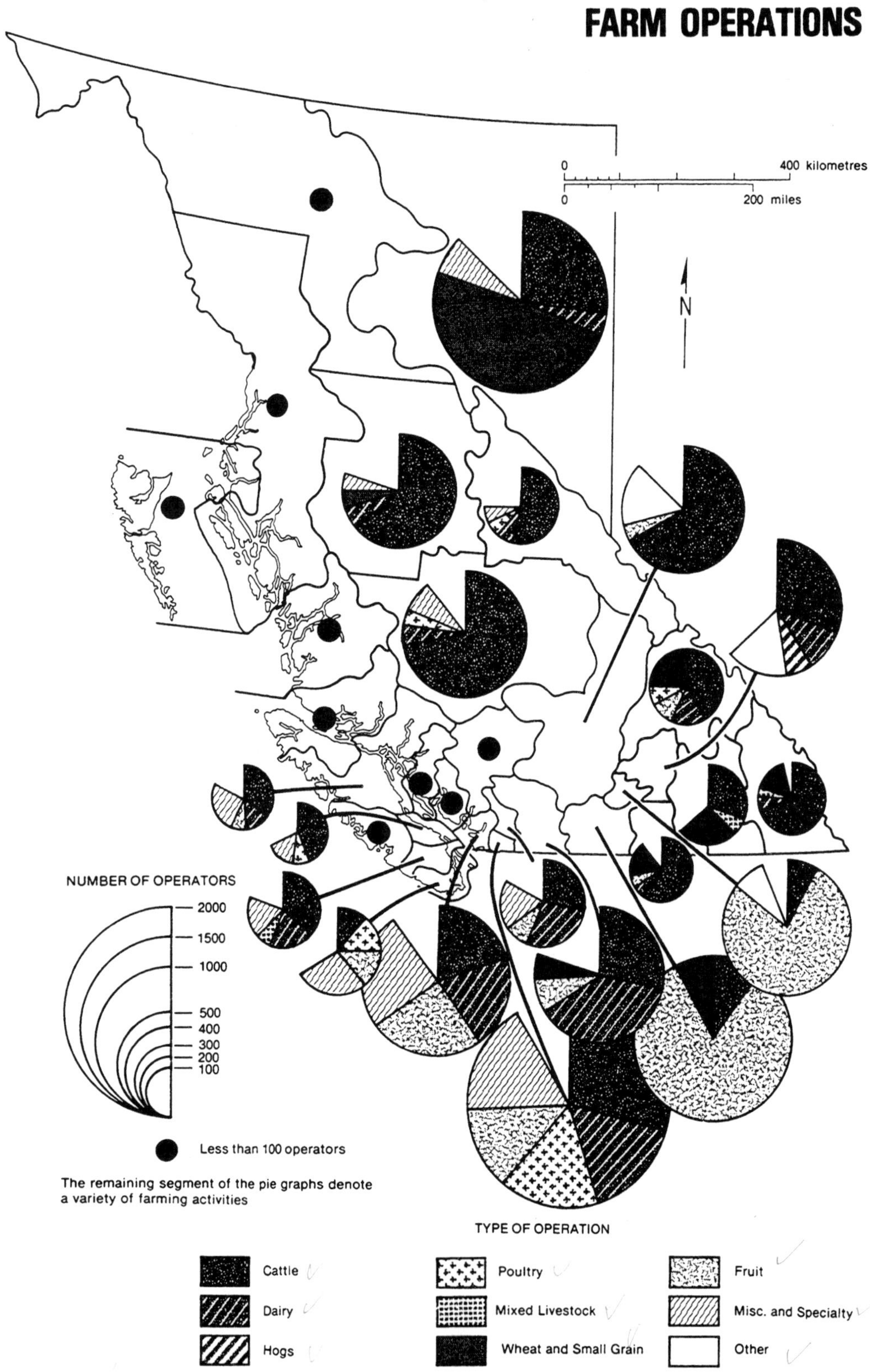

FIGURE 2,7 Number of farm operations and major activities by census division, 1981

Table 3,7 LIVESTOCK NUMBERS IN BRITISH COLUMBIA AND PERCENTAGE OF CANADIAN TOTAL (in brackets), 1961-1983

	1961	*1971*	*1981*	*1983*
CATTLE	461,000 (4)	573,000 (7)	789,000 (6)	748,000 (6)
DAIRY CATTLE (milk cows)	91,000 (3)	80,000 (3)	89,000 (5)	88,000 (5)
PIGS	41,000 (1)	78,000 (1)	254,000 (2)	268,000 (2)
SHEEP	102,000	50,000	66,000	60,000
HENS	5,600,000	7,800,000	9,900,000	no data

SOURCES: *Census of Canada*, 1961, 1971, and 1981; BRITISH COLUMBIA, *Agri-stats*. Victoria: Ministry of Agriculture and Food, Economics Branch, 1984.

that problems were persisting, particularly through the vertical integration of milk sales by supermarket retailing outlets and the decline of door-to-door sales. This has resulted in an industry concentrated generally into larger production units and strictly controlled by the provincial Milk Marketing Board.[14] Entry is restricted according to possession of a quota which belongs to a specific dairy herd, while the actual production is allocated according to the quota. Unfortunately, anomalies persist in the system, since some milk produced on Vancouver Island is shipped to the mainland for packaging and sale, while some milk produced on the mainland is sold on Vancouver Island.[15] Clearly, the industry is not operating at optimum efficiency, reflecting perhaps the structure of the retailing industry rather than producers or creameries. That this sector is still in a state of instability is reflected in the recent mergers of dairy co-operatives and the decision by the British Columbia Marketing Board to withdraw from the national marketing system. The number of operators is holding steady at about 1,130 farmers.

An example of a medium to large dairy operation is illustrated by the Pendray farm near Sidney on Vancouver Island.[16] Originally, the family farm was located closer to Victoria, but urban expansion made farming increasingly difficult, hence the move. The new location necessitated considerable investment in new equipment and buildings. The terrain is generally flat, with a mixture of sandy loam and clay loam soils, enabling large scale operations. Irrigation is practised to ensure and enhance yields. With the surrounding land zoned Agricultural Land Reserve, it is unlikely that they will have to move again through urban pressures.

The dairy industry has been experiencing deteriorating returns in the face of rising costs and relatively stable demand. Only efficient operators with large capital investment are able to survive. Over a 20-year period the total number of dairy animals has shrunk from 91,000 to 89,000, but, as noted earlier, output per animal has increased substantially.

Ranching

Ranching was the first type of farming practised in western Canada on any scale. It grew in response to the meat requirements of both the mining industry and the fledgling coastal cities; furthermore, it produced a product which could be driven to the markets. Ranchers were attracted to the extensive natural grazing available in the Interior grasslands and open forest.[17] The pioneer cattlemen quickly evolved a seasonal form of transhumance between the valley hay meadows, the ranges and the open forest of higher altitudes. Railway construction in the late nineteenth and early twentieth centuries was a further stimulus to growth. In the early years, range management was not well controlled and disputes were frequent. The establishment of cattle overseers in 1896 brought order and smoothed problems during roundup.[18] By the 1900s, virtually all of the natural grassland had been pre-empted for settlement, with many Crown land parcels being held as leaseholds. Herds were comprised of slaughter cattle nearly two years old or older, while the remainder were put out on natural pastures to fend for themselves during the winter. Generally, carrying capacities were low: 20 hectares (49 acres) per animal.

The inexpensive natural grazing inevitably began to deteriorate. Gould quotes G. Dawson of the Geological Survey as noting that as early as 1878 the native bunch grass was badly deteriorating, so that operators turned to producing forage crops to over-winter their stock.[19] Natural grass meadows were tamed, seeded and irrigated. The increased costs led, in turn, to further specialization. Animals were marketed earlier and separate finishing outfits arose. Thus, the beef industry is now generally a two-step process, with "cow-calf" operations which produce "feeder cattle", and a second stage which fattens and finishes them. In some cases both activities may be carried out in the same unit.[20]

Consequently, from the southern provincial boundary northward to the Chilcotin, Cariboo and Bulkley valleys, ranching has evolved as a dominant land use activity in the Interior, being supplanted in importance only by the specialized fruit growing of parts of the Interior valleys and the grain growing of the Peace River region. Cattle numbers gradually increased until the effects of low prices and higher input costs during the early 1980s caused a decline. Input costs obviously play a key role in the industry's viability, of which forage conditions are a major component. With a reliable source of forage, ranchers are able to exert a measure of control over feed costs. However, this in turn has generated other problems. It has meant more investment in equipment during an era of high interest rates. Irrigation for forage has resulted in conflict with the fishery agencies' requirements of specific stream and river levels for maintaining fish spawning environments.

Finishing feeder cattle has relied to a large extent on Prairie grain since the 1950s and because feedlot operations are market oriented many are located in the Lower Mainland. There is generally a close correlation between number of livestock and grain prices which, in turn, interact with beef prices. An important variable in this situation is the freight rate for grain. The recent federal government action of abolishing the Crowsnest freight rate structure (which was related to export grain) may have significant repercussions in the whole system of grain and livestock farming in Western Canada.

In the face of declining returns and foreign competition it is not surprising that by January 1985 total cattle numbers had decreased to 638,000 from 789,000 in 1981. Western beef producers face keen competition from Australia and New Zealand, where grass-fed livestock can be grazed cheaply the year round. Import controls have been instituted since the late 1970s to prevent further underselling by these exporters who tend to compete at the cheaper end of the market.[21]

Neither the government nor the operators have sat idle in the face of such challenges, for several measures have been undertaken to increase the viability of the livestock industry. British Columbia is a beef deficit region, producing only 40 percent of its consumption, so there is a rational basis for developing it, providing that input costs can be kept competitive with those of other producers. The government is improving the resource base by upgrading rangeland through reseeding and knapweed control, breeding improvement, disease eradication and technical information diffusion.[22] Producers have in many instances taken collective and individual actions to enhance the industry. The formation of the British Columbia Livestock Producers Co-operative Association in the 1940s was important in providing a reliable market and competitive pricing for operations. It was instrumental in setting up markets at Williams Lake, Okanagan Falls, Merritt and Kamloops.[23]

Individually, many ranchers have striven to improve their herds through careful breeding. The herd at the Chataway Ranch near Ashcroft is representative of the type of quality herd that has evolved, partly as a response to the necessity of keeping an outfit viable and also due to the owners' enthusiasm. Starting in the early 1950s as a commercial cow-calf operation, the owners gradually concentrated on purebred Herefords. By the 1970s herd size was close to 150 head, and through breeding, the herd gained the accolade in 1977 of having a Grand Champion Bull. Initially, grazing was on the wild hay meadows and Crown rangelands, but investment in irrigation permitted further expansion. By the 1980s the owners were able to purchase an adjoining ranch and co-ordinate the irrigated areas and Crown ranges of each ranch.[24]

Generally, the livestock industry is experiencing difficult times, reflected in the decreasing farm animal population. A farm income insurance plan is meant to protect operations, but returns have been low for several years and, consequently, many smaller, and even a few very large ranches have gone out of business. Both the operators and the provincial government have been opposed to suggestions from the federal government for a national beef plan.

Grain

Cereals are grown mainly in the Peace River region, which is an extension of the northwest Prairie conditions into northeastern British Columbia. Due to the short growing season (Figure 1,7) it is essential that precipitation and growing temperatures occur at the right time, otherwise harvests may be poor. A recent estimate of net returns to Canadian grain producers shows that this region has a productivity potential which is only 50 percent of the best area in Canada, the Red River region.[25]

Despite this regional disadvantage, arable farms do have some short run flexibility compared with livestock operations in that they can switch between the five main small grains and oilseeds, according to rotation and anticipation of price trends. However, since this region constitutes only two percent of the western grain growing area, British Columbia operators have to follow national and international grain price trends. A good indication of producers' sensitivity to world prices for wheat and feed grown is shown in Table 4,7.

Tree Fruit

One of the most distinctive characteristics of British Columbia's agriculture is the growing of fruit (Table 5,7). From the very beginning of

Table 4.7 GRAIN AREA BY TYPE OF GRAIN IN HECTARES, 1979-85

	1979	*1981*	*1983*	*1985*
WHEAT	33,000	50,000	65,000	69,000
BARLEY	54,000	102,000	75,000	81,000
OATS	20,000	no data	28,000	24,000
CANOLA (RAPESEED)	112,000	35,000	81,000	71,000

NOTE: Figures are rounded to the nearest thousand.

SOURCE: British Columbia, *Agri-Stats*. Victoria: Ministry of Agriculture and Food, Economics Branch, Annual.

European settlement, pioneers invariably planted orchards as part of their farm operations, both for self-sufficiency and as cash crops to supply the infant cities, towns and mining centres. More specialized orchards were planted in the late nineteenth century around Victoria, on the Gulf Islands and in the lower Fraser Valley. These were comprised of varieties and types of apples which have virtually disappeared, such as Bellflower and Lemon Pippin.[26] Such peri-urban farms also supplied strawberries and other berries.

The coastal areas produced well into the twentieth century, but because of the high incidence of disease, the storage and marketing problems of the varieties grown and the attractions of the better growing conditions of the dry Interior valleys, the early orchards have almost vanished and the Okanagan Valley dominates output. However, the recent development of dwarf varieties, better pomological practices and "pick your own" sales has seen a small comeback for the original coastal areas, with new plantings of small orchards of two to five hectares (five to ten acres) near the cities.

The Okanagan is synonymous with fruit production in Western Canada. Although the first fruit trees were planted by Father Pandosy at the Oblate Mission near Kelowna in 1862, the extensive modern plantings date from the early 1900s. Several land development schemes drew immigrants to various parts of the region, extolling the attractions of its environment and the financial attractions of fruit farming. Some of the pioneer areas eventually proved to be unsuccessful, due to stressful growing conditions, as in parts of the Kootenays farther east and in the Wallachin area to the west. In the Okanagan, fruit growing boomed as orchards were planted on the frost draining lower valley slopes. Varieties consisted of the Delicious

Table 5,7 FRUIT AND VEGETABLE LAND USE

	Nfld.	P.E.I.	N.S.	N.B.	Que.	Ont.	Man.	Sask.	Alta.	B.C.	CANADA
MAJOR TREE FRUITS (number of trees in millions)											
Apples – 1971	–	–	.71	.10	1.20	1.90	–	–	–	2.60	6.60
1981	–	–	.89	.11	1.50	3.10	–	–	–	3.00	8.80
Pears – 1971	–	–	–	–	.60	.68	–	–	–	.34	1.10
1981	–	–	–	–	.50	.47	–	–	–	.24	.70
Peaches – 1971	–	–	–	–	–	–	–	–	–	.27	1.60
1981	–	–	–	–	–	–	–	–	–	.43	1.50
SMALL FRUIT (thousands of hectares)											
1971	–	.20	2.60	1.40	2.90	10.80	–	–	–	3.90	22.00
1981	–	.40	6.00	3.10	4.40	11.40	–	–	–	5.50	31.00
VEGETABLES (thousands of hectares)											
1971	.40	1.40	2.40	2.00	32.00	51.00	2.10	.30	3.00	7.50	103.00
1981	.50	1.60	3.10	3.40	32.00	61.00	1.80	.60	4.00	7.80	117.00
GREENHOUSE (millions of square metres)											
1981*	.02	.03	.18	.08	1.10	3.60	.07	.05	.35	1.00	6.60

– insignificant

* comparable data for 1971 are not available

SOURCES: *Census of Canada*, 1971 and 1981.

group whose strains now dominate the retail market, while Bartlett and Anjou have become the main pear varieties.

The viability of tree fruit production in a relatively isolated region has been possible through three factors. First, the development of extensive irrigation networks based on a well organized water licence system enabled the main obstacle of summer drought to be overcome. Second, even in the early years of this century it was clear that co-operation between growers was necessary to establish quality control, manage the occurrence of fluctuating production, and co-ordinate marketing.[27] The British Columbia Fruit Growers Association was formed in Vancouver in 1889, but as the locus of tree fruit production moved to the Okanagan, the association moved there also in 1932. Legislation to control production was introduced in 1937. Third, to remain competitive, modern pomology requires extensive ongoing research in all stages of production. The industry has had substantial support from provincial, federal and university research into the newer cultivars, harvesting and storage methods, and marketing techniques.

Producers today are faced with a variety of environmental, managerial and marketing issues, the particular combination depending on the orchard involved. Growing fruit are prone to several problems, notably rain and hail damage on cherries, causing split and rot: over 30 percent of the 1982 crop was damaged in this way. Some producers are experimenting with fibreglass sheeting as protection, but the costs are as high as 25,000 dollars per hectare (10,000 dollars per acre).[28] Other hazards include Little Cherry disease which killed off cherry growing in the Kootenays in the 1940s (related to zinc deficiency and mealybug infestation).[29] Such problems can be solved only by destruction. Spraying against biological hazards is an integral part of modern fruit farming.

The management of fruit production generally entails a long run arrangement of factors of production and appropriate evaluation of input costs because fruit trees take several years to mature and produce. However, the development of dwarf cultivars has meant a shortening of the time span between planting and production of a profitable crop. Many producers grow a range of fruit and different varieties of the same fruit. In this way labor can be used more evenly, since different fruits and varieties ripen at different times. Indeed, there are also environmental benefits associated with this approach.

For example, the Kidson orchard in the Okanagan has over 16 hectares (40 acres) devoted to apples (88 percent) and pears (12 percent). Among the apples, McIntosh constitute 50 percent, Spartan 25, Golden Delicious and Tydeman 15, and Red Delicious 10. The typical investment per hectare has been estimated at land, 18,000 dollars, equipment, 6,000 and accommodation, 6,000.[31] Labor costs are a perennial difficulty, since growers often rely on migrant pickers, and comparable costs are much lower across the border in Washington. The response has been to develop mechanical fruit pickers and smaller dwarf varieties suitable for both machine picking and easier manual picking. Thus, tree fruit acreage has declined during the decade 1971-1981 (Table 5,7), but the numbers of trees per hectare have increased substantially. The labor issue has been exacerbated by confrontations between migrant workers and local trouble makers.

Marketing the produce has always been one of the most complex problems facing producers. High production coupled with strong competition from citrus fruits has resulted in depressed prices for several years, with costs for apples averaging 13 to 14 cents per pound in 1983 and returns varying from 5 to 11 cents.[32] The producers' association and the marketing boards are finding it increasingly difficult to deal with a problem of this magnitude. Understandably, in view of these pressures, many growers have been tempted to give up fruit farming altogether.

Grapes

The Okanagan and Similkameen valleys are also noted for their grape production in one of the most northerly locations in the world. The initial output (early 1900s) was for table grapes, but after the construction of a winery in 1927 production gradually switched to supplying grapes for wine. Several other wineries and a number of cottage wineries have since been built and today 90 percent of the output is geared to wine.[33] There has been a gradual change from Labrusca to Vinifera (French hybrid varieties) to improve grape and wine quality.[34] Vines are grown under irrigation, and to take advantage of soil conditions and cold air drainage vineyards occur in scattered locations in the two valleys. At present 1,400 hectares (3,300 acres) are in grapes, producing about 17,000 tonnes (18,750 tons).[35] Wineries can only import up to 20 percent of their grape requirements, a measure clearly of benefit to local growers. Recent problems have been intense competition from cheaper European wines and a changing consumer preference for light white wines, leaving large stocks of red wine unsold.

Specialty Agriculture

In addition to the dairying and mixed farming activities of southeast Vancouver Island and the lower Fraser Valley, there are a variety of specialty crops grown. These include berries, vegetables, mushrooms, filberts, honey, flowers and ornamental shrubs. Mild winters, high sunshine hours and proximity to markets have encouraged this development.

Strawberry production has gone through several cycles of expansion and decline, responding to changing competition from the United States and Mexico and disease problems. Current production is 25 percent of the Canadian total. Raspberry production requires the special growing conditions found in the province so that British Columbia accounts for 90 percent of national production.[36] Raspberry output has seen periods of prosperity and decline too, with growers recently facing intense competition from Oregon and Washington producers. Cranberry production is, in contrast, very stable, reflecting the restricted area of suitable sites for production and the near North American monopoly of the industry by Ocean Spray, a United States growers' co-operative.[37]

Vegetable production again benefits from the climatic advantages of the region over other parts of Canada, but invariably is struggling against American competition. Mushroom producers appear to be one of the few groups able to make profits consistently. The mushrooms are grown in 12-week cycles, using Californian spawn. Currently, as long as a grower can produce more than 14.6 kilograms per square metre (3 pounds per square foot), a profit is there.[38]

Greenhouse production of fruit, vegetables and ornamentals has increased substantially, not only through expanding acreage, but also as a result of better greenhouse technology and horticultural practices. For example, the Vander Zalm butterhead lettuce enterprise in Surrey municipality near Vancouver is a highly automated operation of 3 hectares (7.4 acres), with a projected output of 3,500,000 heads.[39] At this rate the market clearly has to extend beyond the province.

Floriculture continues to flourish as a specialty of the province. In March, daffodils are air freighted to eastern Canada to herald the arrival of Easter. In December, holly is sent for the Christmas season.

The intensity and high technology of the farming in the southern part of the province contrasts with the experimentation and pioneering spirit of the more remote regions. In some interior locations growers are experimenting with ginseng production. On remote Flores Island one farmer grazes his beef stock (Brown Swiss, Durham Shorthorn and some Welsh Black Cross), specially bred to withstand rain, on the muskeg of a former lake bed drained by missionaries in 1893.[40]

SUMMARY

Agriculture has an important role in the way of life of the province, particularly on the rangelands, in the southern towns and villages and in the lower Fraser Valley. It is a productive, stable, resource industry which also enhances the aesthetic quality of the landscape. The industry is facing difficult times, a situation which is not unique to the province or to Canada. The problems are complex, being a combination of high input costs, especially interest rates which affect investment in machinery, relatively high labor costs and competition from cheap, often subsidized, foreign products. Only the dairy industry has been able to protect itself, partly through locational monopoly and partly through carefully controlled supply. Other agricultural activities, particularly ranching and fruit growing, manage to continue only through government assistance and anti-dumping legislation. Cheap food does not necessarily mean nutritious and uncontaminated products, and agriculture in the province can best survive through concentrating on quality and flavor, features which foreign agribusiness cannot always provide.

REFERENCES

1. BRITISH COLUMBIA, *Agriculture Statistics*. Victoria: Ministry of Agriculture and Food, Release No. 7, December 1983.

2. BRITISH COLUMBIA, *Reports of the Legislature*. Victoria: Legislative Assembly Select Standing Committee on Agriculture, 1979, p. 34.

3. Among some of the useful sources of information on agriculture in British Columbia are: *Census of Canada*, agriculture volumes, annual reports of the British Columbia Ministry of Agriculture, the farming newspaper, *Country Life*, and FARLEY, A.J., *Atlas of British Columbia*. Vancouver: University of British Columbia Press, 1979.

4. CANADA, *The Agricultural Land Reserves of British Columbia: An Impact Analysis*. Ottawa: Environment Canada, Lands Directorate, Land Use in Canada Series, No. 13, 1978.

5. BRITISH COLUMBIA, *Reports to the Legislature*, *op. cit.*, p. 30. It has been estimated that to achieve self-sufficiency all of the agricultural land would have to be used. Without increased productivity this could supply only 70 percent of needs by the end of the century.

6. A wealth of detail on the early agricultural development can be found in local histories, for example, DOUGAN, R.I., *Cowichan My Valley*. Duncan: R.E. Dougan, 1973; WOODS, J.J., *The Agassiz-Harrison Valley*. Sidney, British Columbia: Kent Centennial Committee, 1958; RUFLI, E. (ed.), *History of the District of Rutland, British Columbia, 1858-1958*. Kelowna: Rutland Centennial Committee, 1958. A good general picture is given by FARLEY, A.J., *op. cit.*, reference 3, pp. 78-87.

7. BRITISH COLUMBIA, *Marketing Boards in British Columbia Phase II Research Report*. Victoria: Legislative Assembly Select Standing Committee on Agriculture, Vol. 1, 1979, pp. 5-9.

8. CANADA, *Census of Canada*. Ottawa: Statistics Canada, 1971 and 1981.

9. *Ibid.*, 1981.

10. For a detailed and comprehensive study of regional productivity and self-sufficiency see, BRITISH COLUMBIA, *Land Productivity*. Victoria: Legislative Assembly Select Standing Committee on Agriculture, 1978.

11. BRITISH COLUMBIA, *Agri-Stats*. Victoria: Ministry of Agriculture and Food, Economics Branch, 1983-1985.

12. The Hans Frueh herd of Cobble Hill has achieved yields of 20,000 pounds per cow, as reported in *Country Life*, July 1976.

13. It was a national problem: see CANADA WEST FOUNDATION, *Western Canadian Agriculture to 1990*. Calgary: Canada West Foundation, 1980, p. 139.

14. BRITISH COLUMBIA, *Marketing Boards in British Columbia Phase II Research Report*, *op. cit.*, p. 61. Note that the Milk Board, unlike the other agricultural marketing boards, is not controlled by the British Columbia Marketing Board. For a detailed discussion see *An Assessment of the Rationale for Marketing Boards in British Columbia*. Victoria: Legislative Assembly Select Standing Committee on Agriculture, Vol. IV, January 1979.

15. MOYO, E., *A Study of the Spatio-Temporal Competition in the Dairy Industry, with Particular Reference to Vancouver Island*. Victoria: University of Victoria, Department of Geography, M.A. Thesis, 1978.

16. Personal communication, D. Pendray, August 1985.

17. GOULD, E., *Ranching*. Seattle: Hancock House, 1978, p. 109.

18. *Ibid.*, p. 113.

19. *Ibid.*, p. 115.

20. CANADA WEST FOUNDATION, *op. cit.*, reference 13, p. 99.

21. *Ibid.*, p. 110. A recent problem has been complaints against European Economic Community (Irish) beef dumping.

22. The policies and programs are detailed in BRITISH COLUMBIA, *Annual Report of the Ministry of Food and Agriculture*. Victoria: Legislative Assembly, 1975-1984.

23. *Country Life*, November 1983, pp. 20 and 24.

24. *Country Life*, February 1984, p. 12.

25. CANADA WEST FOUNDATION, *op. cit.*, reference 13, pp. 38-41.

26. UPSHALL, W.K. (ed.), *History of Fruit Growing and Handling in United States and Canada, 1860-1972*. University Park, Pennsylvania: American Pomological Society, 1976.

27. CANADA WEST FOUNDATION, *op. cit.*, p. 172. From 1939 to 1974 British Columbia Tree Fruits was the sole marketing agency. In 1978 independent sales were again permitted.

28. *Country Life*, August 1983, p. 12.

29. *Country Life*, May 1983, p. 7.

30. *Country Life*, June 1983, p. 9.

31. *Ibid.*

32. *Country Life*, July 1983, p. 5.

33. CANADA WEST FOUNDATION, *op. cit.*, p. 176.

34. *Country Life*, October 1984, p. 16.

35. BRITISH COLUMBIA, *Agri-Stats*, *op. cit.*, 1981-1985.

36. CANADA WEST FOUNDATION, *op. cit.*, pp. 179-186.

37. *Ibid.*, p. 183.

38. *Country Life*, April 1982, p. 5.

39. *Country Life*, July 1982, p. 19.

40. *Victoria Times Colonist*, October 20, 1983, p. 14.

PLATE 10 Coal mining and processing installations near Sparwood. ▶
B.C. Government Photo

8 MINING

William M. Ross
University of Victoria

Mining is one of the oldest and most important of British Columbia's resource industries. The industry has had a significant impact on the economic health of the province, the patterns of exploitation and settlement and the frontier nature of the province's lifestyles and politics. There have been periods when mining dominated the economy, such as the gold rush era, but the industry has traditionally been second to forestry among the resource industries.

As is the case with many staple commodities, such as forest products, the prices for British Columbia minerals are affected by prices set on a world market and are often beyond the control of domestic producers. These prices have had impacts on the particular mineral resources that were developed and the life of many mines. The early gold rushes into the Cariboo, for example, could not have been sustained were it not for the high prices associated with this precious metal. Access and transportation costs precluded development of other minerals until an adequate transportation system was in place. When this was achieved it became possible to exploit a wider range of minerals. Since not all mineral prices rise or fall simultaneously, this wider range gave some stability to the industry, but the prices were still beyond the control of local producers.

The fluctuating prices have been particularly important in influencing patterns of exploitation and settlement. The lure of gold between 1858 and 1862 attracted the first large wave of settlers and encouraged clearance of agricultural lands. As gold diminished in importance, a pattern which is now all too familiar began to emerge. Miners directed their efforts to the highest grade sources and when the first signs of depletion and overworking occurred they moved on to new sites. When mines closed communities which had been established on the basis of the single industry suddenly became ghost towns. Some old mines have periodically reopened when mineral prices rose considerably. In these cases improved technology and price increases were usually the central factors. At the turn of the century, for example, ore grades of two percent copper were considered economical, while now grades of 0.4 percent are regarded as commercially viable.

The roots of the provincial mining industry are tied very closely to the United States. Many of the early miners in the province moved here as a result of the gold rush and others came to work the mines in the Kootenays. Even when the ties to the rest of Canada improved with the building of the Canadian Pacific Railway the connections with the United States, and particularly California, remained strong. Some of the technology developed in California was used in British Columbia and ore was sent from the province to San Francisco for smelting. After the turn of the century, and particularly as gold gave way to the development of minerals such as zinc, silver and copper, the connection diminished.

The pattern of mineral dominance continues to shift. Coal was one of the first minerals developed in the province, but was overtaken by gold. At the turn of the century silver, lead, zinc and copper mines were opened and by 1930 these minerals had overtaken gold in importance. Copper and lead were the leading minerals in this period, but in the 1940s and 1950s zinc became more prominent and replaced copper as one of the two dominant minerals. In the 1960s and 1970s copper re-emerged as a leading mineral and the development of the coal fields in the southeast and northeast has led to the rebirth of that mineral, such that in the 1980s coal and copper are now the dominant minerals.

THE MINING INDUSTRY IN PERSPECTIVE

The early history of British Columbia's mining industry was dominated by coal and placer gold, but ever since the Canadian Pacific Railway was completed in the 1880s lode minerals and fossil fuels have contributed the bulk of provincial mining revenues. The earliest mining in the province took place along the coast. Coal was mined at Saquash on the northeast coast of Vancouver Island as early as 1836 and later at Nanaimo in 1852. There are also records of gold being found in a small quartz vein on Moresby Island in the early 1850s, but all of these early developments were overshadowed by the placer gold mining in the lower Fraser in 1858 and upriver in the Cariboo in 1861 and 1862. Coal was an important resource, since it attracted shipping and gave promise of industrial development, but the discovery of gold was more important to the early rapid growth of the region and to the creation of British Columbia as a province within the Canadian Confederation.

By the early 1880s most of the rich deposits of gold had been mined and prospectors began to turn their attention elsewhere. They moved up from the south through the Columbia and Kootenay valleys, concentrating their attention on the areas along and south of the Canadian Pacific Railway route. Many significant discoveries were made, including silver and copper at Nelson in 1886, gold and copper at Rossland in 1889, copper at Boundary Camp and Greenwood in 1891, and silver, lead and zinc at Slocan in 1891 and at Kimberley in 1892. Some silver and lead ore was shipped south to San Francisco for smelting, but between 1896 and 1901 a number of smelters were brought into production at Nelson, Trail, Grand Forks and Greenwood. In addition to the lode minerals noted above, coal production in this period increased significantly. The coastal areas, particularly Vancouver Island, dominated production and new coal mines were opened in the East Kootenay. The product from these mines was used by the railroads for fuel and by the newly built smelters for coking coal.[1]

Between 1900 and 1935 the production of gold continued to decline, while copper production increased substantially. Zinc and silver emerged as important minerals during the period and the volume of lead production was stable. The spatial pattern did not change significantly from the previous period. Coal was important on Vancouver Island and in East Kootenay. Some gold was taken from the northern part of the province in the Cassiar Mining District in the early part of this period and later copper was mined at Anyox, on Observatory Inlet. Silver, lead and zinc production was dominated by the Sullivan Mine at Kimberley.

By the late 1950s there were profound changes in the mining industry. Demand for industrial minerals increased after World War II and substantial amounts of capital became available. Technological improvements in mining led to large-scale production of lower grade ores and the appearance of open-pit mines in the province, where previously most mining was underground. Production from these open-pit mines was often sold to overseas buyers in large quantities under long-term contracts. Mines that were opened during this period include the iron ore mine at Tasu, the asbestos mine at Cassiar and the Craigmont copper mine in the Highland Valley.[2]

The earliest commercial natural gas in British Columbia was discovered at Pouce Coupe in 1948 and oil at Boundary Lake in 1955. These discoveries ushered in an era of unprecedented growth in the oil and gas industry. Most of the oil and gas production is concentrated in the Peace River area. The gas reserves have always exceeded those of oil. British Columbia produces sufficient gas for its own needs and exports large quantities to the United States. The oil production, on the other hand, has seldom met more than 50 percent of the province's needs.

The trends that emerged during the 1950s intensified during the 1960s and 1970s. More and more large low grade open-pit operations replaced the small high grade mines of the past. The Sullivan Mine at Kimberley declined in importance as open-pit mining intensified in the Highland Valley. New mines were opened in the Omineca area of northern British

Columbia. Coal re-emerged as a significant mineral with large-scale operations in southeast British Columbia. Unit trains were used to transport this coal from the mines to Roberts Bank near Vancouver from which the coal was shipped to Japan.

By the late 1970s coal, copper, iron ore, oil and natural gas dominated the export oriented mineral production, while sand, gravel and cement were produced in large quantities to serve the domestic construction industry which grew rapidly between 1950 and 1960.

THE PHYSICAL SETTING

The great variety of geological environments in British Columbia is reflected in a wide variety of mineral and energy deposits. Most of the world's mineral production comes from two types of geological terrain — cordilleran belts and precambrian shields. Over 85 percent of British Columbia is underlain by cordilleran belts of various metamorphic, faulted and folded rocks which range in age from Precambrian to recent.[3]

The province can be divided geologically into six rather distinct regions (Figure 1,8). The Insular Belt, the westernmost region, consists of heavily faulted, unmetamorphosed volcanic and sedimentary rocks which range in age from Tertiary to Paleozoic. The area is known for coal deposits and minerals, such as copper and iron ore. The mountainous Coast Plutonic Belt is a crystalline region comprised of coarse intrusives of the late Cretaceous to early Tertiary age. It consists of granitic rocks and crystalline gneisses. A fairly wide range of metal ores is associated with these rock formations. In the Intermontane Volcanic Belt the formations consist mainly of unmetamorphosed sedimentary and volcanic rocks. Extrusive lava flows of late Tertiary age dominate in the southern and northern extremities. Streams have deeply eroded parts of this region and exposed intrusive complexes, such as copper porphyries. The area includes many fault and contact zones where copper and molybdenum are commonly found. The Omineca Metamorphic Belt consists of very old sedimentary rocks, primarily schists and granitic intrusions. The major lode metal industries in the province are found in this belt, notably the historic Boundary area in the south, well known for copper, lead, silver and zinc deposits. The Rocky Mountain Belt consists mostly of sedimentary carbonate rock. Thrust faulting is evident throughout the region and, unlike the situation in the Omineca Belt, there is very little volcanic rock. The best known mineralization in this region is in the southern and central areas where metallurgical quality coal deposits are found. The sedimentary rocks found in the Interior Plain Belt lie horizontally and have not been folded as in the

GEOLOGY

FIGURE 1,8 Geologic regions

Rocky Mountain Belt. The sandstones and shales are of Cretaceous and Early Tertiary ages and it is in this region where natural gas and petroleum are to be found.

PATTERNS OF MINERAL DEVELOPMENT

Location of Major Mines and Smelters, and Oil and Gas Production

The patterns associated with mineral production in British Columbia vary considerably. Much of the variation can be attributed to sudden changes in ore prices, the qualities of the ore bodies and operating costs. The locations of the major mines and smelters operating in 1985 are shown in Figure 2,8. A number of points are noteworthy about the spatial pattern. First, the mines in the southeast produce mainly coal, with the exception of the Sullivan Mine at Kimberley which produces lead, silver, zinc and iron. Second, there is a concentration of copper mines in the Highland Valley area, including Afton and Lornex. The Highmont and Craigmont mines in this area have been closed in recent years because of depressed copper prices. Third, the development of northeast coal has led to the establishment of the Bullmoose and Quintette mines. Fourth, the Granduc and Endako mines in the Omineca area are now closed, largely because of low copper prices. The Equity Mine near Houston is still operating, producing gold, silver and copper. The Brenda and Bell copper mines were reopened after the Commissioner of Critical Industries in British Columbia was able to renegotiate some of their major operating costs, particularly electricity. The other major mines are scattered. They include the Gibraltar copper and molybdenum mine near Williams Lake, the Cassiar asbestos mine near the Yukon-British Columbia border, the Island Copper molybdenum and copper mine at Rupert Inlet, and the Lynx/Myra copper, zinc, lead and silver mine on Buttle Lake.

There are only two major smelters in British Columbia. The Trail smelter produces lead, zinc and sulphur. The ore for the smelter comes from the Sullivan Mine, noted above, and from a mine at Pine Point near Great Slave Lake in the Northwest Territories. The smelter at Kitimat produces aluminium, the ore being imported largely from mines in the Caribbean.

Most of the oil and gas industry is located to the north of the Peace River and Fort St. John is the service centre for the industry. There has been exploration for oil and gas in the southeastern part of the province and on Vancouver Island, and there have been moves to resume exploration off the north coast of British Columbia, but there are no producing wells in these areas.

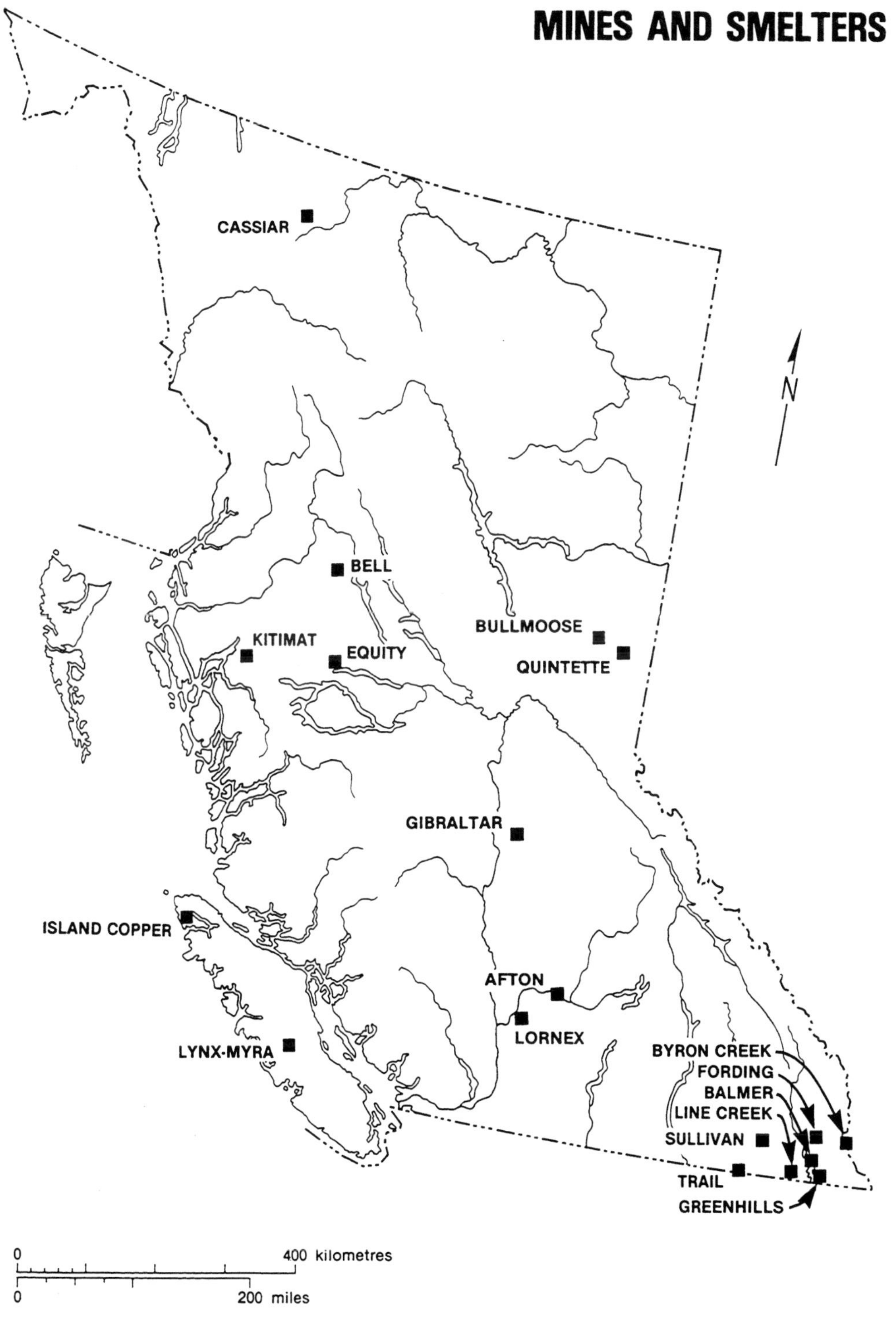

FIGURE 2,8 Location of major mines and smelters, 1985

Value of Mining by Economic Region

The vagaries of international markets and the non-renewable nature of the resource have also affected the value of ore produced in different regions within the province. The net mining revenue, excluding oil and gas production, is shown in Table 1,8. The most important region between 1981 and 1983 was the Kootenays where most of British Columbia's coal was mined. Although mining revenue in the Peace River-Liard region was negligible, the development of northeast coal has greatly increased that region's relative position, but precise figures are not yet available. If petroleum and natural gas production figures were available by economic region and included in Table 1,8 this region would be roughly comparable with the Kootenays in revenue generated. In most regions during this period mining revenue dropped marginally, but in Cariboo-Fort George there was a precipitous decline. The key factors responsible were the depressed prices of copper and molybdenum and the closure of a number of mines, including the Granisle, Endako and Boss Mountain. It is evident from Table 1,8 that the interior of the province is the centre of the mining industry and it is likely to remain so in the immediate future.

Relative Value of Minerals and Oil and Gas

The historical and percentage contribution of various minerals to the British Columbia total is shown in Table 2,8. A number of trends are evident. During most of the nineteenth century only gold and coal were of commercial significance. In the twentieth century the range of minerals exploited was much broader. Gold has continued to decline in importance, although minor recoveries did take place during the worldwide depressions of the 1890s and 1930s. Coal has had a checkered history. In the 1860s, coal was exported. Later, after completion of the railway, it was used for fuel in steam engines, as well as for coking purposes in the early smelters that operated in the province. From 42 percent of the total mineral value produced in 1910, coal declined to only 3 percent in 1960. Since that time it has re-emerged and now accounts for more than 20 percent of the value of mineral production.

Between 1900 and 1950 silver, lead, zinc, copper, gold and coal were the most important minerals. Silver, lead and zinc were often found together, particularly in the spectacular Sullivan Mine at Kimberley. Silver ranged between 5 and 10 percent of mineral value between 1900 and 1950, while lead and zinc varied more widely from a low of 6 percent in 1910 to a high of 63 in 1950. A large proportion of the lead and zinc

Table 1,8 NET MINING REVENUE BY ECONOMIC REGION IN MILLIONS OF DOLLARS, 1981-1983

	1981		*1982*		*1983*	
Economic Region	*Revenue*	*Percent*	*Revenue*	*Percent*	*Revenue*	*Percent*
KOOTENAYS (EAST AND CENTRAL)	962	52	898	55	960	57
OKANAGAN-BOUNDARY	85	5	80	5	72	4
LILLOOET-THOMPSON	272	15	208	13	253	15
LOWER MAINLAND	5	–	4	–	7	1
VANCOUVER ISLAND AND COAST	140	8	129	8	143	9
CARIBOO-FORT GEORGE	273	15	203	12	123	7
PEACE RIVER-LIARD	–	–	–	–	9	1
SKEENA-STIKINE	98	5	114	7	105	6
TOTAL	1835	100	1636	100	1672	100

NOTES: 1. Economic regions are shown in Figure 2,8.

2. Oil and gas revenues are not included.

SOURCE: Price Waterhouse, *The British Columbia Mining Industry in 1983*. Vancouver: Price Waterhouse, Chartered Accountants. Report prepared for the Mining Association of British Columbia, 1984.

was smelted at Trail and then shipped out of the province for fabrication. Copper production fluctuated considerably as well. The invention of electrical power in the late nineteenth century was instrumental in boosting the industry, as transmission lines, generators and batteries devoured copper. Toward the end of the period the cost of producing copper in British Columbia increased relative to that of foreign competitors and copper production declined. Gold declined in the period, except during the depression when prices were very high and old claims were reworked. Most of the best mines had been depleted of their best ore and there were no large discoveries which warranted new mines being opened. Coal production also

Table 2,8 VALUE OF MINERAL PRODUCTION IN PERCENTAGE OF THE TOTAL, 1860 to 1980

	1860	*1870*	*1880*	*1890*	*1900*	*1910*	*1920*	*1930*	*1939*	*1950*	*1960*	*1970*	*1980*
Gold	98	92	56	19	29	23	7	6	34	8	4	1	6
Silver	–	–	–	3	14	5	9	8	7	5	4	2	5
Copper	–	–	–	–	10	18	22	21	11	6	5	25	22
Lead	–	–	–	–	16	5	8	23	18	30	21	7	2
Zinc	–	–	–	–	–	1	9	16	13	33	28	9	2
Coal and coal products	2	8	44	78	29	42	38	17	10	7	3	4	15
Iron	–	–	–	–	–	–	–	–	–	–	6	4	–
Asbestos	–	–	–	–	–	–	–	–	–	–	7	3	3
Sulphur	–	–	–	–	–	–	–	–	–	–	2	1	1
Cadmium	–	–	–	–	–	–	–	–	–	–	–	1	–
Molybdenum	–	–	–	–	–	–	–	–	–	–	–	11	9
Nickel	–	–	–	–	–	–	–	–	–	–	–	1	–
Oil	–	–	–	–	–	–	–	–	–	–	1	13	7
Gas	–	–	–	–	–	–	–	–	–	–	4	6	20
Structural Materials	–	–	–	–	–	–	–	7	3	7	10	9	8
Other	–	–	–	–	2	6	7	2	4	4	5	3	–
Total	100	100	100	100	100	100	100	100	100	100	100	100	100
TOTAL VALUE (in dollars)	2,285,531	1,456,328	1,816,612	2,598,803	16,344,751	26,377,066	35,543,084	55,391,993	65,681,547	148,155,060	179,849,790	485,233,614	3,077,049,327

SOURCE: British Columbia, Ministry of Mines, annual reports.

declined, as alternative heating fuels replaced coal and diesel locomotives took over from coal-fired steam engines.

Between 1950 and 1980 the range of minerals produced expanded and there were changes in the most important minerals. In 1950 seven minerals reached the level of one percent or more of the total value; the comparable figures for 1960, 1970 and 1980 were 12, 15 and 12. The increase is linked to the industrial expansion of the late 1960s and 1970s and the resulting demand for industrial minerals. In 1950 lead and zinc dominated mineral production, but as the quality and amount of ore at the Sullivan Mine declined, Cominco secured more of its ore from the Pine Point Mine. By 1980 copper, molybdenum, coal, oil and gas were the leading minerals. Coal and copper re-emerged as important minerals, partly because open-pit mines and a vastly improved transportation system decreased costs of production. Molybdenum, often a by-product of copper mining operations, was used mainly as an alloying agent to strengthen steel and as a catalyst in petroleum refining. Oil and gas were not produced commercially until the middle 1950s, but by 1980 they constituted 27 percent of the value of mineral production. Most of the oil is consumed in Canada, while a significant portion of the natural gas is exported to the United States.

THE MINING INDUSTRY IN THE 1980s

The mining industry established record levels of value and production as British Columbia entered the 1980s, but these rosy figures have not been matched since and by the middle of the decade the industry was quite depressed. Still, it remains important to the national and provincial economies: British Columbia is the leading provincial producer of copper, molybdenum, lead and coal, and mining is second only to forestry among the resource industries in terms of its contribution to the gross provincial product.[4] Solid minerals have decreased in production and value, while petroleum and natural gas have increased. In 1982 coal replaced copper as the leading mineral in terms of value. Natural gas ranked second, copper third and crude oil was fourth. If these trends continue the value of oil and natural gas production may soon outrank the value of metal production in the province.

Mining companies did not respond well to the worldwide economic depression which developed in 1981. They believed that the depression would be temporary, and only later cut production and closed mines. These actions did not, however, lead to significant recovery in the industry.

Since 1982 many third world producers, particularly in Africa and South America, have continued to glut the market. They have maintained production without regard to cost, in order to obtain foreign exchange and protect domestic employment. In addition, the mining industry has faced increased costs, including higher taxes, power rates, transport charges and wages. In 1985 the provincial government recognized some of the special problems faced by all resource industries in a period of economic decline and established the Commissioner of Critical Industries whose job it is to help bring marginal plants back into production.

A significant change in recent years has been the rise of coal to first position among the solid minerals. There has been considerable expansion in production capacity, with the opening of the northeast coal fields. Most of this production is going to Japan through a new coal terminal built on Ridley Island near Prince Rupert. Attempts have been made to diversify markets, with shipments going to Korea, Brazil, Spain, Denmark and Italy, but Japan remains the single most important market. The Japanese have encouraged coal mine development in many areas, partly because they overestimated their needs and partly because they wanted to reduce their reliance on any one supplier, as well as to control production and price. The result has been that coal prices and production have declined in southeastern British Columbia at the same time that northeast coal has been brought into production. Even in the northeast the Japanese have attempted to renegotiate the prices that they pay for the coal.

The worldwide depression of the 1980s also affected oil and natural gas production. This portion of the industry reached record levels in 1980, but has slumped since, as world prices for petroleum products have declined. British Columbia has a large surplus of natural gas. Much of this gas is exported to the United States, but exports have declined recently because of the high Canadian border price for natural gas relative to that of other gas sources, such as Mexico. Government initiatives, such as the National Energy Program, were also cited as reasons for cutbacks, since wellhead pricing and federal tax regimes discouraged investment. Governments and industry have made attempts to look for other markets. These initiatives include the possibility of liquid natural gas exports to Japan, construction of a major fertilizer plant and construction of a natural gas pipeline to Vancouver Island. Oil contrasts sharply with natural gas, as all of British Columbia's production is consumed within the province. However, this amounts to only 20 percent of the total provincial oil consumption. Some recent oil discoveries, particularly in the Fort Nelson area, may lead to production increases, but it is likely that the province will remain largely dependent on outside supplies.

CONCLUSION

The mining industry in British Columbia is the second most important resource industry in the province. It is not as unstable as it was in the last century, largely because the range of minerals exploited is greater, but it remains an industry subject to price fluctuations on a world scale. In the last 30 years energy minerals, such as coal, oil and gas, have become dominant. Most minerals, with the exception of oil and structural materials, are exported. When prices are depressed export markets suffer, mine production decreases and communities experience rapid declines in population. Regionally, the non-metropolitan areas of the province face the greatest difficulties, since most of the mines are located in these regions.

During periods of economic recovery and high prices for minerals boom conditions tend to heighten optimism about the future of the mining industry. When new mines are opened towns are often developed to serve the mines. Other industries, such as sawmills, are sometimes drawn into the region and the local economy experiences a boom until the mine runs out of ore or prices decline and the mine closes. In either case the bust cycle appears and the decline begins.

The mining industry in British Columbia has often experienced these extremes and the mid-1980s are no exception. The province faces depressed economic conditions and declines in the value of mineral production. Some have suggested that the industry is a sunset industry in the sense that it is declining in importance in terms of value and employment. Others note that the industry has experienced bad times in the past and that the sun will rise again for the industry. If there is to be a strong recovery in the mineral industry prices for commodities will have to rise significantly, markets will have to be diversified and production costs will have to fall. Recovery would also be aided by greater processing within the province. Without some of these developments the mining industry is likely to remain stagnant and depressed for much of this decade.

REFERENCES

1. HEDLEY, M.S., *The Mineral Industry of British Columbia*. Victoria: Department of Mines and Petroleum Resources, 1969, pp. 3-4.

2. FARLEY, A.L., *Atlas of British Columbia*. Vancouver: The University of British Columbia Press, 1979, p. 72.

3. BARKER, M.L., *Natural Resources of British Columbia and the Yukon*. Vancouver: Douglas, David and Charles, 1977, pp. 16-17; British Columbia, *Inventory of the Natural Resources of British Columbia*. Victoria: British Columbia Natural Resources Conference, 1964, pp. 383-387.

4. PRICE WATERHOUSE, *The British Columbia Mining Industry*. Vancouver: Price Waterhouse, Chartered Accountants. Reports prepared for the Mining Association of British Columbia, 1975-1984.

PLATE 11 West coast seiner. *R.C. Dixon Photo* ▶

PLATE 12 Fishing fleet at Prince Rupert ▶
Prince Rupert Convention and Visitors Bureau Photo

ZAPORA

9 FISHERIES

William M. Ross
University of Victoria

The fishing industry in British Columbia faces a dilemma, not so much of survival, but of scale, direction and character. Sol Sinclair and Peter Pearse,[1] in two recent reports on the fishery, have argued for policies to make the industry more efficient, yet these policies portend declines in industry employment and particular hardships for many isolated communities which depend on the industry. Attempts to reduce capitalization in the primary sector may affect people who have little opportunity for alternative employment without relocation. These same policies threaten traditional lifestyles, the number of people in the industry and aboriginal involvement in the fishery.

The resource base has traditionally been very rich. Salmon dominates the industry, yet herring, halibut and other demersal species are also important. No one area along the coast dominates fish production, but certain fisheries statistical areas are the most important: 23 (Barkley Sound), 2E (the eastern Queen Charlotte Islands), 7 (Central), 12 (Johnstone Strait), and 4 (Skeena) (Figure 1,9). In most areas salmon dominate, but herring are prominent on the central west coast of Vancouver Island and the eastern shore of the Queen Charlotte Islands. The Strait of Georgia area is of modest importance, but much of the shrimp and oyster production comes from this region.[2]

The resource base is increasingly threatened by larger numbers of fishermen with sophisticated equipment and by urban and industrial development. Stocks have diminished considerably since the turn of the century. The exact causes of such declines are not known, but overfishing and loss of habitat are thought to be prime factors. The role of the fishery in the economic and social life of the province has changed considerably since Captain Cook first started a fishery trade in the 1770s and the first commercial cannery was opened on the Fraser in 1870. At one time during the 1880s and 1890s the industry dominated life in British Columbia, yet today the industry contributes less than five percent of the gross provincial product.[3] During its most prominent period the industry occupied many sites along the coast, but has now consolidated primarily in two locations

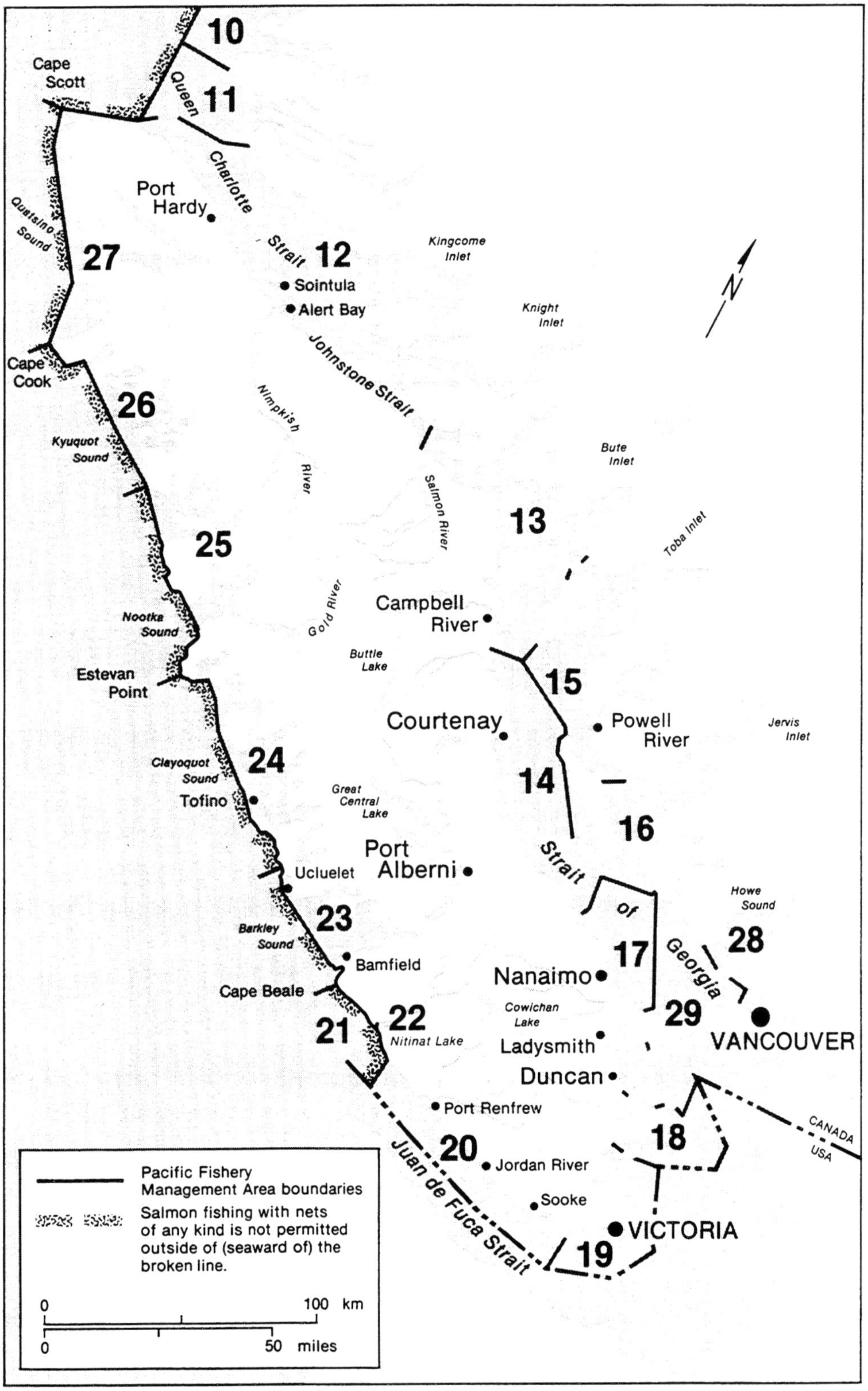

FIGURE 1,9 South coast fisheries statistical areas

— Prince Rupert and Vancouver. Although the industry remains a critical element of life at certain locations along the coast its relative importance in the province has diminished.[4]

Compared with that of Atlantic Canada, the industry in British Columbia is younger, much more advanced, but far less important to the region. The Pacific fishery is more capital intensive and the processing sector has been consolidated. There are other employment alternatives in many communities, unlike the case in much of the Atlantic region. The fishery in British Columbia operates as a total system from Dixon Entrance to the Strait of Juan de Fuca and is no longer characterized by isolated fishing communities dependent on local species.

The industry in British Columbia has evolved in response to both long-standing and relatively new pressures. Too many fishermen are pursuing too few fish. Processing plants remain idle for significant portions of the year. There is pressure on stocks from other industries, from within the industry and from native peoples. Fishery products which were once basic foods that competed with meat and poultry have become luxury items, and are highly subject to fluctuating markets. Finally, the extended fisheries jurisdiction associated with the 200 mile (321.9 kilometre) economic zone presents new challenges for an industry based largely on near-shore fishing.

THE FISHERY IN AN INTERNATIONAL CONTEXT

The earliest fishing in British Columbia was located near shore, particularly in the river mouths. There was little pressure on stocks because the technology of the time did not permit large scale offshore fishing. As vessels were improved, with adoption of gas engines and other improved technology, they moved farther and farther offshore. After World War II other nations began to fish off the coast of British Columbia and interception of fish by American fishermen began to have detrimental impacts on stocks.

In 1964, after years of negotiations, Canada unilaterally declared a nine-mile (14.5 kilometre) exclusive fishing zone beyond the three-mile (4.8 kilometre) territorial sea. In 1970 Hecate Strait and Queen Charlotte Sound were declared to be exclusive fishing zones (Figure 2,9). This action took place at a time when there was general dissatisfaction with the existing law of the sea. By the time the Law of the Sea Conference was convened in 1973 Canada had already negotiated bilateral arrangements for access to stocks surplus to Canadian needs, particularly groundfish. In 1977 Canada declared a 200-mile (321.9 kilometre) economic zone within which the country asserted exclusive fishing rights. These actions

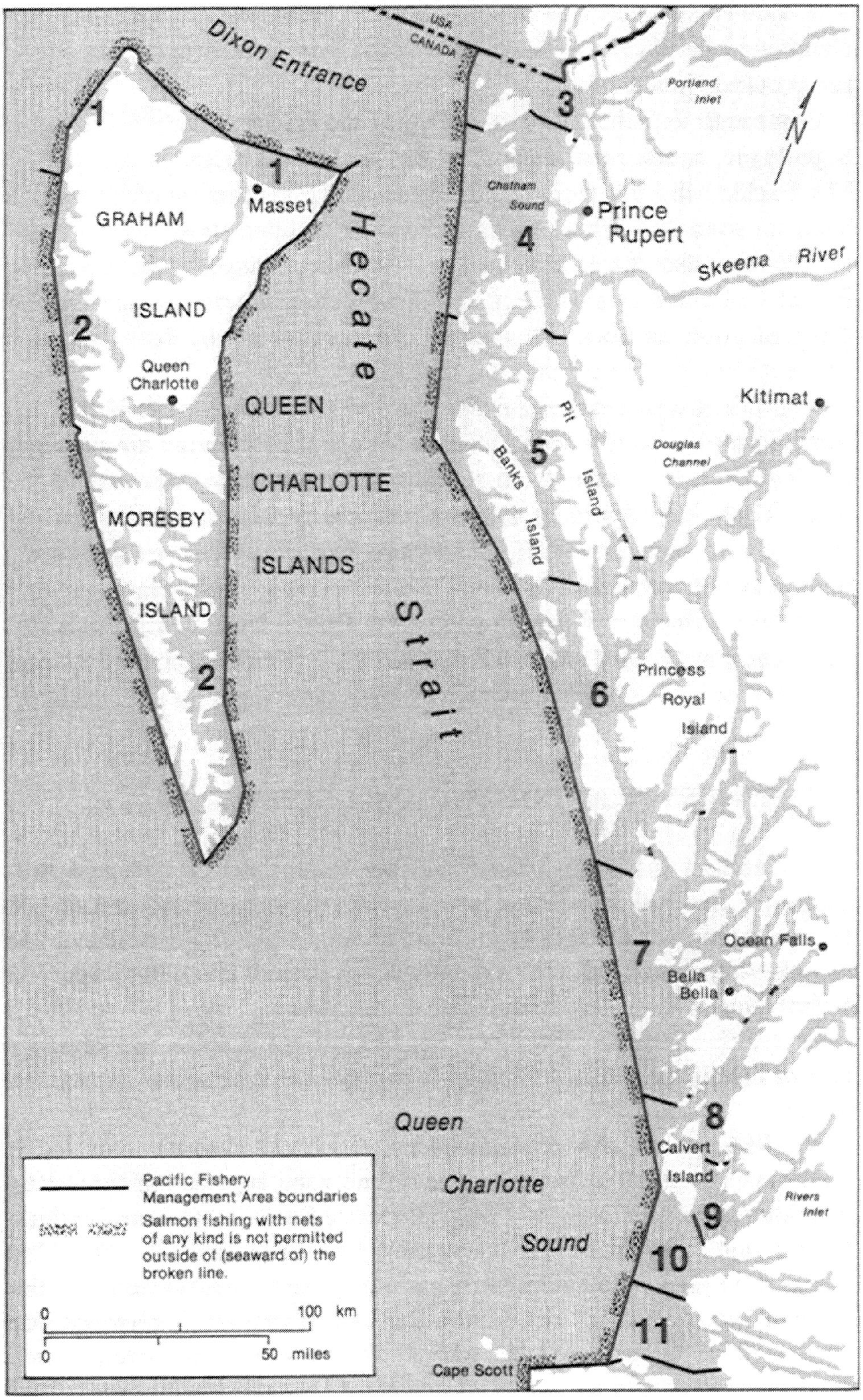

FIGURE 2,9 North coast fisheries statistical areas

permitted Canada to exclude foreign fishing for halibut and other species in Hecate Strait, for example, but they did not eliminate the problem of foreign impingement on Canadian stocks.

Salmon which spawn in the rivers of British Columbia range over large areas of the Pacific Ocean and can be caught when they venture outside the economic zone. Salmon destined to four of British Columbia's rivers are subject to significant interception by American fishermen in the economic zone. Fish bound for the Nass and Skeena must pass the American seine fleet at Noyse Island before entering Dixon Entrance. Salmon bound for the Fraser migrate through Johnstone Strait and the Strait of Juan de Fuca. Those passing through Johnstone Strait are not intercepted by American fishermen, but in the Strait of Juan de Fuca there is considerable interception. On the Stikine most of the spawning areas are in Canada, but the prime area for catching salmon is in the Alaska panhandle.[5] It is only in the last decade that Canada has established a viable commercial fishery on the Stikine.

Canadian fishermen also intercept salmon bound for American territory, particularly coho heading for Washington waters. In addition, Alaskan fishermen have complained about Canadian overfishing of chinook stocks which migrate all along the west coast. Interception of stocks has led to much dissatisfaction on both sides of the border. Only sockeye and pink salmon entering the Fraser after passing through the Strait of Juan de Fuca are subject to joint management by Canada and the United States. An agreement to manage Pacific salmon along the whole coast was initialled in January of 1983. The proposed United States-Canada Salmon Treaty set out levels of chinook interception by Canada. As a result, little progress has been made towards ratification and the positions of both countries seem solidly entrenched.

Access to stocks may also be altered, depending on the final boundaries for the 200-mile (321.9 kilometre) economic zone. The positions of Canada and the United States in the Strait of Juan de Fuca vary only slightly. In Dixon Entrance the positions are quite divergent. The United States claims an equidistant line through Dixon Entrance. Such a boundary would enhance United States' interception of Nass and Skeena stocks. The Canadian government argues for the existing A-B line through Dixon Entrance and then an equidistant line to the 200-mile (321.9 kilometre) limit. British Columbia asserts claim to much of the sea off the coast of southeastern Alaska, claiming that the predominant land mass behind the panhandle is Canadian.[6] If this claim were upheld in the final resolution of the dispute American interception would be reduced and valuable habitat grounds would come under Canadian jurisdiction.

MANAGEMENT AND THE LEGAL FRAMEWORK

The fisheries in British Columbia have highlighted some of the larger conflicts between federal and provincial governments over management of natural resources. The Constitution Act (formerly the British North America Act), 1867, provides the parameters under which the fisheries have been managed for over a century. Section 91 of the Act gives the federal government authority to legislate with respect to seacoast and inland fisheries, navigation and undertakings which are inter-provincial or international in nature. Section 92 gives the province a proprietory (ownership) right to the resource, but in the case of most tidal fisheries the right to fish in public waters is a public right, not a proprietory right. As a result, the management of the fishery is a federal responsibility until such time as the fish are landed. Once landed they are capable of being processed and hence are then subject to provincial control and regulation.

Interactions between the federal and provincial governments over fishery management are extremely complex.[7] The federal government can issue regulations pursuant to the Fisheries Act which governs such items as gear, seasonal openings and fishing effort. Some of these are developed by the Department of Fisheries and Oceans. In other cases, such as trout management, the regulations are prepared by the British Columbia Fish and Wildlife Branch and then forwarded to Ottawa where they are consolidated with tidal regulations and published. While the arrangements are complex, de facto control of the fishery lies with the federal government which administers several acts, including the Fisheries Act, the Fisheries Price Support Act and the Fisheries Inspection Act. Of these the Fisheries Act is the most important. Amendments passed in 1977 give greater authority to the Department of Fisheries and Oceans to prosecute violators in incidents involving adverse impacts of mining, logging and industrial activities.

These amendments heightened provincial suspicions that federal initiatives are sometimes disguised attempts to influence the management of other resources. Four cases are pertinent. In the case of Riley Creek, Queen Charlotte Timber was charged with causing damage to a salmon stream through poor logging practices. In another case Dan Fowler was charged under Section 33(3) of the Fisheries Act with dumping debris into a stream. Charges against Queen Charlotte Timber were subsequently dropped, but in the Fowler case the Supreme Court of Canada ruled that the use of this section was an attempt by the federal government to manage forest resources and was *ultra vires* the federal Parliament.[8] In two other cases the courts have sustained the powers in the 1977 Fisheries Act. One case involved water flow in the Nechako River. The Aluminium Company of Canada wanted to reduce the flow in the river to augment power production at Kemano. The court ruled that flows had to be maintained at a level which would not harm salmon survival. In the other case, Will Paulik, a private Richmond citizen, used the Fisheries Act to lay charges against Richmond Landfill for dumping leachate into the Fraser River.[9] The company pleaded guilty on six counts. There appears to be a clear pattern in these decisions. Where it can be shown that there is a specific scientific relationship between the action and damage the Fisheries Act can be used to help control the actions of companies and individuals. The evidence concerning the impact on salmon of increased siltation and debris in a stream is not as clear as that concerning the effects of water levels, temperature and leachate. As a result, the Act has not been used very successfully to alter forestry practices. The number of incidents and conflicts between the two levels of government is likely to increase, especially since much of the industrial development in the interior of the province is in areas frequented by spawning salmon.

INTERNAL SOURCES OF CONFLICT

Beyond the conflicts associated with federal-provincial clashes other internal problems face the industry. Traditionally, fishery management has been directed to physical rather than economic and cultural relationships. This reliance on biological models for management arose from the expectation that maintenance of physical stock was a sufficient condition for a healthy fishery. One problem is that physical stocks are threatened by overfishing in both the sport and commercial fisheries. Aboriginal peoples claim that the fishery is important to their culture and have laid claims to special access, even for commercial purposes. The claims of the Nishgas to the resources of the Nass River valley illustrate this problem well.

At the heart of almost all the problems associated with the industry is the common property and seasonal nature of the resource. The excess capacity of fishing fleets and processing operations is endemic to the industry. The right to fish traditionally has been regarded as a right of free access to fish, yet to protect stocks many restrictions have had to be placed on fishermen. The fishery is open for a few months of the year and capital invested in vessels and processing plants is idle during the off season. There appears to be a growing awareness among managers of the resource that the essential problems involve questions of choice, allocation and economic efficiency. Reconciling the competing interests of seine and gillnet or sport and commercial fishermen involves complex choices.

THE COMMERCIAL FISHERY

The fishing industry in British Columbia has evolved into a highly capitalized industry dominated by a few major companies, such as British Columbia Packers. It is characterized by a stable but technologically well-equipped fleet and focused on catching salmon and herring (Table 1,9). Salmon are by far the most important species (Figure 3,9). The Fraser, Skeena and Nass rivers, as well as Rivers Inlet, are the prime locations for the fishery. Herring are found all along the coast in sheltered bays and inlets. Traditionally, most salmon were canned, but more and more now are marketed as fresh fish. Herring continues to be reduced to oil and meal, but in the last decade herring have been valued more for the roe.

Salmon

The commercial industry in British Columbia emerged with the first cannery on the Fraser River in 1870. At this time the industry consisted largely of small independent firms strategically situated on the major rivers. There was a low level of plant concentration and a high degree of local proprietorship and financing, especially by Victoria merchants. By 1900, the emergence of limited companies led to a large increase in the number of plants and high concentrations on the major rivers. These companies anticipated large profits from the industry and sought to restrict competition by establishing fish plants in all major fishing areas along the coast. The number of commercial canneries reached its zenith in the 1920s when there were over 120 operating and since then the number has declined to less than 10. Initially most of the smaller independent canneries disappeared and later even many of the large and medium sized companies, such as the Canadian Fishing Company, Anglo-British Columbia Packing Company and British Columbia Packers, began to consolidate their operations. Most of these closoures were caused by high seasonal operating costs at the remote locations and when the ability to transfer fish over larger distances without seriously diminishing quality was achieved, centralized locations became attractive. During the 1960s and 1970s those companies which failed to diversify into fresh fish and other products, or depended too greatly on one product, such as canned salmon, found it difficult to survive. Companies such as Nelson Brothers were acquired by others. This resulted in more cannery closures and greater consolidation of production in Prince Rupert and Vancouver. Many communities, particularly those in isolated areas along the coast, such as Namu, lost important sources of local employment. A few new plants were opened in the late 1970s, primarily co-operatives, but their production is quite small.

Table 1,9 VOLUME AND LANDED VALUE OF THE MAJOR COMMERCIAL SPECIES IN BRITISH COLUMBIA

	Landings in 1981		*Approximate price in 1981*
	volume *(metric tons)*	*value* *(millions of dollars)*	*(dollars per pound)*
SALMON	74,476	154.3	0.56 to 2.20
HERRING (roe)	29,300	32.9	0.51
HERRING (food & bait)	8,663	2.2	0.12
HALIBUT	2,159	6.4	1.35
SABLEFISH	3,720	4.9	0.88
GREY COD	5,154	2.7	0.24
LING COD	1,780	1.6	0.41
ROCKFISH	9,029	3.0	0.15
SOLE	4,282	2.2	0.23
POLLOCK	1,106	0.2	0.11
TURBOT AND FLOUNDER	1,149	0.2	0.09
DOGFISH	755	0.1	0.08
TUNA	200	0.4	0.91
OTHER PELAGIC AND ESTUARIAL SPECIES	261	0.8	1.46
CRAB	1,191	2.6	0.99
SHRIMP AND PRAWNS	839	2.6	1.41
GEODUCKS	2,620	2.1	0.36
ABALONE	73	0.7	4.25
CLAMS AND HORSE CLAMS	754	0.7	0.41
HERRING SPAWN-ON-KELP	186	3.0	7.33
HAKE	6,200	0.9	0.06

SOURCE: Pearse, P., *Turning the Tide – A New Policy for Canada's Pacific Fisheries.* Vancouver: Commission on Pacific Fisheries Policy, Final Report, 1982, p. 119.

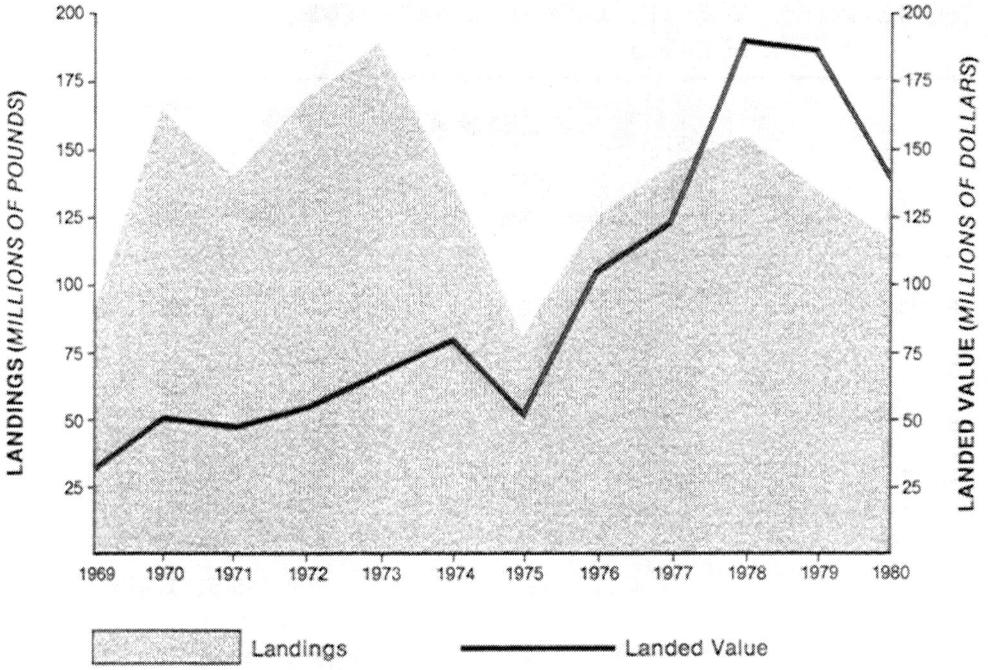

FIGURE 3.9 Salmon landings and landed values, 1969-1980

The commercial fishery entered the 1980s facing additional economic challenges. The price for salmon and other species rose dramatically in the 1970s, to such a level that products became luxury items and consumer resistance began to develop. The value of the Canadian dollar increased in relation to other currencies, further impeding the ability of processors to market products overseas. In addition, two fatalities in Europe, caused by botulism in Alaskan canned salmon, made marketing even more difficult. These events occurred at a time when interest rates were at their highest levels in decades and competition from large Alaskan catches was substantial. Processors had to finance large inventories. This placed added burdens on an already shaky industry. Profits of major companies, such as British Columbia Packers, declined seriously and two of British Columbia's oldest fishing companies, Cassiar Packing and Frances Millard, went into receivership. There has been great instability in the processing sector and consolidation of production in fewer and fewer plants appears to be continuing.

While there has been some rationalization of processing capacity in the industry, the same cannot be said for the fleet. As early as 1880 there was anxiety about provincial stocks because fishing practices similar to those used in British Columbia caused declines in the salmon catches on the Sacramento and Columbia rivers. In the 1880s and again in the early 1900s attempts were made to limit the number of vessels by allocating a particular number to a specific cannery. Neither of these attempts worked, as the lure of high profits led to the demise of each scheme. In 1958 the federal government appointed an economist, Sol Sinclair, to investigate the excess capacity in the salmon and halibut fisheries. Sinclair proposed a system of restricted vessel licences and levies on catches to dampen incentives for over-investment.[10] The report was not acted on directly, but did form the basis of the 1968 Davis Plan, named for the Minister of Fisheries at the time.

The Davis Plan was designed to increase fishermen's earning power and to reduce the size of the fleet. It eventually became a four-phase program, but in the initial stages the intent was to licence only those vessels that could show a significant dependence on salmon. Phases introduced later were designed to reduce the size of the fleet by purchasing vessels, by improving the seaworthiness of vessels and by providing a rational management scheme for the fishery. At best, the results of the 1968 scheme were mixed. The number of vessels in the fleet was reduced. However, the capacity of the fleet increased, as gillnetters and trollers were replaced by seiners, which have a greater per ton capacity to catch fish than the smaller vessels. The seaworthiness of the fleet improved, but the level of capital investment in the industry increased substantially to a point where the capacity of the fleet is estimated to have doubled or perhaps trebled. The lessons to be learned were articulated well in the Pearse Report:

> . . . fishermen and vesselowners will try to expand their fishing power whenever they compete with each other for an unspecified share of the catch of a common-property resource, even if the fleet's capacity is already excessive. When the value of the catch rises and governments fail to capture the extra returns from fishing, the fishermen will reinvest them in expanded fishing capacity until the higher costs absorb the higher revenues. Restrictions on the number or dimensions of vessels will not prevent this investment in greater fishing power.[11]

Dissatisfaction with the Davis Plan led the federal government to commission Peter Pearse to undertake a study of the west coast fishery in 1981. Pearse made numerous recommendations. The most contentious points were proposals to auction 10-year licences to the highest bidders and to impose royalty taxes on fish landings. Fishermen voiced a large number of concerns over the Pearse Report. They argued that bidding for licences would eliminate most independent fishermen. Some of these might be from small coastal communities, such as Sointula, that rely heavily on the resource for economic survival. Pearse's major proposals were rejected by the Minister of Fisheries and Oceans who invited the industry to submit alternatives. No changes have been made to the regulations as of January, 1984.

Other Species

Compared with salmon the value of other species landed in British Columbia is relatively small. Among the various industries dependent on these species herring, particularly the roe-herring industry, is the most important. Prior to 1970, herring was reduced to oil and meal and some was used for bait. Landings varied, but consistently were over 90,800 tonnes (100,000 tons). In 1967, landings fell to 52,664 tonnes (58,000 tons) and the fishery was closed, except for a limited food and bait fishery (Figure 4,9). The industry reemerged in the 1970's after the herring stocks partially recovered from the collapse of the late 1960's and a demand for herring-roe developed in Japan. The main roe-herring fishery takes place from late February to early April when the roe is ripe and the fish are about to spawn. This sector is much more volatile than the salmon industry. Overfishing and the reliance on a single market are the major causes of the volatility in the industry. Attempts have been made to control the size of the fleet, as was the case in the salmon fishery, but these have not prevented the capacity of the fleet from increasing.

The halibut fishery is one of the oldest on the coast. Most of the fishing grounds are in northern British Columbian and Alaskan waters. By 1920, it was apparent that stocks were being severely depleted by overfishing. Because of the trans-boundary nature of the stocks Canada and the United States signed a convention to jointly manage stocks in 1923. The International Pacific Halibut Commission established by the convention recommends catch quotas to both Canada and the United States. The quotas are for particular areas of the Pacific Ocean and are not assigned to any one country. The Commission has no power to regulate entry and, as a result, the fishing season has progressively been shortened as stocks declined. The short fishing season compounded the processing problem, since large quantities were delivered in short periods of time and much of the catch had to be frozen instead of serving the high priced fresh market. During the late 1960s and 1970s halibut stocks declined precipitously (Figure 5,9), primarily because of incidental

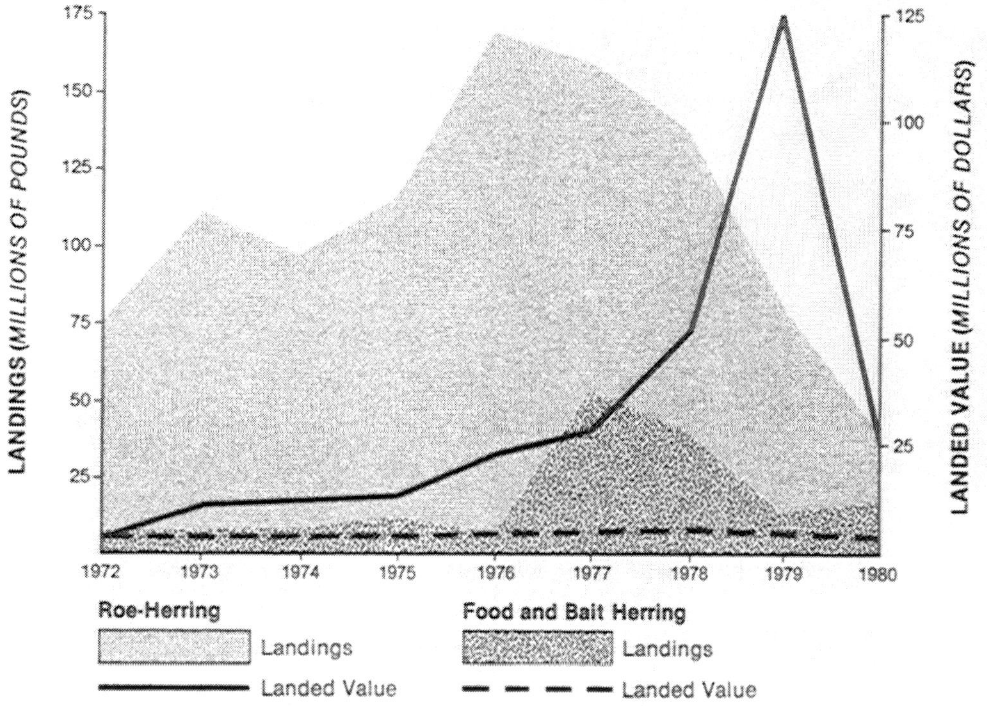

FIGURE 4.9 Herring landings and landed values, 1972-1980

halibut catches by foreign high seas trawl fleets. The declaration of 200-mile (321.9 kilometre) fishing zones by Canada and the United States cut deeply into Canada's production as well. Prior to this event two-thirds of the Canadian catch had been in American waters, but after the declaration Canadian fishermen were excluded. This put added pressure on Canadian fishing grounds because the fleet was not reduced proportionately. During the 1970s the value of halibut rose, but it declined substantially in 1980 and 1981. Prices since then have remained relatively stable in a period when fishermen have faced increased costs for fuel, gear and provisions.

Unlike the case in the major commercial fisheries, there have been modest increases in volume and landed value in many of the minor fisheries. Stocks which have not been exploited greatly in the past, including Pacific cod, sole and hake, show large increases in the volume of fish caught. For others, such as shrimp, prawns and crab, volume has remained constant, but the price has increased significantly. Most of the groundfish are taken offshore, while exploitation of other species is focused on internal waters, particularly the Strait of Georgia. Most of

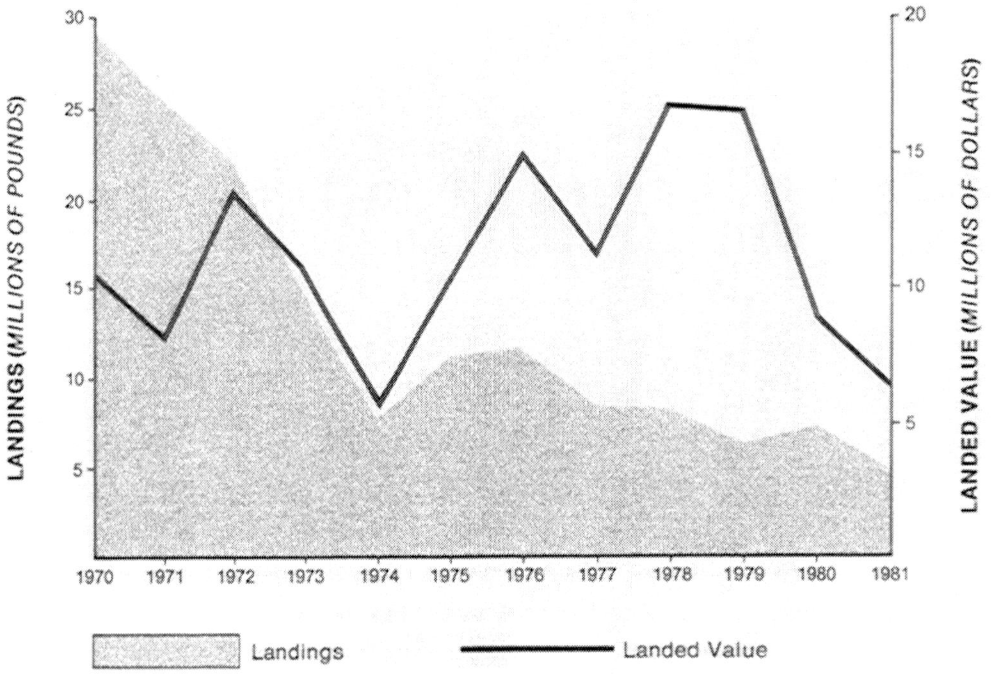

FIGURE 5.9 Halibut landings and landed values, 1970-1981

the minor species are now utilized to their maximum sustained level. Although there is some opportunity to enhance production through mariculture, they will remain a minor element of the fishery compared with salmon.

THE SPORT FISHERY

The sport fishery emerged after World War II in an era when stocks available for commercial exploitation were becoming more limited. The competition between the commercial and sport fishermen has grown more intense over time. The sport fishery attracts fishermen from all over the world and generates a great deal of commercial activity. The opportunities for sport fishermen range from trophy-fishing for chinook salmon to river fishing for steelhead. Estimates of the number of sport fishermen vary from 320,000 to 467,000 for both tidal and non-tidal fishermen. The total catch attributed to these fishermen is highly debatable. It is known that the largest portion of the sport fishery catch is concentrated between

Johnstone Strait and the Strait of Juan de Fuca. There is a small sport fishery on the north coast based in centres such as Rivers Inlet where camps have been established at some of the old canneries. Pearse estimated that the sport catch was approximately 1,000,000 fish in 1982. In terms of species, coho accounted for 65 percent, chinook 35 and pink 2. In some areas, particularly the Strait of Georgia, such figures represent 50 percent or more of the total catch of some species. Since the figures and the method of collection are subject to considerable error, they must be treated cautiously, but they do indicate that the sport fishery has significant impact on stocks.[12]

The impact of the sport fishery on the province and small communities cannot be ignored. Estimates of the total economic value of the sport fishery are quite crude. The capital value of vessels engaged in sport fishing throughout the province is estimated to be 800 to 900 million dollars, or about equal to the value of the commercial fishing fleet. Spending related to salt water sport fishing approximates 100 million dollars annually. The value of local industries that have grown around the sport fishery, such as small custom canneries, marinas and tourist accommodations, has not been determined. For particular places, such as Campbell River, the sport fishery is one of the most important elements of the community's economic base.

As the sport fishery has grown, so has the conflict with the commercial sector. Each blames the other for declines in stocks. The sport fishermen argue that the federal fisheries department has traditional ties with the commercial sector and has failed to take significant action to regulate overfishing. Commercial fishermen point to areas along the coast, such as Howe Sound, that have been closed permanently for commercial fishermen and to the abuses of some sport fishermen who sell salmon. Such arguments are indicative of the numerous interest groups in the fishery that tend to blame each other for depletion of salmon runs.

THE INDIAN FISHERY

The Indian fishery is more than just a fishery. Fish, especially salmon, were an early source of food for Indians, but they also played an important role in the art, myths, ceremonies and feasts. Estimates suggest that pre-colonial Indian consumption of fish amounted to 31.5 - 312.5 kilograms (70 - 700 pounds) per capita annually. Today diets are much more varied and fewer Indians depend on fish as their prime source of food. However, fish remains central to Indian folklore and non-Indians tend to underestimate its importance to the cultural identity of Indians.

In the last decade Indian catches have risen, but they remain less than five percent of the annual catch and much below the levels prior to European settlement. Indians complain that stringent regulations concerning permits and fishing times are an unfair interference with their historical rights and traditions. The tighter regulations have led to stronger enforcement and to federal prosecution. The federal government has been concerned that fish designated for Indian consumption has been sold on the commercial market. Indians argue that, since they fish primarily on rivers and lakes, sport and commercial fishermen take their allotment and federal authorities unduly restrict Indian catches in order to allow sufficient escapement. Intertwined with this conflict is the slow resolution of Indian aboriginal rights and land claims, which tends to sour relations between the Indians and government.

Indian participation in the commercial fishery has declined appreciably since the early years of the century. Prior to the consolidation of processing operations Indian fishermen fished in waters near the canneries. Usually they rented boats from the companies. As the canneries closed Indian participation in the fishing sector fell as well. By 1970 the fishing and processing sectors employed less than half the number of Indians employed in 1950. Pearse commented:

> The erosion of employment in fishing and related occupations has had a devastating impact on dozens of Indian communities that offered no other employment opportunities and where unemployment was already chronic.[13]

The 1968 Davis Plan initially accelerated the decline in Indian participation in the fishery, but the federal government moved to arrest this decline through programs designed to maintain Indian participation. In 1982 the government helped the Northern Native Fishery Corporation purchase over 240 vessels from British Columbia Packers. The Corporation has sold most of these vessels to Indian fishermen. These moves have prevented a further decline in Indian participation in the fishery, but it still remains below 1950 levels.

CONCLUSION

There are pressures on the British Columbia fishery which are altering the scale and character of the industry. The pressures — excess capital in the industry, overfishing, threats to habitat, exclusive access for particular uses, highly competitive international markets and too many fishermen —

have been present for some time. Some segments of the industry have responded more quickly than others. The processing sector has continued to consolidate firms and plants since the 1920s. The traditional romantic nature of life associated with small canneries in isolated areas along the coast has virtually disappeared. In the commercial sector, the number of fishermen has declined, but there is still a substantial number of fishermen who fish only one or two days per week during the short fishing season. If the industry is to become more economically efficient consolidation in both the fishing and processing sectors will have to continue.

Consolidation and the pressures for greater economic efficiency threaten lifestyles and particular groups of fishermen. Stocks cannot be exploited at higher levels to satisfy the wants of all three groups, commercial, Indian and sport fishermen. If, for example, the recommendations in the Pearse Report to enhance aboriginal participation are accepted and the closure of more areas to commercial fishing continues it is evident that the burden of reducing fishing effort will fall most heavily on the commercial sector.

The British Columbia fishery is still a vital element in the social and economic life of the province. In many respects, the direction the industry takes in the next decade will depend on its ability to resolve its internal problems. Against this background, however, is the added complication that the industry has little direct control over the resource itself and the habitat needed to sustain the resource. If habitat is not protected well, and if stocks are not maintained or enhanced by various levels of government, then industry decisions will mean little. This is the dilemma with which the industry and government must contend in the next decade.

REFERENCES

1. SINCLAIR, S., *A Licensing and Fee System for the Coastal Fisheries of British Columbia*. Vancouver: Department of Fisheries and Oceans, 1978; PEARSE, P., *Turning the Tide — A New Policy for Canada's Pacific Fisheries*. Vancouver: Commission on Pacific Fisheries Policy, Final Report, 1982.

2. FARLEY, A., *Atlas of British Columbia*. Vancouver: University of British Columbia Press, 1979, pp. 88-93.

3. BRITISH COLUMBIA, *British Columbia Financial and Economic Review, 43*. Victoria: Ministry of Finance, 1983, pp. 70-71.

4. SINCLAIR, W., *The Importance of the Commercial Fishing Industry to Selected Remote Coastal Communities of British Columbia*. Vancouver: Department of Environment, 1971.

5. LOGAN, R., "Geography and Salmon: The Noyes Island Conflict, 1957-1967", *Journal of the West*, 8, 1969, pp. 438-446.

6. BRITISH COLUMBIA, *Submission of the Province of British Columbia on West Coast Maritime Boundaries Between Canada and the United States*. Victoria: Queen's Printer, 1977.

7. PEARSE, P., *op. cit.*, reference 1, pp. 19-34.

8. DORCEY, A., MCPHEE, M. and SYDNEYSMITH, S., *Salmon Protection and the B.C. Coastal Forest Industry: Environmental Regulation as a Bargaining Process*. Vancouver: Westwater Research Centre, 1980, pp. 71-92.

9. "Quiet Habitat Crusader Wins Round", *The Fishermen*, 48, January 28, 1983, p. 7.

10. SINCLAIR, S., *Licence Limitation — British Columbia: A Method of Economic Fisheries Management*. Ottawa: Department of Fisheries, 1960.

11. PEARSE, P., *op. cit.*, reference 1, p. 79.

12. PEARSE, P., *op. cit.*, reference 1, pp. 187-200.

13. PEARSE, P., *op. cit.*, reference 1, p. 152.

PLATE 13 The wading pool in Beacon Hill Park, Victoria. *Time-Colonist Photo* ▶

PLATE 14 A coastal stream. *G. Gorley Photo* ▶

10 WATER RESOURCES[1]

W.R. Derrick Sewell
University of Victoria

Introduction

More than almost any other natural resource, water acts as a powerful conditioning factor in economic, social and political affairs. In Canada, in particular, its availability has played a critical role in shaping settlement patterns, the structure of economic activity, relationships among various levels of government, and, to an increasing degree, the quality of life.[2] In this respect British Columbians are extremely fortunate. Collectively they have access to more than one-third of Canada's runoff, which itself amounts to almost one-tenth of the world total.[3] And because British Columbia has a relatively small population for an area its size, its water endowment is even more generous.

With such a wealth of resources, one might justifiably conclude that the province has no problems with respect to water, either now or in the foreseeable future. Such, however, is not the case. Nature is niggardly, as well as generous, and does not bestow her bounty in a uniform way. Water availability, in fact, varies considerably from one part of the province to another. Besides this, there are many possible uses for water, some of which can be developed harmoniously and others of which result in serious conflicts. In recent years such conflicts have multiplied enormously in British Columbia. There are other difficulties, too, which result from the division of responsibilities for water management between the federal and provincial levels of government, the fact that some rivers cross international or inter-provincial boundaries, and the fact that social values attached to water resources may change over time. A consequence has been that British Columbian water management issues occupy an increasingly significant place on the political agendas in the legislatures in Victoria and in Ottawa.

In this chapter the nature and distribution of the province's water resources are described first. Then the various uses to which these assets have been put and the problems that have emerged as a result are discussed. The development of water resources, of course, has taken place within an institutional framework of laws, public policies, and administrative agencies. The gradual evolution of this framework is outlined, focussing particularly upon the changes that have been made to it in the past decade or so. The chapter is brought to a conclusion with comments on the challenges that British Columbia's water managers will have to face in the years ahead, and especially in the next few years.

BRITISH COLUMBIA'S WATER WEALTH

Anyone flying across British Columbia cannot fail to be impressed by three major features of the environment: the massive, rugged mountain ranges; the numerous narrow steepsided valleys; and the myriad rivers which in many cases flow for hundreds of miles from their headwaters to the ocean. These features, combined with the province's large area (586,008 square kilometres or 366,255 square miles) and its location on the Pacific coast, are responsible for its huge water wealth.

As indicated in Table 1,10, several of the province's rivers are very large, in terms of their drainage area, length or discharge. In addition, there are numerous large lakes, some of them well over 386 square kilometres (100 square miles) in area. The locations of the major streams and lakes are shown in Figure 1,10. Collectively, British Columbia's rivers discharge some 781 trillion cubic metres (633 million acre feet) per year. This is enough water to cover all of Vancouver Island to a depth of 23 metres. Expressed in another way, it is sufficient to meet the annual needs of a city the present size of Vancouver for more than 37 years without any recycling. Every British Columbian has available to him (or her) some 1 billion litres (2.2 million gallons) of water. This is more than enough to fill 780 million bathtubs!

This is obviously an extremely favorable endowment. But it needs to be interpreted with caution. Water tends to be rather uneven in its spatial distribution and it varies considerably in its availability over time. There may, in fact, be poverty in the midst of plenty, or, alternatively, so much water that it becomes an embarrassment. Both features are present in British Columbia. They arise in part from the nature of the province's terrain and in part from seasonal variations in precipitation.

As noted in Chapter 3, about 75 percent of British Columbia's land area is over 937 metres (3,000 feet) above sea level. More than half of it is over 1,280 metres (4,200 feet). As shown in Figure 2,10, high mountains lead to major differences in precipitation, and hence runoff, across the province. The Northern Cordillera acts as a barrier to moisture-bearing winds from the Pacific. Precipitation is often extremely heavy at

Table 1,10 DRAINAGE AREA, LENGTH, AND DISCHARGE OF MAJOR RIVERS

River	*Drainage Area*		*Length*		*Average Annual Discharge*
	square miles	square kilometres	miles	kilometres	cubic metres per second
FRASER	89,310	231,312	850	1,368.5	3,600
LIARD	55,374	143,418	300	483.0	1,300
PEACE	49,595	128,451	175	281.7	1,400
COLUMBIA	39,770	103,004	466	750.3	2,800
SKEENA	21,038	54,488	360	579.6	1,000
STIKINE	19,445	50,362	335	539.3	1,000
NASS	8,046	20,839	235	378.3	900

SOURCE: Farley, A.L., *Atlas of British Columbia*. Vancouver: University of British Columbia Press, 1979, p. 38.

the coast, sometimes in excess of 750 centimetres (295 inches), but declines rapidly as one moves from west to east. It is especially heavy on the western slopes of the mountain ranges, but much lower on the eastern slopes which create a rain shadow effect. There are several inter-montane areas where precipitation is less than 50 centimetres (20 inches) per annum and a few places where it is as low as 12.5 (5). Spatial variations in precipitation in the province are depicted in Figure 4 of Chapter 4.

Not only are there important spatial differences in precipitation, but there are seasonal variations too. Broadly speaking, in the coastal regions winters tend to be very wet and the summers relatively dry. In much of the interior of the province precipitation is fairly evenly distributed throughout the year, although the spring is often fairly dry. In the coastal regions part of the winter precipitation is in the form of snow and part in the form of rain, especially at the lower elevations. In the Interior much of the winter precipitation is in the form of snow. These factors of terrain and climate result in strikingly different seasonal stream flow patterns across the province, as illustrated by the hydrographs for the Nanaimo River, Capilano Creek, the Fraser River and the Chilko River portrayed in Figure 3,10.

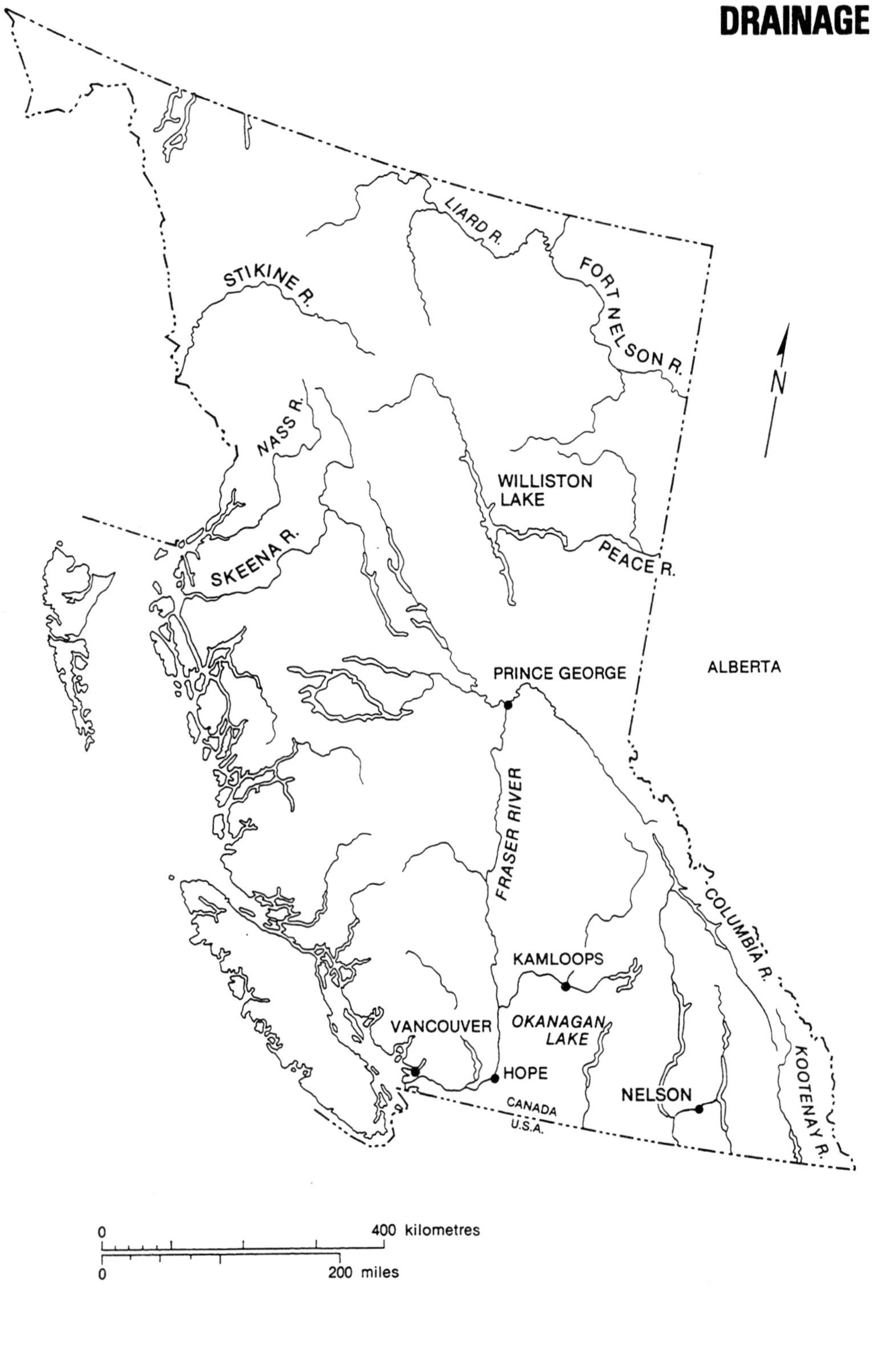

FIGURE 1,10 Major rivers and lakes

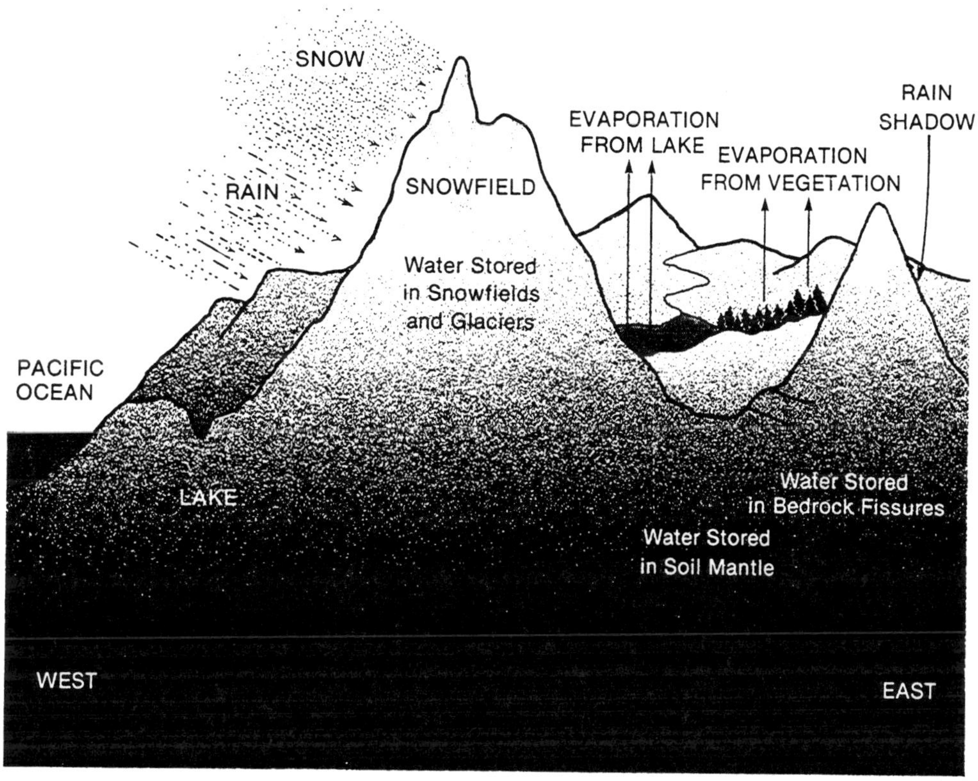

FIGURE 2,10 The hydrologic cycle in British Columbia

Besides the seasonal fluctuations, there are often major variations from year to year. As an example, the maximum recorded flow of the Fraser River at Hope was 15,200 cubic metres per second (536,560 cubic feet per second) on May 31, 1948, while the lowest flow was 340 cubic metres per second (10,200 cubic feet per second) on January 8, 1916. Similarly, the highest flow recorded on the Skeena River at Usk was 9,340 cubic metres per second (329,702 cubic feet per second) on May 26, 1948, while the minimum flow was 51.8 (1,828) on March 1, 1950.[4]

These major spatial and temporal variations in runoff have several implications for human activity. They may mean that water may not be where it is needed when it is required. A consequence may be to limit the economic growth of a particular region, either because the water is not there or it is too expensive to bring it in from elsewhere. Another implication may be that on occasion rivers overflow their banks, causing major losses of property and possibly life as well. A major challenge in British Columbia, therefore, has been to cope with seasonal and other variations in streamflow.

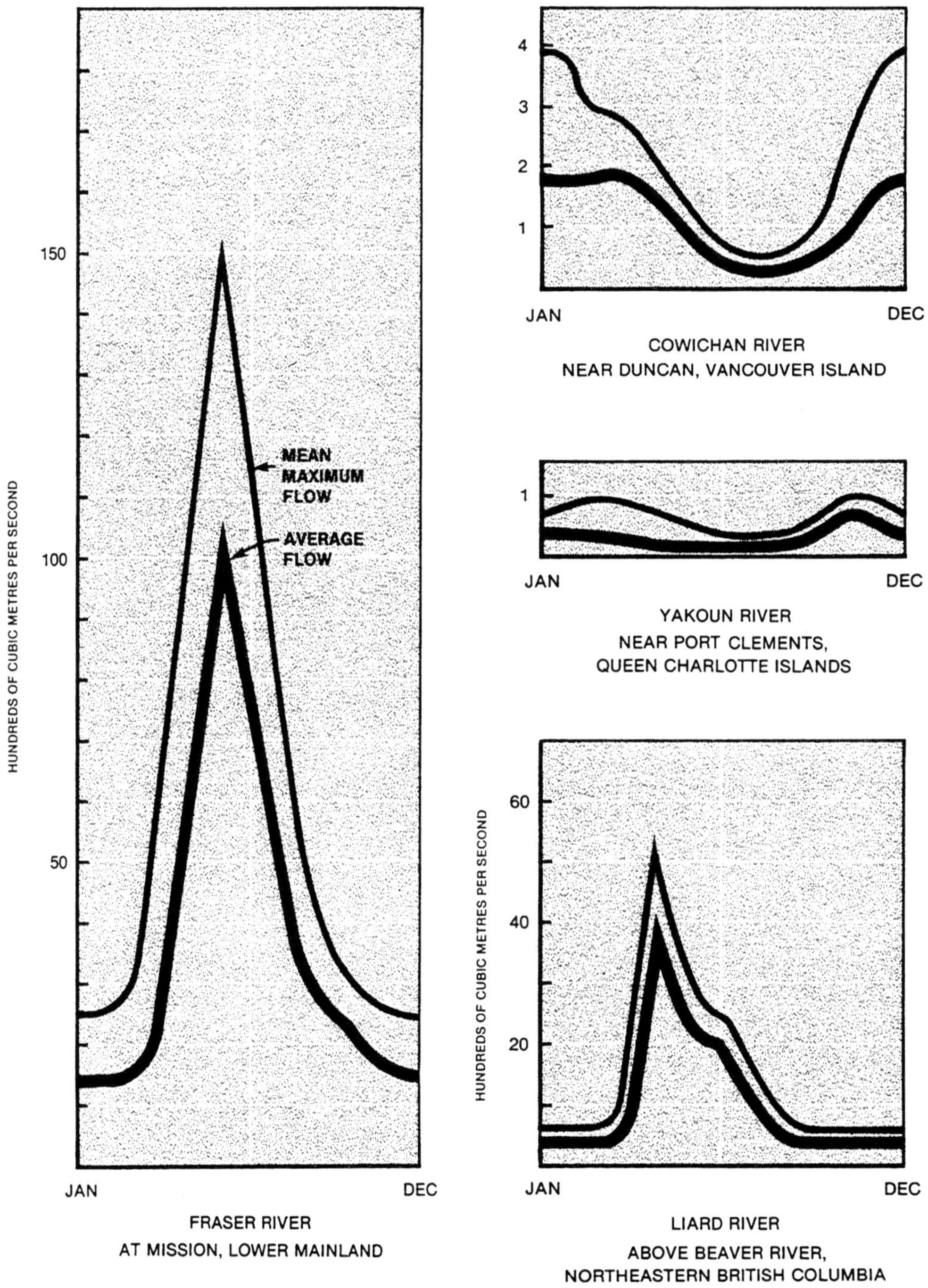

FIGURE 3,10 Hydrographs of selected rivers

The relative abundance of perennial streams and the generally easy access to many of them has resulted in most of the water withdrawals being made from surface sources. While some regions have drilled wells for their supplies, groundwater is not a major contributor to overall supplies. As yet, there are only rough estimates of the magnitude of the province's groundwater resources.

WATER USES

Compared with people in many other parts of the world, British Columbians are heavy water users. In the economically less developed regions of the world, such as in much of Africa, per capita use of water for domestic purposes may be as little as 20 litres (4.39 gallons) per day. In contrast, in Vancouver it now exceeds 950 litres (190 gallons) per day. In addition, there are withdrawals for manufacturing, mining and the cooling of thermal power stations. When all of these are totalled, the average British Columbian accounts for the withdrawal of some 3,407 litres (749 gallons) per day.[5] This is well below the figures for Canada as a whole (7,100 litres or 1,562 gallons per capita per day) and for Ontario, but well above that for Quebec.[6]

For convenience, water uses can be classified into withdrawal, consumption, and in-stream categories. As the term implies, withdrawal uses are those which involve the abstraction of water from a water body, such as for use in a household or factory, or on a farm. Part of this withdrawal may be 'consumed' in the sense that it does not return directly to the water body, either because it is incorporated physically into the product (such as in beer or soft drinks) or because it is evaporated into the atmosphere. In a few cases water may be transferred to another watershed. It should be noted that water withdrawals do not always equal total water use or demands, since in some instances a given volume of water is 'recycled' or used more than once. This has been increasingly the case in certain manufacturing and mining industries. Thus, for Canada as a whole, a total of 11,705 million litres (24,572 million gallons) per day are withdrawn to serve a demand of 173,900 (38,253).[7] Precise figures are not available as to the extent of recycling in British Columbia. In-stream uses are those which take advantage of the water where it is, as in navigation, migration and spawning of fish, hydro-electric power generation, or in a wide variety of recreational pursuits.

Data for water withdrawals and consumption for British Columbia and other parts of Canada are set out in Table 2,10. These data show that the manufacturing sector is by far the largest water user in the province, reflect-

Table 2.10 WITHDRAWAL AND CONSUMPTION OF WATER IN CANADA AND MAJOR REGIONS, 1980

	Millions of Litres Per Day					
	Municipal and Rural	Manufacturing	Mining	Agriculture	Thermal	Total
Withdrawal						
ATLANTIC	1,443	4,078	677	68	8,627	14,893
QUEBEC	4,510	7,900	337	84	2,744	15,575
ONTARIO	3,615	19,676	704	698	30,685	55,378
PRAIRIE PROVINCES	1,739	860	2,339	5,898	5,557	16,373
BRITISH COLUMBIA	1,076	5,642	277	1,568	787	9,350
YUKON AND N.W.T.	27	–	109	–	–	136
CANADA	12,410	39,156	4,443	8,296	48,400	111,705
Consumption						
ATLANTIC	155	118	82	68	65	488
QUEBEC	637	306	45	83	21	1,092
ONTARIO	522	762	46	580	230	2,140
PRAIRIE PROVINCES	470	154	465	2,941	42	4,072
BRITISH COLUMBIA	282	227	109	737	6	1,361
YUKON AND N.W.T.	9	–	45	–	–	54
CANADA	2,075	1,567	792	4,409	364	9,207

SOURCE: Canada, *Canada Water Year Book, 1981-1982*. Ottawa: Environment Canada, 1983.

ing the huge quantities that are withdrawn by the forest products industry, and especially by pulp and paper plants. Some 150 cubic metres (5,300 cubic feet) are needed for the production of a ton of pulp, and about 125 (4,415) for a ton of paper. The second largest user is agriculture, in which irrigation has become increasingly important, while municipal and rural uses lie just slightly behind in third place. In terms of consumption, the order is quite different, with agriculture being by far the most important of the various categories, accounting for over half the total.

The pattern of withdrawals differs considerably from that of the nation as a whole, reflecting, in particular, the relatively minor role that cooling of thermal power stations plays in British Columbia compared with Ontario or the Atlantic Provinces. In terms of consumption, however, the province is much more like Canada at large.

Lack of up-to-date information prevents a precise description of regional water withdrawals across the province. A few inferences, however, can be drawn. As might be expected, the heaviest abstractions for municipal and rural use are concentrated mainly in the regions of the major urban agglomerations: Vancouver, Victoria, Kamloops, and the Okanagan Valley.[8] Withdrawals for manufacturing are focussed on several regions, notably those dealing with forest products, processing of minerals, agricultural products, and fish. In this respect, the Vancouver area, Vancouver Island, and several points in the Interior stand out. Concerning agricultural use, the Kootenays and the Okanagan Valley stand out as the major areas.

Among the most important uses in the province are those for which no withdrawal is made, the in-stream uses. Unfortunately, there are only a few statistics on such uses. Recent figures indicate that Crown corporations, private power utilities and various industries have licences which allow them collectively to divert some 19,139 cubic metres per second (675,894 cubic feet per second) for the generation of hydro-electric power.[9] Beyond this, however, few data are available. It is well known, nevertheless, that considerable use is made of the province's streams for the driving of logs and for barge transportation.

As Chapter 9 has shown, many of the major rivers support runs of anadromous fish. These and many inland lakes furnish a habitat for other species of fish as well. In addition, water-based recreation has expanded extremely rapidly in popularity in recent years, both for British Columbians and for a growing tourist industry. In 1983, the provincial government sold some 418,000 fishing licences to resident and non-resident sportsmen.[10] Lakes and streams furnish recreational opportunities for more than 15,000 pleasure craft that are used beyond the coastal regions of the province.[11] All of these activities generate millions of dollars in income for residents in the areas where the recreation takes place and along the routes leading to them.

EMERGING PROBLEMS AND ISSUES

British Columbia's water resources have clearly played an important role in the province's economic development and in the evolution of a quality of life of which its citizens are justly proud. Over the years, however, several problems and issues have emerged, some of which are reaching critical proportions. These difficulties stem largely from:

1. the spatial and temporal distribution of the water resource; and
2. the rapid expansion of population, urbanization and industrialization, leading particularly to declining water quality, together with changing social values.

Droughts and Floods

As noted earlier, water in British Columbia is not always available when and where it is wanted. This is especially so in regions which receive on average below 80 centimetres (20 inches) of precipitation per year, but which, for various reasons, have developed activities that have become dependent upon at least that amount, such as irrigated agriculture or salmon fishing. From time to time the Okanagan, the Cariboo, and the Kootenays have suffered long periods of below average runoff and agricultural production has suffered as a result.[12] Sometimes the parks and the gardens of Vancouver and Victoria have withered as a consequence of prolonged drought.[13] And occasionally, salmon have been unable to reach their spawning grounds because water levels have been too low, as in the case of the Goldstream River in Victoria.[14] Compared with certain other regions of Canada, notably the Prairies, however, British Columbia has only minor problems with respect to drought, particularly of a long term nature. Among the various regions, only the Okanagan seems to be in serious trouble.[15]

A much more pervasive problem is that of floods. A study by Burton in 1965 showed that in Canada as a whole there were more than 200 urban places that were subject to major floods.[16] Of these towns, about 50 were in British Columbia. Another investigation undertaken by Sewell about the same time confirmed these results.[17] The latter indicated that these flood-prone locations were concentrated particularly in the basins of the Fraser and Columbia rivers, and at various locations on Vancouver Island. The problem was especially severe in the lower Fraser Valley. In the years since then the flood loss potential has risen considerably, particularly as a result of population growth and industrial expansion. This is emphasized by the fact that during the past decade there have been several floods that have caused losses of well over 10 million dollars in property damages alone.[18] To this would need to be added losses in income and costs of emergency measures and rehabilitation.

The biggest single problem is in the lower Fraser Valley. Some 1,250,000 people now have homes, work and recreate there.[19] A substantial part of the province's manufacturing industry is concentrated in the triangular area between Vancouver and Hope, as well as much of its horticulture and dairy farming. When the last major flood occurred in 1948 it cost some 20 million dollars in damages that were compensated, mostly for property losses, rehabilitation and repairs. Actual losses were much larger. If a similar flood occurred today and protective works did not hold, losses could be many times that amount, perhaps as much as 300 million dollars.[20]

Water Quality

Declining water quality has become an increasingly pervasive issue. It results both from expanding urbanization and industrialization, and especially from an attitude that the use of water for the disposal of wastes should be regarded as free. This presented few difficulties until about two decades ago. The major polluters were generally located far away from the larger established centres of population. There were no major outbreaks of water-borne, vector-related diseases nor fish kills of a significant magnitude. There seemed to be little to worry about. Things have changed dramatically, however, since the mid-1960s, in part as a result of the explosive growth of population and polluting industries, and in part as a consequence of increasing awareness and concern by the public at large about the environmental deterioration.

Today, most British Columbians are shocked to learn that only 46 percent of the province's urban population is served by sewage treatment systems, compared with 63 percent for the country as a whole, and 100 percent in Prince Edward Island, Ontario, Manitoba, the Yukon and Northwest Territories.[21] They are certain to be worried when they hear that the complex ecological web of the lower Fraser River is in jeopardy, as a result of the persistent use of its waters for the disposal of toxic and other wastes.[22] They reacted adversely when it was disclosed that mining companies are dumping thousands of tonnes of tailings into lakes, rivers or estuaries, such as Alice Arm on Vancouver Island.[23] They were also alarmed when they observed the impact of a Eurasian milfoil weed infestation, as in Okanagan Lake.[24]

While the root of some of the concern has been the impact upon human health, the actual or potential effects on the preservation of fisheries, and especially anadromous species, have been the focus of some of the sharpest conflicts in water use. There are numerous examples. They include proposals to use certain fish migration streams for the development of hydro-electric power, as in the case of a further diversion of the Nechako River into the Kemano River to enable an increase in power output on the latter,[25] or the harnessing of certain parts of the Fraser River system for hydro-electric power generation and flood control.[26] They also embrace cases where forest companies wish to use such streams for log driving, log storage, or processing of products.[27] Fisheries interests claim that these uses may either physically impede the migration of fish or destroy habitat required for spawning or feeding, as in the case of logging on Riley Creek in the Queen Charlotte Islands, or log storage in Cowichan Bay on Vancouver Island.[28]

Other Conflicts

An increasing number of conflicts involve outdoor recreation pursuits, notably where wilderness is an important part of the experience. Thus, hydro-electric power development schemes which threaten to impair sports fishing or canoeing are bitterly opposed by recreationists, as was the case with the proposed McGregor Diversion on the Fraser River,[29] and the Site C project on the Peace River.[30] Opposition in the latter instance was especially acute in the light of the adverse effects that the construction of the W.A.C. Bennett dam had had on the Peace-Athabaska delta.[31] Filling of the reservoir behind the dam resulted in the drying up of large areas of habitat for birds and other wildlife, and in the reduction of income and food supplies for native peoples in the delta area below the dam.

In short, although British Columbia possesses a vast wealth of water resources, there are rapidly growing problems to be faced in their management. While some of these are rooted in the nature of the resource itself, an increasing number stem from conflicts among users and among different levels of authority. Ultimately the solutions must be found in a flexible and sensitive institutional framework.

THE INSTITUTIONAL FRAMEWORK

Over the past century a complex set of institutions has evolved for the management of British Columbia's water resources. These consist of laws, policies and administrative arrangements that are intended to allocate water among competing users and to furnish various water-related goods and services. These institutions have been modified over time, in response to the emergence of new problems, the improvement of technologies, and the introduction of new concepts and ideas. Changes have been particularly profound in the past 20 years.

Federal Government Activities

Jurisdiction over British Columbia's water resources is shared between the federal and provincial governments. This division of authority has had an important influence over the manner in which the resources have been developed. In broad terms, the federal authority has helped shape water management in British Columbia in five major ways — namely, by:

1. carrying out programs designed to improve navigation or enhance fishery stocks;
2. imposing regulations either to maintain or improve water quality, especially where fishery stocks are involved, or to ensure that the national interest is maintained in the development of international or inter-provincial streams;
3. undertaking co-operative data collection programs with respect to water quantity and quality;
4. participating in water resources planning; and
5. co-operating in various schemes of development, notably those concerning flood control or the treatment of municipal wastes.

Generally, the federal government has exerted a leadership role in those areas of water management for which it is given clear responsibility in the Constitution (formerly under the British North America Act).[32] Thus, it has carried out programs of harbor development, as in Vancouver and New Westminster, and fisheries enhancement, as in the Salmonid Enhancement Program. It has vigorously opposed developments which might harm the fisheries and on occasion has taken offenders to court.[33] Where it has believed that the national interest is at stake it has enacted new legislation or it has brought existing regulations or negotiation procedures to bear. The passage of the International River Improvements Act in 1955, for example, was designed to ensure that development of the Columbia River would be in Canada's interest. It enlisted the good offices of the International Joint Commission when disputes arose over water supply for Point Roberts in 1973,[34] and the raising of the Ross Dam on the Skagit River in 1980.[35]

Where the federal government's role is less well defined it has usually waited for signals from provincial authorities for it to become involved. Sometimes this has occurred in response to a crisis, such as in the declaration of a national disaster. The floods on the Fraser River in 1948 and in Port Alberni in 1964 are cases in point. In these instances the federal government provided funds to assist in the relief and rehabilitation of the flood victims. Beyond this, with respect to the Fraser River, the federal government co-operated with the province in a 14-year investigation of the flood problem and the search for a viable solution, notably through the studies of the Fraser River Board[36] and its successors, especially the Fraser River Joint Advisory Board established in 1968.[37] In other instances the federal government has participated on an on-going basis in co-operative ventures with the provincial government, such as in the collection, analysis, and dissemination of hydrologic data, and the undertaking of river basin studies. The Okanagan Basin Study is an outstanding case in point.[38]

The federal government also has provided financial assistance for a number of water management schemes. For example, it has shared with the provincial government on the costs of a 120 million dollar program for improved flood control in the Fraser River basin.[39] It has undertaken projects to improve flood protection along the Capilano River.[40]

The principal responsibilities for the management of water resources at the federal level are in the hands of Environment Canada, through its Inlands Water Directorate. It derives its mandate mainly through the Canada Water Act, 1970 (Table 3,10). There are, however, at least 19 other federal agencies which have specific interests in water management, administering a myriad of different Acts.[41] Experience has shown that co-ordinating these different interests and perspectives is a formidable challenge.[42]

Provincial Government Activities

While the federal government exercises some control over water use and assists materially in overcoming several problems, the major direction for water management in British Columbia comes from the provincial government. The provincial authority allocates water among competing users, draws up plans for development, sets priorities, and does most of the regulation to ensure that quality is maintained. The province also relates water management to the development of other resources and assesses its environmental impacts and other consequences.

The major provincial agency dealing with water matters is the Ministry of Environment (Table 4,10). Established in 1977, its various divisions now cover a wide range of purposes of water management, including water

Table 3,10 FEDERAL GOVERNMENT ADMINISTRATIVE AGENCIES AND LEGISLATION CONCERNED WITH WATER IN BRITISH COLUMBIA

Agency	*Legislation*
ENVIRONMENT CANADA	Department of Environment Act, SC, 1970
	Canada Water Act, RSC, 1970, c. 5, 1st supp.
	International Rivers Improvement Act
	Migratory Birds Convention Act, RSC, 1970, c. F-16
	Navigable Waters Protection Act, RSC, 1970, c. N-19
	International Boundary Waters Treaty Act, RSC, 1970, c. I-20
	Environmental Contaminants Act, SC, 1974-1975, c. 72
	National Parks Act, RSC, 1970, c. N-13
FISHERIES AND OCEANS CANADA	Department of Fisheries and Oceans Act
	Fisheries Act, RSC 1970, c. F-14, as amended
	Ocean Dumping Control Act, SC, 1974-1975-1976, c. 55
TRANSPORT CANADA	Canada Shipping Act, RSC, 1970, c. S-9
	Navigable Waters Protection Act, RSC, 1970, c. N-19
NATIONAL HARBORS BOARD	National Harbors Board Act, RSC, 1970, c. N-8, as amended
	Harbor Commissions Act, RSC, 1970, c. H-1
NATIONAL ENERGY BOARD	National Energy Board Act, RSC, 1970
DEPARTMENT OF NATIONAL HEALTH AND WELFARE	Department of National Health and Welfare Act, RSC, 1970, c. N-9
DEPARTMENT OF INDIAN AFFAIRS AND NORTHERN DEVELOPMENT	Department of Indian Affairs and Northern Development Act, RSC, 1970, c. I-7
	Northern Inland Waters Act, RSC, 1970, c. 28, 1st supp.

Table 4,10 BRITISH COLUMBIA PROVINCIAL GOVERNMENT ADMINISTRATIVE AGENCIES AND LEGISLATION CONCERNED WITH WATER MANAGEMENT

Agency	*Legislation*
MINISTRY OF ENVIRONMENT	Ministry of Environment Act, SBC, 1980, c. 30
	Water Act, RSBC, 1979, c. 429
	Waste Management Act, SBC, 1982, c. 41
	Environment Management Act, SBC, 1981, c. 14
	Environment and Land Use Act, RSBC, 1979, c. 110
	B.C. Fisheries Act, RSBC, 1979, c. 137
	Pesticide Control Act, RSBC, 1979, c. 322
	B.C. Wildlife Act, SBC, 1982, c. 57
MINISTRY OF LANDS, PARKS AND HOUSING	Land Act, RSBC, 1979, c. 214
	Park Act, RSBC, 1979, c. 309
MINISTRY OF FORESTS	Forest Act, RSBC, 1979, c. 140
MINISTRY OF HEALTH	Health Act, RSBC, 1979, c. 161

supply, water quality control, and the maintenance of fisheries. It gathers, analyses and disseminates information about streamflow, snowpacks, water quality and, to a more limited extent, water use in the province. Under the provisions of the Water Act, it issues licences for water withdrawal. Those wishing to obtain such a licence make application to the Comptroller of Water Rights, indicating the timing and quantity of the intended abstraction. Providing there is no objection from other licence holders, a licence is granted on payment of a fee. Such fees vary considerably in magnitude. An irrigator in the Okanagan, for example, withdrawing some .01 cubic metres per second (100 acre feet) might pay about 20 dollars for this amount. In contrast, British Columbia Hydro which requires a flow of 150 million acre feet must pay a fee of 183 million dollars.[43] Presently there are some 51, 775 licences issued in the province, permitting a total withdrawal of some 15,653 cubic metres per second (552,786 cubic feet per second). Fees collected in 1985 amounted to over 205 million dollars.[44]

The Ministry also carries out responsibilities with respect to water quality management. Firstly, it calls together various interested parties from time to time to obtain advice on standards that should be set in this connection. Potential polluters, as well as representatives of the public, are involved in these discussions. Secondly, it issues permits to those who intend to discharge wastes into water bodies. These permits specify the conditions under which the discharge may be made, including the volume that may be released. In 1984 there were some 3,983 permits issued to individuals or companies for disposal of wastes into water bodies.[45] A third function of the Ministry in this regard is the prosecution of those who do not meet the requirements of the permit. On the whole, there seems to have been a reluctance to take offenders to court, either because there is insufficient evidence or there is a belief that other kinds of pressure may be more effective.[46] Whatever the case, the total number of successful prosecutions has been small and the fines levied extremely low. For example, over the past three years, hundreds of violations of the Waste Management Act have been reported and investigated. In any given year, however, prosecutions have never exceeded a dozen and total fines have been less than 20,000 dollars.[47]

There are several other provincial agencies involved in various aspects of water management. These include the Ministry of Agriculture, which is concerned particularly with irrigation; the British Columbia Forest Service, which monitors impacts in watersheds within which logging, log transport or forest products manufacturing takes place; and the Department of Lands, Parks and Housing, which exercises various planning and regulation functions, particularly with respect to the use of water bodies in parks and the use of the foreshore for log handling and storage.

Water management in British Columbia is clearly complex. It involves two levels of government and myriad agencies, each with a set of purposes and functions which on occasion may duplicate or conflict with those being pursued by other bodies. To overcome these difficulties various co-ordinating mechanisms have been established. Some are permanent, others are "ad hoc", and many are informal. They are, indeed, numerous and a few examples will suffice. They include mechanisms for bringing the federal and provincial governments together to discuss programs they undertake jointly on a continuous basis, as in the case of streamflow measurement, or to carry out shorter term co-operative ventures, such as river basin planning as was done in connection with the Fraser, the

Okanagan, and the Mackenzie basins. At the provincial level they include devices whereby programs proposed in one ministry that may have impacts on those of other ministries can be considered by the latter before implementation. Various "referral" systems are now in place in this connection, facilitated in part by the Environment and Land Use Committee, and in part by formal procedures, such as those instituted by the British Columbia Forest Service.[48]

THE CHALLENGES AHEAD

Most visitors to British Columbia would doubtless conclude that British Columbians have managed their water resources well. And definitely in comparison with certain other parts of North America this would be a reasonable conclusion. The large majority of British Columbians have access to cheap, safe, and potable water supplies. Water-related epidemics are very rare and fish kills from the disposal of industrial wastes are seldom heard of in the province. British Columbia has attracted large energy-using industries to take advantage of low cost hydro-electric power. And the province has become renowned for its outstanding water-based recreation opportunities.

While British Columbians may be justly proud of this record, they can clearly not afford to be self-satisfied or complacent. As noted earlier in this chapter, there are already some major problems to be faced and others are certain to appear in the very near future. The question arises, how well prepared is the institutional framework to deal with the emerging challenges?

Important changes have been made to legislation, policies and administrative structures relating to resources management in British Columbia in recent years, notably that concerned with water. These modifications appear to have been in response to the increasing severity of certain issues, on the one hand, and to the emergence of new concepts and various social movements on the other. Rising flood loss potentials and rapidly declining water quality in some regions emphasized the need for change. The Fraser River flood in 1948, for example, made it clear that the conventional approach to building dikes, evacuating flood plain dwellers, and compensating victims was inadequate. Similar events elsewhere, notably in the United States,[49] showed that construction of flood control dams was not a satisfactory solution either. A more effective strategy seemed to be that of controlling flood plain occupance, together with the provision of a higher degree of flood control. These concepts were embodied in the agreements drawn up between British Columbia and the federal government in 1968.[50]

The dikes have been repaired and raised to withstand a flood of the magnitude of that experienced in 1894. At the same time the province has embarked upon a program of flood plain mapping, and has introduced legislation to limit the future occupance of the flood plain to those who are willing and able to cover the cost of losses that may occur from a flood event.[51] Attempts have also been made to deal with the problem of sewage disposal. There are now no instances of communities dumping raw sewage into fresh water bodies. Primary treatment is a widespread practice and secondary treatment is being undertaken in many places. A number of communities now have tertiary treatment as well.

Two social movements, in the mid-1960s and early 1970s, concerned with the environment, on the one hand, and with the involvement of the public in planning and policy-making, on the other, also had an important impact.[52] These movements made it clear that the social context of resources management was changing and that its institutions would have to be altered too. More attention would need to be paid to environmental effects of resource development, and a more direct role would need to be given to the public at large in shaping policies. The enactment of federal and provincial legislation with respect to pollution and environmental protection, the reorganization of agencies dealing with water so that they became a key part of Environment Canada at the federal level and the British Columbia Ministry of the Environment at the provincial, the provision of specific opportunities for public involvement at both levels, and the establishment of an Environmental Land Use Committee Secretariat in British Columbia were all illustrations of an enlightened approach to resources management, and especially that concerned with water. This was given concrete expression in the planning undertaken with respect to the Okanagan River basin, a joint federal-provincial venture which lasted from 1969 to 1974.[53] This particular process has been applauded, not only in other parts of Canada, but also in other countries. Since then a new type of planning has been introduced, called Strategic Planning. Designed to streamline the process and to foster local initiatives, it has already been applied in six regions. The plans which have emerged have enabled improved information systems to be put in place and suggestions for water allocations to be made.[54]

The mid-1970s was an era of intense institutional change in water management in British Columbia at both the federal and provincial levels. The pace, however, seemed to slow down in the latter part of the decade and a period of consolidation followed. It continues today. To some it may be seen, perhaps, as a period of reflection, but to others it has been viewed as a time when momentum was lost and when retardation began.[55] Whatever the case, it seems pertinent to pose the question, where do we go from here?

Assuming that the goal is the attainment of a progressively higher level of sophistication in water management, one would do well to:

1. examine experience elsewhere with respect to the adoption of particular institutional innovations; and
2. consider the extent to which present institutions in British Columbia are "resilient" and the ways in which they might be modified to enable them to cope with major change in external variables.

Recent research has shown that seven major themes characterize contemporary water management in countries that are regarded as "advanced" in this connection. While there may be some disagreement as to their applicability to particular situations, they do offer a measure of sophistication of modern management. These themes have been discussed in some detail elsewhere.[56] For convenience they may be summarized as follows:

1. the adoption of a progressively broader perspective;
2. the search for as wide a range of choice as possible;
3. the recognition of water as an economic good;
4. the adoption of the river basin as a basic unit in water planning;
5. the provision of a direct role for the public in planning and policy-making;
6. the inclusion of environmental considerations as an integral dimension in decisions; and
7. the recognition of social impacts as a critical consideration in planning and policy-making.

As Table 5,10 indicates, each of the themes has been adopted to some degree in British Columbia, but none has been fully implemented. While there has been no thoroughgoing review of the extent of adoption in Canada as a whole, there is evidence that British Columbia has made as much progress in this regard as any of the more "advanced" of the provinces.[57] Nevertheless, there is still a long way to go with respect to some of the themes in British Columbia, notably those concerning the adoption of economic principles, widening the range of choice, and social impact assessment.

Another measure of the progress yet to be made is that of the resilience of existing water management institutions. A major mark of sophistication is the extent to which laws, policies and administrative structures can cope with massive change brought about, particularly, by "external forces", such as shifts in interest rates or in world commodity prices, the introduction of new technologies, alterations in political relationships, the appearance of new ideas, or major changes in social values. Those institutional frameworks that foster and facilitate adaptations are resilient. Those that resist change may not only act as barriers to progress, but may also break down in the process.[58]

The record to date suggests that, on the whole, British Columbian water institutions have been amenable to change. Recently, however, there have been signs that all may not be well.[59] As the various problems outlined earlier have suggested, existing institutions are finding it more and more difficult to cope with diverse value systems, to adjust to major shifts in external variables, such as interest rates, or to generate policies with short lead times or projects that are incremental, rather than massive in scale. The challenge to British Columbia's water managers is both urgent and clear.

Table 5,10 ADOPTION OF THEMES OF CONTEMPORARY WATER MANAGEMENT IN BRITISH COLUMBIA

Theme	*Degree of Adoption*
1. BROADENING PERSPECTIVE	
– Consolidation of purpose	Moderate at provincial and federal levels
– Integration of functions	Moderate at provincial and federal levels
2. WIDENING THE RANGE OF CHOICE	Major advances with flood management, less with drought and water quality
3. WATER AS AN ECONOMIC GOOD	Increasing recognition in water allocation among uses, and in charges for water-related services
4. RIVER BASIN AS AN AREAL UNIT UNIT IN MANAGEMENT	High recognition in data collection and planning, much less in allocations or programs of development
5. PUBLIC INVOLVEMENT	Much higher level than in the past, but mostly in project or plan appraisal, little involvement in policy review, mainly "ad hoc"
6. ENVIRONMENTAL IMPACT ASSESSMENT	Some pioneering developments in process and in guidelines for evaluation, based on general policy guidelines and administrative discretion, rather than legal requirement
7. SOCIAL IMPACT ASSESSMENT	In early stages of evolution

REFERENCES

1. The author wishes to offer his sincere thanks to the following individuals who provided information or who commented upon various drafts of the chapter: Anthony Dorcey, Robert Durie, Chad Day, Harold Foster, Jon O'Riordan, Linda Hannah and Murray Rankin.

2. FOSTER, H.D. and SEWELL, W.R.D., *Water: The Emerging Crisis in Canada*. Toronto: James Lorimer, 1981.

3. CANADA, *Canada Water Year Book*. Ottawa: Environment Canada, 1983.

4. Environment Canada, Water Survey of Canada data, personal communication.

5. CANADA, *Water is a Mainstream Issue*. Ottawa: Inquiry on Federal Water Policy, 1984, p. 8.

6. *Ibid.*, p. 14.

7. *Ibid.*

8. British Columbia Natural Resources Conference, *British Columbia Atlas of Resources*. Victoria, 1956.

9. British Columbia Ministry of Environment, personal communication.

10. REID, R., *The Value and Characteristics of Freshwater Angling in B.C.*. Victoria: Ministry of Environment, 1984.

11. BRITISH COLUMBIA, *Park Data Handbook*. Victoria: Ministry of Lands, Parks and Housing, Parks Program Branch, 1985.

12. O'RIORDAN, T. and O'RIORDAN, J., *Okanagan Water Decisions*. Victoria: University of Victoria, Department of Geography, Western Geographical Series, Vol. 4, 1972.

13. BANCROFT, C., *Strategies for Urban Water Supply Shortages*. Victoria: University of Victoria, Department of Geography, Unpublished M.A. Thesis, 1974.

14. WOOD, C.J.B., "Conflict in Resource Management and the Use of Threat: The Goldstream Controversy", *Natural Resources Journal*, Vol. 16, No. 1, 1976, pp. 137-158.

15. FOSTER, H.D. and SEWELL, W.R.D., *op. cit.*

16. BURTON, I., "Flood Damage Reduction in Canada", *Geographical Bulletin*, Vol. 7, No. 3, 1965, pp. 161-186.

17. SEWELL, W.R.D., *Water Management and Floods in the Fraser River Basin*. Chicago: University of Chicago, Department of Geography, Research Paper No. 100, 1965.

18. British Columbia Ministry of Environment, personal communication.

19. BRITISH COLUMBIA, *The Fraser River: Protecting a Resource*. Victoria: Ministry of Environment, 1982.

20. Canada-British Columbia, Fraser River Joint Advisory Board, *Fraser River Storage Review Report*. Victoria, December, 1976.

21. CANADA, *Water is a Mainstream Issue*, *op. cit.*

22. DORCEY, A.H.J. (ed.), *The Uncertain Future of the Lower Fraser*. Vancouver: University of British Columbia, Westwater Research Centre, 1976.

23. SEWELL, W.R.D., "Water Management", in FORWARD, C.N. (ed.), *Vancouver Island: Land of Contrasts*. Victoria: University of Victoria, Department of Geography, Western Geographical Series, Vol. 17, 1979, pp. 195-214.

24. DEARDEN, P., "Public Perception of a Technological Hazard: A Case Study of the Use of 2,4-D to Control Eurasian Milfoil in the Okanagan Valley, British Columbia", *Canadian Geographer* (in press).

25. SEWELL, W.R.D., "Large Scale Water Transfers in North America: Solution or Disaster?", in MILES, E.D. (ed.), *Natural Resources Policy and Management: Essays in Honor of James A. Crutchfield*. Seattle: University of Washington Press (in press).

26. SEWELL, W.R.D., *Water Management and Floods in the Fraser River Basin*, *op. cit.*

27. DORCEY, A.H.J., McPHEE, W. and SYDNEYSMITH, S., *Salmon Protection and the B.C. Coastal Forest Industry: Environmental Regulation as a Bargaining Process*. Vancouver: University of British Columbia, Westwater Research Centre, 1980.

28. SEWELL, W.R.D., "Water Management", *op. cit.*, reference 22.

29. SEWELL, W.R.D., "Large Scale Water Transfers in North America: Solution or Disaster?", *op. cit.*, reference 24.

30. HENRY, K., *Site C Hearings: Final Report*. Victoria: British Columbia Utilities Commission, 1983.

31. PEACE-ATHABASKA DELTA PROJECT, *The Peace-Athabaska Delta: A Canadian Resource*. Ottawa: Peace-Athabaska Delta Project, Summary Report, 1972.

32. SEWELL, W.R.D., "The Administrative Framework for Water Management in Canada: Challenges and Responses", in CANADIAN COUNCIL OF RESOURCE MINISTERS, *The Administration of Water Resources in Canada*. Montreal: Canadian Council of Resource Ministers, 1968, pp. 17-23.

33. DORCEY, A.H.J., McPHEE, M.W. and SYDNEYSMITH, S., *op. cit.*, reference 26.

34. SEWELL, W.R.D., "Large Scale Water Transfers in North America: Solution or Disaster?", *op. cit.*, reference 24.

35. GIBSON, J.A., "The Evolution of the High Ross Dam Settlement", *Northwest Environment Journal*, Vol. 1, No. 2, Spring 1985, pp. 1-18.

36. FRASER RIVER JOINT ADVISORY BOARD, *Fraser River Upstream Storage Review Report*. Victoria: Fraser River Joint Advisory Board, December, 1976.

37. SIMMONS, G.E., "Approaches to Flood Control on the Fraser River", in SEWELL, W.R.D. and BARKER, M.L., *Water Problems and Policies*. Victoria: University of Victoria, Department of Geography, Cornett Occasional Papers, No. 1, 1980, pp. 35-45.

38. CANADA - BRITISH COLUMBIA OKANAGAN BASIN COMMITTEE, *Final Report*. Victoria: Canada-British Columbia Okanagan Basin Committee, 1974.

39. CANADA and BRITISH COLUMBIA, *The Fraser River Flood Control Program*. Victoria: Ministry of Environment, (no date).

40. Environment Canada, personal communication.

41. PEARSE, P.H., BERTRAND, F. and MacLAREN, J.W., *Currents of Change: Final Report of Inquiry on Federal Water Policy*. Ottawa: Environment Canada, 1985, p. 158.

42. MITCHELL, B. (ed.), *Institutional Arrangements for Water Management: Canadian Experiences*. Waterloo: University of Waterloo, Department of Geography, Publication Series No. 5, 1975.

43. BRITISH COLUMBIA, *Annual Resource Report for the Calendar Year 1985*. Victoria: Ministry of Environment, Water Management Branch, 1986.

44. *Ibid.*

45. British Columbia Ministry of Environment, personal communication.

46. DORCEY, A.H.J., McPHEE, M.W. and SYDNEYSMITH, S., *op. cit.*, reference 26.

47. British Columbia Ministry of Environment, personal communication.

48. DORCEY, A.H.J., McPHEE, M.W. and SYDNEYSMITH, S., *op. cit.*, reference 26.

49. WHITE, G.F., *Choice of Adjustment to Floods*. Chicago: University of Chicago, Department of Geography, Research Paper No. 93, 1964.

50. BRITISH COLUMBIA, *The Fraser River: Protecting a Resource*. Victoria: Ministry of Environment, 1982.

51. SIMMONS, G.E., *op. cit.*

52. O'RIORDAN, T. and SEWELL, W.R.D. (eds.), *Project Analysis and Policy Review*. Chichester: John Wiley, 1981.

53. SHANKS, G.R., "Environmental Policy Implementation: The Case of Federal-Provincial Co-operation in the Okanagan Basin", in

MITCHELL, B. and GARDNER, J.S., *River Basin Management: Canadian Experiences*. Waterloo: University of Waterloo, Department of Geography, Publication Series No. 20, 1983, pp. 189-202; O'RIORDAN, J., "The Public Involvement Program in the Okanagan Basin Study". *Natural Resources Journal*, Vol. 16, No. 1, 1976, pp. 177-196.

54. O'RIORDAN, J., "Strategic Planning and River Basin Management", in SADLER, B. (ed.), *Water Policy for Western Canada: The Issues of the Eighties*. Calgary: University of Calgary Press, 1983, pp. 77-85.

55. DORCEY, A.H.J., "The Management of Super-Natural B.C.", in STANBURY, W.T., *Political Conflict and Public Bodies in British Columbia*, (in press).

56. SEWELL, W.R.D., "Institutional Innovations in Water Management: An Appraisal of Scottish Experience", in SEWELL, W.R.D. and BARKER, M.L., *Water Problems and Policies*. Victoria: University of Victoria, Department of Geography, Cornett Occasional Papers, No. 1, 1980, pp. 108-124.

57. MITCHELL, B. and GARDNER, J. (eds.), *River Basin Management: Canadian Experiences*. Waterloo: University of Waterloo, Department of Geography, Publication Series No. 20, 1983.

58. SEWELL, W.R.D. and FOSTER, H.D., *Energy Policy Formulation for an Uncertain Future*. Ottawa: Energy, Mines and Resources, February, 1983.

59. SADLER, B. (ed.), *Water Policy for Western Canada: The Issues of the Eighties*. Calgary: University of Calgary Press, 1983.

PLATE 15 Natural gas processing plant in the Peace River area. ▶
B.C. Government Photo

11 ENERGY[1]

W.R. Derrick Sewell
University of Victoria

THE CHANGING SCENE

For more than 30 years energy has been big news in British Columbia. Hardly a month has gone by in that period when the media have not given a high profile to proposals, policies or problems in this connection. In many ways this is not surprising. It has long been believed that those regions which have the highest levels of energy consumption are also those with the highest standards of living,[2] and that regions which have large energy resources are especially fortunate.[3] Consequently, any news in British Columbia which suggested that a new source was about to be unlocked was greeted with great enthusiasm, while signs that the pace of development might be slowing down were interpreted, by some at least, as portents of impending disaster.

For the most part the news has been good. Year by year new projects have been announced, beginning with oil and natural gas developments in the mid-1950s, and continuing with the massive hydro-power schemes on the Columbia and Peace rivers in the 1960s, 1970s and early 1980s. Various large-scale coal development schemes have been introduced throughout the period. As each new project appeared, the scope and scale of development seemed to increase. By the early 1980s several multi-billion dollar schemes were underway or were at the advanced planning stage. While other parts of the country, notably Eastern Canada, reeled under the shock of the crisis caused by the Organization of Petroleum Exporting Countries, British Columbians remained optimistic and mostly unconcerned. For them the problem was not that of energy scarcity, but, rather, how to unlock more of the province's resources fast enough to meet new demands.

In the mid-1980s, however, the news has not been so consistently good. True, British Columbia Hydro has been able to find a market for some of its surplus electrical energy in the United States,[4] and sales of natural gas to that country have begun to increase once more.[5] In addition, interest in the province's oil and natural gas resources remains high. In 1985 sales of rights for exploration totalled more than 16,500,000 dollars.[6] But there has been some bad news too. Several mega-projects have been cancelled or postponed, including the Site C dam on the Peace River,[7] the Kemano II project in northwestern British Columbia,[8] and a natural gas pipeline from the mainland to Vancouver Island.[9] In addition, there are signs that the huge coal project in northeastern British Columbia may be in trouble, following the announcement of Denison Mines that it was going to write off its 240,700,000 dollar investment in the Quintette mine which is part of the scheme.[10] Meanwhile, Japanese companies have succeeded in renegotiating a 10 percent reduction in prices for the coal they are importing from the province under long term contracts already signed.[11] Besides this, British Columbia Hydro, which faces interest charges in excess of 950 million dollars a year, has had to increase the price it charges for electricity.[12]

Some observers believe that the present situation is only temporary and that the traditional pattern of constant growth in energy supply and demand will soon reappear. Others, however, feel that fundamental changes are in prospect, and that the energy news in the future will be very different from that in the past. They emphasize that this does not mean that it will be bad. Rather, they suggest, energy is more likely to be a follower rather than a leader in the province's economic and social evolution.[13]

The energy situation in British Columbia in the mid-1980s, therefore, is one of flux. Clearly, its future cannot be predicted with great confidence. As of mid-March, 1986, oil prices have reached their lowest level in over a decade, with the current price of a barrel of crude set at 13 United States dollars. Obviously, this has important implications for not only the oil industry in the province, but also for other aspects of energy resource development.[14]

In this chapter the nature and magnitude of the province's energy resource base, and the extent to which it has been developed to date are examined first. Then the various uses to which these resources have been put are described. There follows a discussion of the factors which have helped shape energy development and utilization in British Columbia. The chapter concludes with a review of some of the emerging problems and issues facing the province in this field.

ENERGY RESOURCES: POTENTIALS AND DEVELOPMENTS

British Columbia is one of Canada's most richly endowed provinces with respect to energy resources, both in terms of volume and variety. Not only does it have large reserves of non-renewable resources, such as coal and natural gas, but it has considerable renewable energy resources too, notably hydro-electric power. Just how big the energy resource potential is, no one is certain, largely because only superficial surveys have been undertaken of many of the sources. Reappraisals are constantly being made as new knowledge is accumulated, as new technologies increase development opportunities, and as costs and prices change. In broad terms, however, British Columbia's energy resource potential now seems even larger than that which was estimated three decades ago.

Although this energy wealth is enormous, it is not evenly distributed. For the most part, the major sources are located far from the main centres of population and industry. Figure 1,11 indicates the distribution of the various sources of oil, natural gas, and coal. Figure 2,11 sets out the major developed and potential hydro-electric power sites, and the main opportunities for large scale thermal-electric power development. The present situation with respect to the various types of energy resource in the province may be summarized as follows.

Coal

The province has very substantial coal reserves, currently estimated at some 2,740 million tonnes (3,020 million tons).[15] They represent about 60 percent of Canada's metallurgical coal (used in steel production) and some 10 percent of the country's thermal coal (used mainly in electric power stations). Current production in British Columbia is about 20 million tonnes (22 million tons) per annum. While coal deposits occur quite widely in British Columbia, there are currently four regions of major interest; the southwest (particularly in the Crowsnest area), the Interior (especially at Hat Creek), the northeast (presently focussed upon Tumbler Ridge) and the northwest (in the Groundhog area, north of Hazelton).

With such large reserves, it is clear that coal can make an important contribution to satisfying the province's energy requirements for possibly hundreds of years to come. As indicated in Chapter 8, there are a number of problems, however, that currently limit development. Some of these are related to costs of mining, on the one hand, and costs of transportation, on the other. There are also difficulties concerning potential impacts on the environment of coal extraction and the burning of coal. There were major concerns, for example, when large scale open cast mining began in the Crowsnest area in the 1960s. Legislation was passed, however, to ensure that the overburden was replaced and vegetation reestablished. Guidelines have now been introduced for coal exploration and development.[16]

More problematical, perhaps, is whether large scale thermal power production, such as that contemplated for Hat Creek or in the East Kootenay, will eventually take place. While it had been confidently expected by

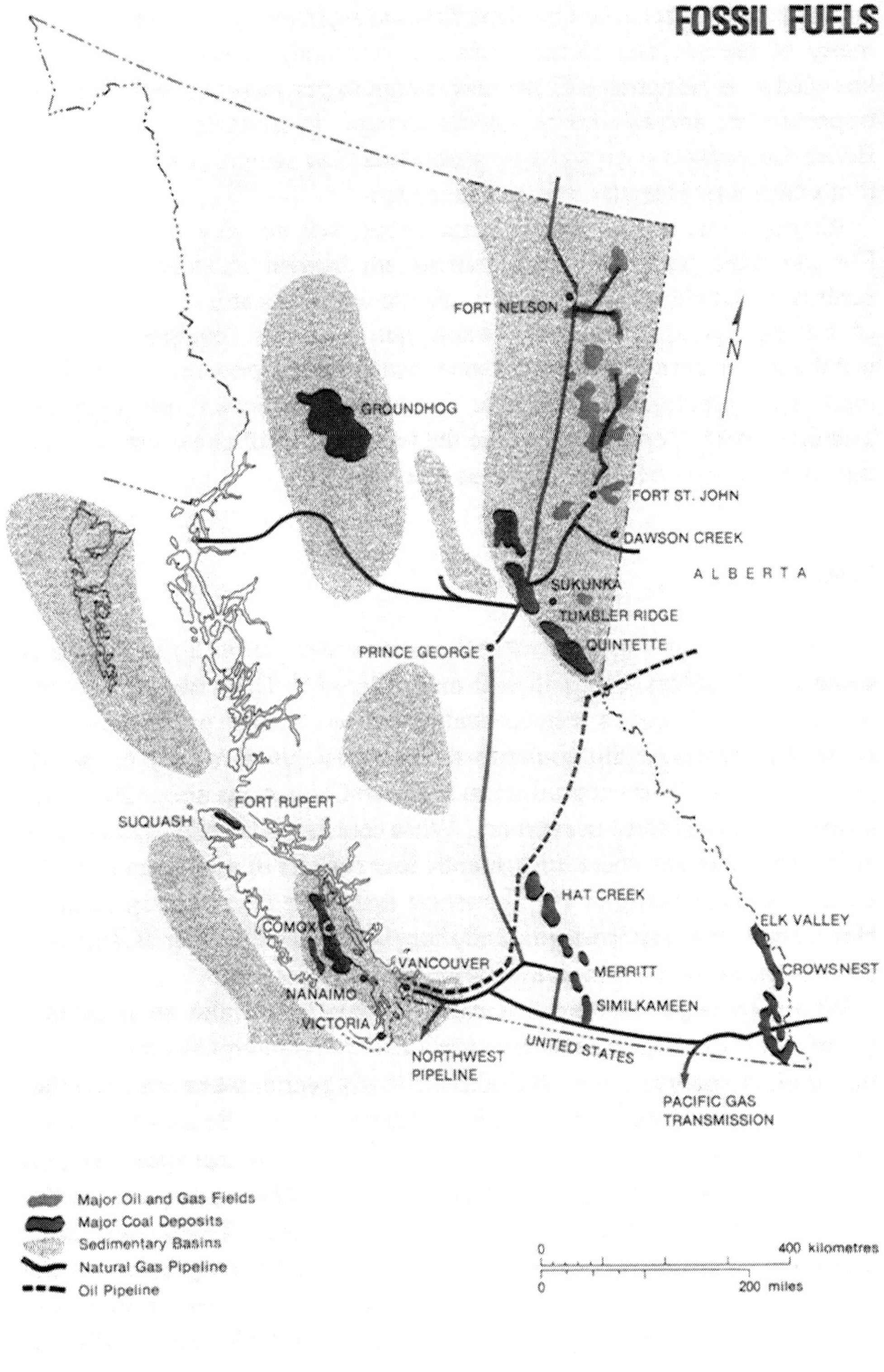

FIGURE 1,11 Distribution of oil, natural gas and coal resources

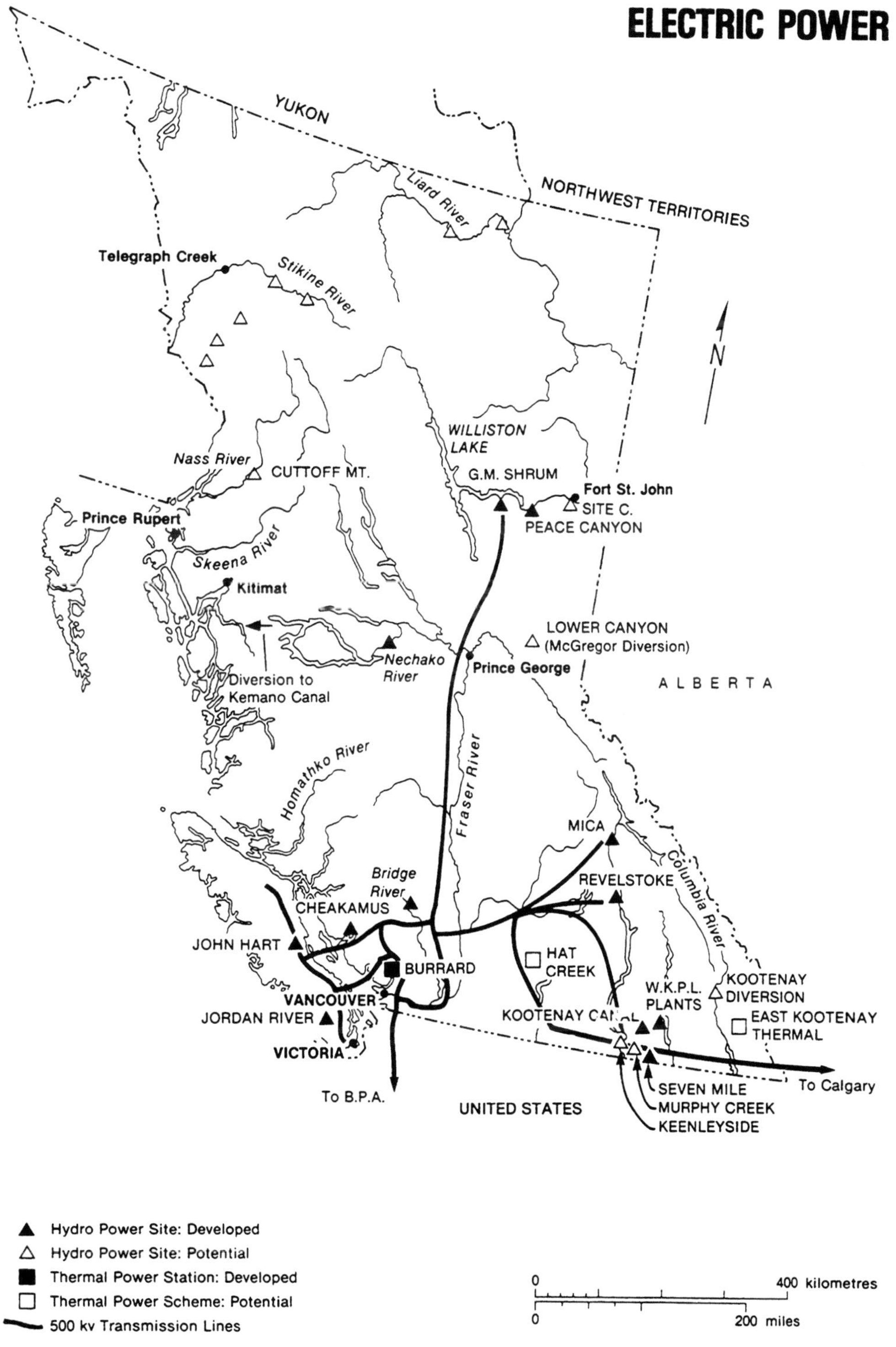

FIGURE 2,11 Developed and potential electric power projects

British Columbia Hydro that such development would certainly occur in the former, the decline in the demand for energy, coupled with rising costs, has resulted in a downplay of such prospects by British Columbia Hydro and others.

Wood

Wood was an important source of energy in the early days of the province, among the original inhabitants and then with the first settlers. It provided the main fuel for home heating and for steam engines, both on the railways and in industry. Eventually, however, coal replaced wood and was itself superseded by oil and natural gas.

In recent years, however, wood has experienced a dramatic revival, principally through the use of two by-products of the forest industry, 'hog fuel' and 'black liquor'. The raw material for hog fuel is combustible waste products — sawdust, bark, shavings, and small chips — from lumber, shingle, pulp and paper mills. When shredded and dried, this material is burned to provide heat in the production of pulp, paper, plywood and lumber.[17] Presently the forest industry jetisons about 10 million tonnes (11 million tons) of wood waste from sawmills alone. Only about half of this, however, is presently used. There are other sources of wood waste, too, most of which remain unutilized. This is especially the case with material left over from logging operations or recovered from hydro-electric power reservoirs.

Black liquor is another type of wood waste. It is a by-product of the kraft pulp process and can be used in boilers which generate steam for electricity or process heat. Considerable use is made of black liquor by forest product companies in several parts of the province. But, again, there are many other opportunities.

Recent estimates suggest that there are presently available 12 million gravity packed units of hog fuel, together with 6.9 million of forest residuals.[18] A gravity packed unit is a volumetric measure used for hog fuel: it consists of 5.7 cubic metres (200 cubic feet) of loosely packed material. Besides this, there is a considerable volume of biomass available in under utilized species of trees.[19]

The distribution of available wood waste, of course, follows closely the distribution of the forest industry. The two major concentrations are (1) Vancouver Island and the southern coastal region (from Vancouver to Knight Inlet), and (2) the central Interior (from Nelson to Vanderhoof). However, there is a major difference between the two regions in terms of utilization. Whereas almost all of the available wood waste in the first region is utilized, only about one-third of that in the central Interior is put to use. The importance of this is emphasized by the fact that even raw wood waste accounts for some 17 percent of all the energy used in British Columbia.

Wood burning stoves have become a popular means of home heating in several areas of the province, in part as a result of incentives provided by the federal government, such as its Forest Industries Renewable Energy Program.

Natural Gas

Natural gas was discovered in northeastern British Columbia in the late 1940s and soon became the focus of large scale programs of exploration and development. The Westcoast Transmission Company built a major trunk line to transport the gas from the Fort Nelson-Fort St. John region to markets in the Vancouver area and in the Pacific Northwest of the United States. The Inland Natural Gas Company constructed another pipeline, linked to the Westcoast line near Kamloops, to distribute gas in the Interior. Subsequently, a total of seven natural gas utilities were established to serve various regions of British Columbia.

Exploration and development continue, mostly in the Peace River region. As indicated in Figure 1,11, there are several other large sedimentary basins. Except for the Queen Charlotte and Tofino basins, however, they have attracted little attention on the part of gas companies so far.

The present estimate of the ultimate natural gas potential of the northeastern region is some 900 billion cubic metres (31,783 billion cubic feet), while that for the Queen Charlotte and Tofino basins has been set at 100 billion (3,531 billion).[20] Experts believe that this is sufficient to provide the province with enough gas to meet its present domestic and external commitments, as well as provide for needs up to 30 years hence.[21]

Oil

Oil is frequently discovered in association with natural gas, as has often been the case in nearby Alberta. In British Columbia, however, the oil story has been much less exciting than that with respect to natural gas. While gas exploration continues to yield developable reserves, the news with respect to oil has not been sensational.

Oil has been discovered in the northeastern portion of the province and wells in that region now produce a total output of about 14 million barrels a year.[22] A pipeline, roughly paralleling the Westcoast Transmission line, now carries the oil southward to the Vancouver area and northwest Washington. But it is not anticipated that much more oil will be found and already the government and oil users have begun to look elsewhere for supplies to satisfy provincial needs when domestic sources have run out. Even so, there has been a substantial response to the provincial government's incentive program for oil exploration and development introduced in 1985.

Hydro-electric Power

British Columbia has long been known to have a very large hydro-electric power potential. Estimates have altered, of course, over time, particularly as knowledge has been accumulated about streamflow and geological conditions. The most recent assessment suggests that there are opportunities to harness at least 222,240 gigawatt hours per annum on the province's major streams.[23] A gigawatt hour is equivalent to one million kilowatts generated for one hour. Generating capacity is calculated on the basis of average water conditions during a specified period.

Table 1,11 shows that there are opportunities for hydro-electric power development in almost every region. In some cases, very large amounts of power could be harnessed. Potentially, for example, some 26,090 gigawatt hours per annum could be developed on the Liard River and about 15,180 on the Stikine-Iskut rivers.[24]

Hydro-electric power development began in Nelson in 1896, with the installation of a 55 kilowatt plant on the Kootenay River. Development proceeded steadily throughout the following half century. A major change began, however, during the 1950s when the demand for electricity suddenly accelerated and for almost 30 years showed little sign of slackening. In the period 1953 to 1973, for example, demands increased at an average rate of 11 percent per annum.[25] Population, meanwhile, grew at one-third of that rate or less. Industries, such as Alcan and Cominco, and electric power utilities added large amounts of capacity, but there never seemed to be enough. And so the era of the mega-project was born. From the 1960s onward projects increased rapidly in size from tens of thousands of kilowatts to millions. No longer did engineers limit their horizons to single projects to serve the next five years. Schemes for harnessing entire rivers with a time perspective of 20 years or more became the prescribed pattern.

The process was actively encouraged by the governments of the day. Provincial politicians, in particular, were convinced that the key to British Columbia's future economic success was the rapid development of energy resources, and especially hydro-electric power. The

Table 1,11 HYDRO-ELECTRIC ENERGY POTENTIAL

River Basin	*Average Energy Capability (gigawatt hours per annum*)*					
	Total	*Developed*	*Under Construction*	*Undeveloped*	*Development Probable†*	*Development Improbable†*
PEACE	23,990	16,520	0	7,470	4,610	2,860
SKEENA	6,470	50	0	6,420	6,420	0
FRASER	54,790	4,460	0	50,330	1,900	48,430
VANCOUVER ISLAND	3,740	2,090	0	1,650	270	1,380
YUKON-TAKU	22,960	0	0	22,960	0	22,960
LIARD	30,590	0	0	30,590	26,090	4,500
STIKINE-ISKUT	22,950	0	0	22,950	15,180	7,770
COLUMBIA	32,560	19,080	7,890	5,590	4,060	1,530
SOUTHWEST COASTAL	24,190	7,880	0	16,310	13,600	2,710
BRITISH COLUMBIA	22,240	50,080	7,890	164,270	72,130	92,140
PERCENT OF TOTAL BY CATEGORIES	100	23	3	74	33	41

* One gigawatt hour per annum equals one million kilowatt hours per annum.

† Projects classified as "probable" and "improbable" have been artibrarily selected using preliminary information, and may be reclassified as more information becomes available.

SOURCE: British Columbia Hydro and Power Authority data.

British Columbia government made several moves to back up its views in this regard, notably the signing of an agreement with the Wenner-Gren interests of Sweden to explore the various resource potentials of the north-eastern part of the province, and especially those of the Peace River, the takeover of the largest electric power utility in the province, the British Columbia Electric Company, to amalgamate it with the British Columbia Power Commission in order to form the British Columbia Hydro and Power Authority in 1961, and the implementation of the Columbia River Treaty concurrently with the development of the Peace River.[26]

The overall result was a more than threefold increase in the province's hydro-electric generating capacity in a period of less than 25 years. In 1961 that capacity stood at three million kilowatts.[27] By 1984 it had increased to 11 million kilowatts (Table 2,11).[28] As indicated in Table 3,11, almost 70 percent of the province's generating capacity now is provided by projects with installations in excess of 500,000 kilowatts. Another feature has been the dominant role assumed by British Columbia Hydro in

Table 2,11 PRINCIPAL PRODUCERS OF ELECTRIC POWER IN BRITISH COLUMBIA

	Installed Capacity on March 31, 1984 (kilowatts)		
	Hydro	*Thermal*	*Combined*
UTILITIES			
British Columbia Hydro and Power Authority	9,571,002	1,131,140	10,702,142
Kitasoo Hydro	320	0	320
The Corporation of the City of Nelson	8,995	0	8,995
Tsawataineuk Band Council	75	0	75
West Kootenay Power and Light	184,875	0	184,875
SUB-TOTALS (UTILITIES)	9,765,267	1,131,140	10,896,407
FOREST PRODUCTS COMPANIES			
British Columbia Forest Products	0	36,300	36,300
Canadian Cellulose	0	44,500	44,500
Canadian Forest Products	0	4,500	4,500
Cariboo Pulp and Paper	0	28,000	28,000
Crestbrook Forest Industries	0	15,000	15,000
Crown Zellerbach	0	23,420	23,420
Evans Products	0	7,500	7,500
MacMillan Bloedel	74,980	119,360	194,340
Northwood Pulp	0	31,500	31,500
Ocean Falls Corporation	13,320	0	13,320
Tahsis	0	8,000	8,000
Weldwood of Canada	0	6,500	6,500
Western Forest Industries	0	1,700	1,700
Western Pulp	4,587	39,800	44,387
SUB-TOTALS (FOREST PRODUCTS)	92,887	366,080	458,967
MINING COMPANIES			
Aluminum Company of Canada	812,800	5,000	817,800
Cassiar Asbestos	0	14,970	14,970
Cominco	468,800	0	468,800
Dickenson Mines	200	0	200
Endako Mines	0	2,250	2,250
Granduc Operating Company	0	32,300	32,300
Wesfrob Mines	0	10,440	10,440
Westmin Resources	3,060	0	3,060
SUB-TOTALS (MINING)	1,284,860	64,960	1,349,820
OTHER INDUSTRIES			
British Columbia Sugar Refining	0	5,500	5,500
Pacific Petroleums	0	7,500	7,500
Walden North Enterprises	450	0	450
SUB-TOTALS (OTHER INDUSTRIES)	450	13,000	13,450
MISCELLANEOUS MICRO-HYDRO	1,413		1,413
TOTALS	11,144,877	1,575,180	12,720,057

SOURCE: British Columbia Ministry of Environment, Water Management Branch records.

electric power development. Its projects now account for some 84 percent of the installed capacity (Table 2,11).[29]

The trend toward large scale hydro-power development has slowed down, temporarily at least. What the future mix will be remains an open question. All of the projects that have been under active discussion in the past few years would have installed capacities of at least 400 megawatts, and most of them several times that figure (Table 4,11). The Yukon-Taku and the Liard schemes, for example, would each be more than twice the size of the most recently completed hydro-power project in the province, Revelstoke on the Columbia River. Recently, however, there has been pressure for a slower, more incremental approach, bringing in small, as well as large schemes, as the load expands.

Other Energy Sources

There are several other potential sources of energy in the province. These include solar, geothermal, tides, waves, wind and uranium. The federal and provincial governments, industry, universities, and various individuals have undertaken a variety of studies of such possibilities. There has been a particularly active interest in solar energy, especially through the efforts of the Solar Energy Society of Canada which now has chapters in several parts of the province. A study undertaken for the British Columbia Ministry of Energy, Mines and Petroleum Resources indicates that even in areas, such as Vancouver, that experience a great deal of cloudy weather, there may be important possibilities for the use of solar technology.[30] There has also been quiet enthusiasm about geothermal energy and tidal power potential, following studies by British Columbia Hydro.[31] Interest has focussed, particularly, upon possibilities in Portland Inlet, the mouth of the Skeena River, Sechelt and Howe Sound. In addition, there have been investigations of the possibilities of developing the province's uranium resources for nuclear power, both through British Columbia Hydro and through the British Columbia Ministry of Energy, Mines and Petroleum Resources. There is little support, however, within government or in the public at large, for nuclear power development.

There has been interest in the possibilities of making better use of energy, notably through improved building design, retrofitting, improved engine efficiency, alterations in behavior patterns, and changes in land use regulations.

Despite all of this, however, the attention to alternative sources of energy or to energy conservation has been minute in comparison to the effort expended on conventional sources.[32]

Table 3,11 THE TEN LARGEST POWER PLANTS

Name	*River*	*Installed Capacity (in megawatts)*	*Owner*
1. G.M. SHRUM	PEACE	2,730	B.C. HYDRO
2. REVELSTOKE	COLUMBIA	1,843	B.C. HYDRO
3. MICA	COLUMBIA	1,736	B.C. HYDRO
4. KEMANO	NECHAKO	813	ALCAN
5. PEACE CANYON	PEACE	700	B.C. HYDRO
6. KOOTENAY CANAL	KOOTENAY	529	B.C. HYDRO
7. SEVEN MILE	PEND OREILLE	525	B.C. HYDRO
8. BRIDGE RIVER	BRIDGE	438	B.C. HYDRO
9. WANETA	PEND OREILLE	300	COMINCO
10. JORDAN RIVER	JORDAN	130	B.C. HYDRO

SOURCE: British Columbia Hydro, Personal Communication.

PRESENT PRODUCTION AND CONSUMPTION

Over the past 30 years British Columbia has spent billions of dollars in developing its energy resources. The overall result has been to put in place a highly diverse energy supply system that has led to the province becoming an overall net exporter of energy. Not only has it been able to attract energy intensive industries, such as aluminium production, and provide the citizenry with abundant, relatively low cost supplies on the domestic scene, but it has made an important contribution to the earnings of the province beyond its own borders.

There are significant differences, of course, in the relative contributions of the various forms of energy to overall production, consumption and trade in that commodity in the province. To enable an assessment of such contributions it is necessary to convert the measure of each energy type into a common unit. In the past the British thermal unit was typically used for this purpose. Today the petajoule is more commonly employed. Table 5,11 presents the various petajoule equivalents.

The present energy production and consumption scene in British Columbia is depicted in Figure 3,11. Essentially, it is a flow chart, consisting of three main stages. It begins with primary production, such as the extraction of natural gas from a well. In British Columbia most of this production moves on to the next stage, secondary production. Part of it, however, is

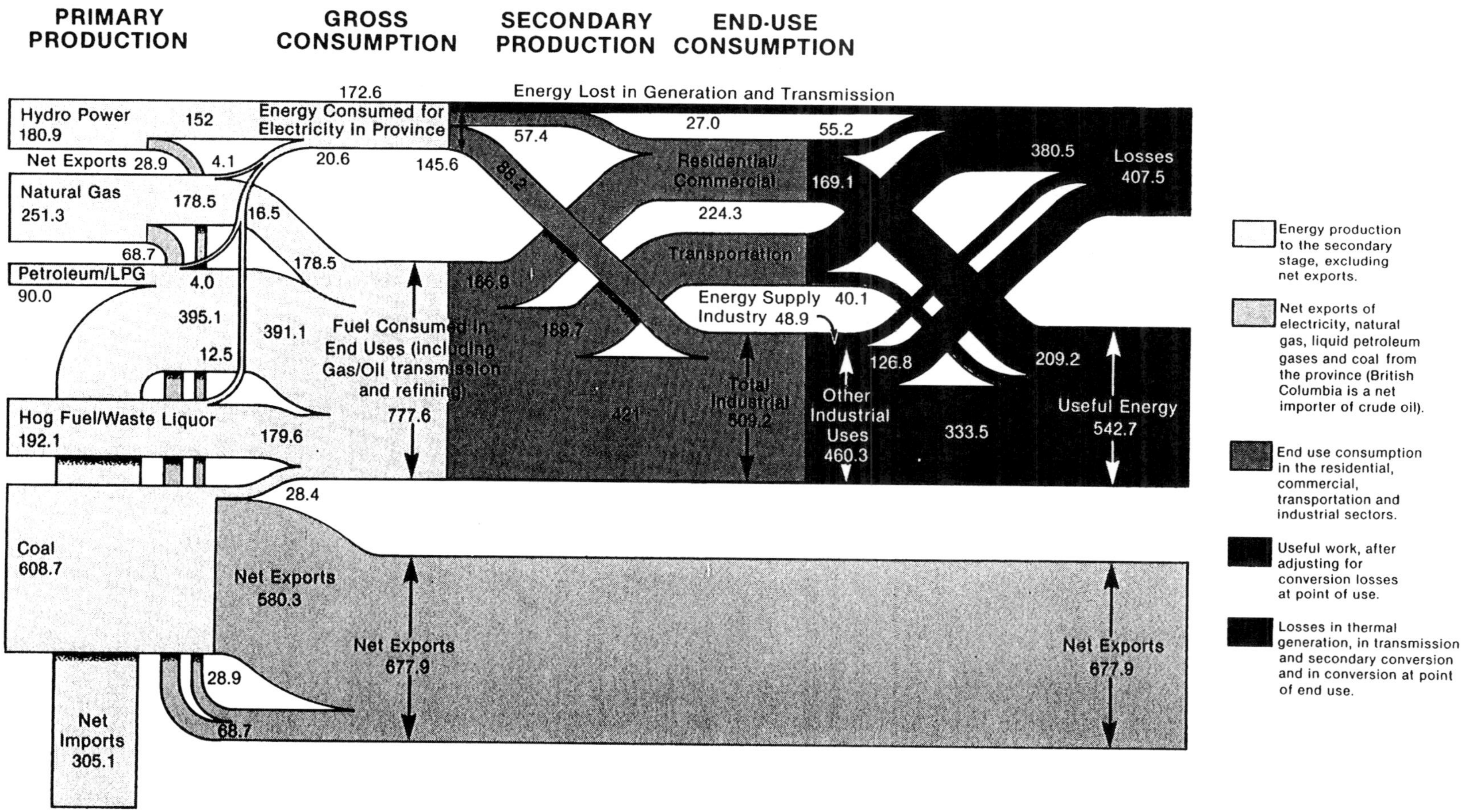

FIGURE 3,11 Energy flows, 1985, in petajoules

Table 4,11 POTENTIAL HYDRO-ELECTRIC POWER AND THERMAL POWER PROJECTS

Project	*Installed Capacity (megawatts)*	*Average Annual Energy (gigawatt hours)*
HYDRO-ELECTRIC POWER		
Murphy Creek	400	1,870
Site C	900	4,610
Iskut	980	440
Stikine	1,920	9,080
Yukon-Taku	3,692	22,960
Liard	4,770	26,090
THERMAL POWER		
Hat Creek	2,000	11,400
East Kootenay	600	3,680

SOURCES: British Columbia Hydro, *Electric System Plans, 1981-1985; 1981-1995*, Vancouver: British Columbia Hydro and Power Authority, December 1980; British Columbia Hydro, *Stikine-Iskut Development*, Vancouver: British Columbia Hydro and Power Authority, December 1980.

exported, notably in the case of natural gas and coal. At the same time, some energy, particularly petroleum, is imported into the province. This, together with the remainder of primary production left after exports, constitutes gross consumption. It is the feedstock for the next stage, secondary production. Here the "energy mix" is converted into forms suitable for end uses, such as in houses, factories and transportation. Natural gas, oil or petroleum, for example, might be burned in a thermal power station to produce electricity. A petroleum refinery produces a wide variety of products for use in houses, industry and transportation.

The conversion of energy from one form to another inevitably involves losses. Sometimes these are small, but mostly they are large. As the diagram shows, while end use consumption accounts for 550.3 petajoules in British Columbia, losses are responsible for 408.5 petajoules. To those concerned with making the most efficient use of resources, or with providing for the future, the message is clear: ways must be found to improve the efficiency of energy utilization in British Columbia!

The relative importance of the various energy forms varies from one part of the province to another, as does the significance of the different uses for such energy. This is indicated in Figure 4,11 which

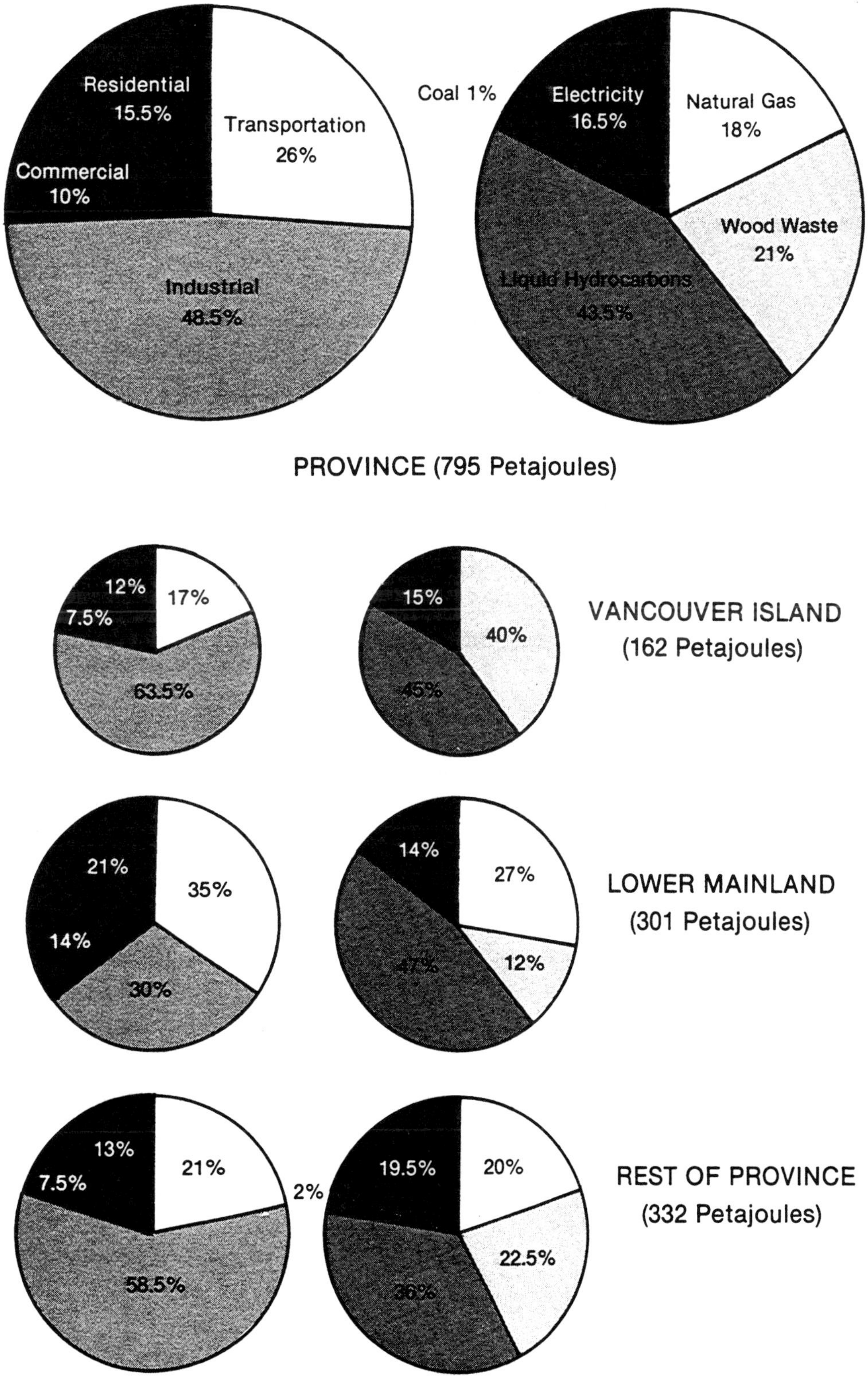

FIGURE 4,11 Energy consumption by end use, fuel type and region, 1978

Table 5,11 ENERGY EQUIVALENTS

1 Petajoule	=	160,000 barrels of oil *(or 25,426 m^3)*
	=	50,000 short tons of thermal coal *(or 45,000 metric tonnes)*
	=	950,000 cubic feet of natural gas *(or 27 × 10^6 m^3)*
	=	280 gigawatt hours of electricity *(1 gigawatt hour = 1 million kilowatt hours)*

is based on 1978 data. The situation, however, has not substantially changed. The diagram shows that, while liquid hydrocarbons dominate everywhere, wood waste is extremely important on Vancouver Island. Natural gas assumes particular importance in the Lower Mainland. From the point of view of uses, industry accounts for almost half and transportation about one-quarter. Again, the relative importance of the uses alters from one region to another. Industrial uses are especially dominant on Vancouver Island, while transportation is a particularly large use in the Lower Mainland.

FACTORS INFLUENCING ENERGY DECISIONS

The present pattern of energy development and use is the outcome of a complex process of decision-making. This process involves interactions between several groups of "actors" who seek to influence each other in a variety of ways which, for convenience, may be described as "strategies". The latter may consist of fairly simple actions, such as the provision of information, or they may extend to more forceful activities, like the provision of incentives or the imposition of sanctions. Different actors, of course, employ different kinds of strategies, reflecting the extent to which they can wield power over other actors. Figure 5,11 presents in very simplified form the various groups of actors that are involved in energy decisions in British Columbia, and a selection of the strategies that they tend to use in trying to influence each other.

Governments and Government Agencies

Because energy is so important to the country's economic and social well-being, the federal government plays a major role in setting policy guidelines for development and use of energy resources. Responding to

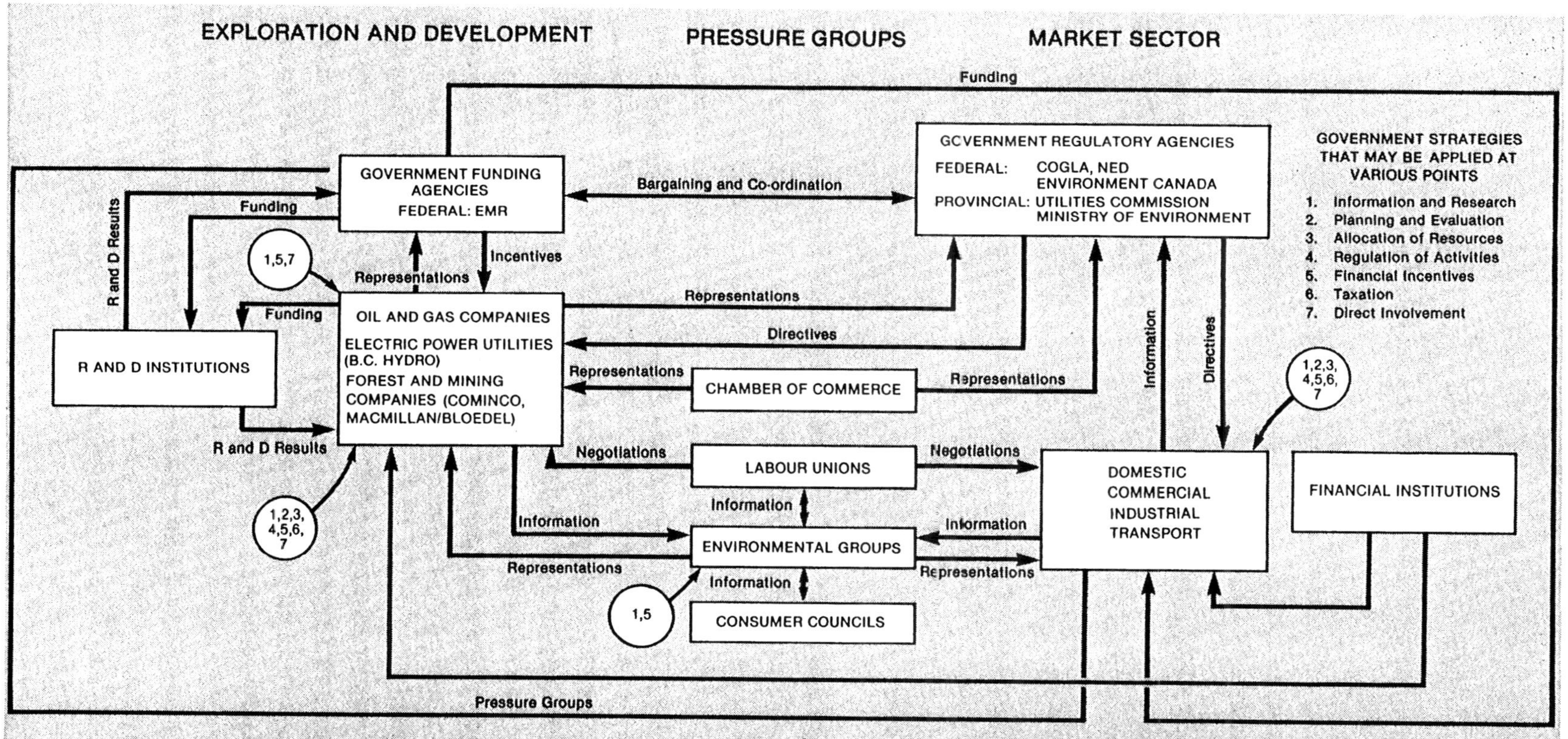

NOTE: *Only a selection of the actors and types of influence is shown here. Also only a selection of possible interactions among actors is indicated.*

FIGURE 5,11 Energy supply and demand systems: actors and influences the energy crisis that followed the oil embargoes of the middle and late 1970s, the federal government drew up a National Energy Program in 1980, based upon three main policy goals:

1. attainment of Canadian independence of the world market for oil within a decade;
2. provision for greater Canadian participation in the oil industry, through increased ownership, and for sharing the benefits from the expansion of this sector; and
3. establishment of a petroleum pricing and revenue sharing regime which would more evenly distribute the wealth generated from Canada's energy resources.[33]

The federal government (Liberal) introduced several policies in 1980 and in the succeeding years in an attempt to implement these goals. They included measures designed to stimulate exploration by Canadian owned oil companies, particularly in the North, strategies to encourage energy users to switch to fuels other than oil, and programs to encourage energy conservation. Many of the government's policies had direct relevance to the energy scene in British Columbia, particularly with respect to the export of natural gas and the development of offshore hydrocarbon resources.

These policies were the focus of much debate at both the federal and provincial levels of administration. The oil companies reacted adversely and the exploration for new sources slowed appreciably. An alienation developed between Ottawa and the western provinces. Energy policy became a major issue in the election campaign in 1984. The newly-elected (Conservative) government has set about dismantling the previous policies and has replaced them with new ones designed to reduce government involvement in the industry, placing greater emphasis upon market forces to determine prices of oil and natural gas and sources of supply. Perhaps the most important of the policy changes affecting British Columbia has been the Western Accord, signed in March 1985 by the federal government and the four western provinces.[34] Under this agreement several controversial federal levies and regulations were removed, and a new natural gas pricing regime was introduced. The Petroleum Incentives Program (P.I.P.) which provided cash incentives to petroleum companies exploring for oil and gas is being phased out, and replaced by a new policy focussing upon frontier regions.

Meanwhile, the federal government joined the provincial authority in a review of the possible implications of exploration of offshore hydrocarbon resources in the vicinity of the Queen Charlotte Islands. The hearings began in Alert Bay on September 9, 1985 and were completed in Victoria on October 26, 1985. A report is expected in mid-April, 1986.

The major agency at the federal level concerned with energy matters is the Department of Energy, Mines and Resources. It is responsible for drawing up, co-ordinating and implementing energy policies for the country as a whole. It carries out a wide spectrum of functions, ranging from the collection of information to the financing of development. Under its mandate several policies have been introduced which have been influential in the development of energy resources in British Columbia, as well as in the use of energy in the province.

Another federal agency is the National Energy Board, which was established in 1959 to advise the government on the development of energy resources, and to regulate specific aspects of the oil and gas industry and the electric utilities. One of the most demanding of its tasks has been to determine the extent to which Canada should export oil, natural gas, or electricity to the United States, and whether various proposals for the construction of pipelines or transmission lines that cross provincial or national boundaries are in the national interest. To assist it in this function, the Board holds hearings which involve proponents, such as oil companies, and other interested parties, like energy agencies or environmental interest groups. The Board has reviewed many proposals concerning British Columbia energy resources.

The British Columbia government and its various agencies are also part of the first group of actors. Like its federal counterpart, the provincial government announced the broad outline of an energy policy in 1980.[35] Its two basic aims were to promote security of energy supply, on the one hand, and to take advantage of the province's comparative advantage in the availability of particular sources of energy, on the other. Several specific strategies were to be used to attain these objectives, including:

1. the improvement of government organization;
2. the assignment of British Columbia Hydro as the government's major instrument to implement policy relating to electric power;
3. the promotion of exports of surplus electric energy, natural gas and coal;
4. the reduction of consumption of oil, particularly through the substitution of natural gas, coal and wood waste, and through the construction of a natural gas pipeline to Vancouver Island;
5. the promotion of improved efficiency in the use of energy, through demonstration programs and more realistic energy pricing schedules;

6. the intensification of research and development; and
7. the streamlining of the evaluation of energy projects in order to sort out more quickly those which have major problems attached to them and to accelerate the development of those which do not.

The Ministry of Energy, Mines and Petroleum Resources which had been established in December 1978 was to continue as the province's principal agency in the energy field. It would have broad powers to gather information as to the resource base, energy use and development opportunities. It was to recommend charges to be made by the provincial government with respect to the harnessing of energy resources, as well as policies that might be pursued through subsidies, regulations and other measures. The program of the provincial government's two energy-related Crown corporations, the British Columbia Hydro and Power Authority and the British Columbia Petroleum Corporation, would be co-ordinated by an Energy Policy Committee chaired by the minister.

A new agency was established, the British Columbia Utilities Commission. It was to assume regulation of the electricity and gas sections of British Columbia Hydro, and undertake regulating functions formerly carried out by the British Columbia Energy Commission which was thereby abolished. The non-regulatory functions of the latter were to be assumed by the Ministry of Energy, Mines and Petroleum Resources.

The Utilities Commission has been extremely busy ever since its establishment. It has reviewed several applications of British Columbia Hydro for rate increases, and has established review boards for a number of energy resource development schemes, notably Site C on the Peace River,[36] and an inquiry into the merits of the disposal of surplus natural gas outside the province.[37]

Another agency with major interests in energy-related developments, particularly with respect to their likely environmental effects, is the Ministry of Environment. It carries out its responsibilities in this regard in part through its own legislation. The review of proposed exploration and development of hydrocarbons in the offshore region adjacent to the Queen Charlotte Islands is in this latter category. The Ministry also exerts control in the hydro-electric power field through its responsibilities for the issuing of water licences.

Like the federal government, the provincial authority has played an important role in guiding the development of energy resources in British Columbia, especially in the past decade. Concerned, for example, that there was insufficient oil exploration activity in the province, it introduced a royalty holiday for wells drilled between the summer of 1985 and the summer of 1989. It estimated that this program would lead to the drilling of 600 more wells than would otherwise have been the case.[38] Worried about the large amount of surplus generating capacity in the British Columbia Hydro system, and the one billion dollar a year interest cost, much of which is associated with that capacity, the government has encouraged the utility to seek markets south of the border.[39] Recently British Columbia Hydro has been selling energy worth more than one million dollars a day in the Pacific Northwest and is now exploring the possibilities of selling electricity to California.[40] Wishing to encourage the development of a larger petro-chemical industry, the government has also introduced a program of incentives for that purpose.[41] In addition, it has tried to speed up the process of review of projects for energy resource development through the introduction of new guidelines.[42]

Developers, Producers and Manufacturers

The second group of actors is comprised of developers, producers and manufacturers. They include electric power utilities, oil and gas companies, coal companies, and manufacturers of equipment that is used in the development and utilization of energy. The British Columbia Hydro and Power Authority is especially important in this connection. Though considerably trimmed in size over the past year or so, due to economy measures, it employs more than 6,500 workers in professional and other capacities, and has capital expenditures of over one billion dollars a year. During the past two decades it has wielded considerable influence in planning and policy-making with respect to the harnessing of hydro-electric resources in the province.

The oil and natural gas companies are also major players in the energy game in British Columbia. The multi-national corporations, including Standard Oil, Gulf and Esso, have long been involved in the province, first as sellers of products and more recently as explorers for potential resources. They have been joined by various Canadian companies, such as Home Oil and, more recently, Petro-Canada. Some of them have focussed mainly upon the search for resources. Others, like Westcoast Transmission and Inland Natural Gas, have been more concerned with the transportation and distribution of the products. While most of the external interest in British Columbia oil and natural gas resources has come from the United States, that in coal has originated in Japan. Japanese capital and Japanese markets, in fact, now dominate the British Columbia coal industry.

Consumers

The electric power utilities, the oil and natural gas companies, and the coal development interests put pressure on the government from a production point of view. They wish to satisfy demands at the lowest possible costs and, in the case of the private sector, to yield a profit for investors. Lined up on the other side are the energy consumers who wish to ensure that they have sufficient energy to meet their needs at a price that they can afford. Domestic consumers, industries, transport, and other uses share this concern. They put pressure on the Utilities Commission to make sure that electric power and natural gas rates stay as low as possible. They can be relied upon to resist increases in such prices.

Interest Groups

In between the producers and the consumers are other interest groups, with a wide spectrum of concerns, ranging from the rights of minority groups to the long term impact of energy projects on the environment. They bring pressure to bear in a variety of ways, including presentations at public hearings, letters to the editor, and protest marches. Several of these groups have been very active on the British Columbia energy scene, notably the Society Promoting Environmental Conservation (S.P.E.C.) and the Sierra Club.

Thirty years ago the development of energy resources seemed a rather simple matter. Everyone seemed to be in favor of it, whatever it meant. The scale at that time, however, was relatively small. There was little concern about environmental effects, and there were no apparent threats to the rights of native peoples or minority groups. The administrative machinery was simple, readily understood, and relatively easy to make one's way through. The scene today stands in sharp contrast. Proposals for energy resource development almost always engender conflict and confrontation. Decision-making is often extremely slow and drawn out.

The province enters the second half of the 1980s with the knowledge that it has a very generous endowment of energy resources, but also that it faces some nagging issues about their development and utilization.

EMERGING ISSUES

There are three major issues that confront energy planners and policymakers as of late 1985:

1. How much energy will be needed in future in British Columbia?
2. What are the most efficient means of satisfying energy demands?
3. How can the views of the public be sought most efficiently and incorporated most effectively in decision-making with respect to energy?

Energy Demand Forecasting

A major requirement in modern energy resource management is the provision of accurate forecasts of future energy demands. This is especially so where long lead times and massive expenditures are involved in the provision of new capacity. The costs of being wrong have been amply demonstrated in British Columbia where British Columbia Hydro is now in the difficult position of having far more generating capacity than it needs and having to pay out hundreds of millions of dollars in interest payments on projects which seem unlikely to yield a return for many years to come. The origins of the utility's difficulties are not recent. For more than a decade it has consistently over-forecast electric power demands by a considerable margin.[43] Reviews have shown that on average since 1973 British Columbia Hydro has over-forecast by more than 18 percent within five years from the time when the forecast was made. The errors have been especially large in the bulk sector (the major industrial consumers) which accounts for about one-third of the utility's sales. In this instance, it seems, forecasts were out by more than 40 percent within six years of the forecast date.

Various explanations might be offered for the consistent inaccuracy of the forecasts. A major one, of course, is the uncertainty of future economic conditions, especially in a province which is so vitally affected by external factors. More fundamentally, however, it seems that British Columbia Hydro has used rather simplistic forecasting methodologies, and has assumed that electric power demands are generally inelastic; that is, they are unaffected by a change in the price of electricity. It has long been recognized by utilities elsewhere, however, that this is a false assumption.[44] During the past decade the price of electricity in British Columbia has climbed steeply relative to that of other goods and services in the province. Naturally, this has had effects on electricity purchases in the various sectors of demand, both domestic and industrial.

A potential consequence of over-forecasting of future demands may be to place British Columbia Hydro on a "utility death spiral", similar to that which occurred in the Pacific Northwest a few years ago.[45] In this process a utility finds itself with more capacity than is required and is unable to generate sufficient financial returns to cover the costs of servicing the debt associated with the provision of that capacity. Its natural reaction is to raise rates for electricity. The consequence, however, is to dampen demand even further. In the case of British Columbia Hydro this process has several additional implications, since the utility must now borrow hundreds of millions of dollars on money markets in the United States and elsewhere. Inability to repay huge debts already incurred poses the possibility that these markets will be less willing to loan money to the province for other, and perhaps even more necessary, purposes. Thus far, the province seems to have escaped the "death spiral", in part because of its "spot" sales of energy to the Pacific Northwest, and in part because its rate increases have not been so large as to seriously alter demand.

Selection of Optimal Strategies

A second problem relates to the selection of the most efficient means of satisfying future demands. Critics of energy planning in British Columbia, and electric power planning in particular, have suggested that there are several major weaknesses in the present approach.[46] Firstly, they claim that, because of its overwhelming dependence on mega-projects, energy resource management in the province is not resilient to changes in economic circumstances or in social values. The shelving of the Site C project, Kemano II, and the natural gas pipeline to Vancouver Island are given as illustrations. Secondly, they suggest that the provincial government's commitment to energy conservation has amounted to tokenism. For example, they point to the fact that only one percent of British Columbia Hydro's multi-million dollar budget is devoted to this means of dealing with the energy problem. Thirdly, the critics suggest that British Columbia Hydro's policies tend to foster waste, since they are based upon a declining block rate structure. This means that the initial increment of power that is purchased is more expensive than subsequent increments. Research in many parts of the world has shown that such structures not only encourage waste, but also lead to serious social inequities.[47]

An examination of British Columbia Hydro's present rate structure and revenues derived therefrom (Table 6,11) shows why such criticisms have been made. While residential consumers furnish 35 percent of the revenues, they pay almost twice as much for their electricity as the bulk consumers. Although it might be argued that industrial consumption to some extent makes domestic use possible, the margin between the rates charged seems very large. Beyond this is the fact that, even within the domestic rate structure, there are possible social inequities. This is because

Table 6.11 BRITISH COLUMBIA HYDRO ELECTRIC POWER RATES AND REVENUES

Customer Class	*Percent of Total Sales**	*Percent of Revenues**	*Average Rates (cents per kilowatt hour)*
BULK†	32	19.5	2.5
GENERAL‡	34	33.2	4.0
RESIDENTIAL	31	35.0	4.7

* These do not add to 100 percent because they do not include irrigation, street lighting and exports.

† Bulk consumers number about 61, who receive service directly from high voltage transmission lines.

‡ General refers to all commercial and industrial consumers not included in the bulk class.

SOURCE: British Columbia Hydro, Personal Communication.

of declining block rates: additional increments of consumption cost progressively less. This may mean, as some have suggested, that low income families bear relatively more of the cost burden than their more affluent counterparts.

A third criticism is that British Columbia Hydro appears to downplay the possibilities of co-generation as a means of meeting part of its energy demands. Present policies, in fact, seem to discourage that option. Evidence presented at the Site C Hearings suggested that a potential exists to generate almost as much energy from wood waste in British Columbia as would have been provided by the project on the Peace River and at a cost competitive with other options available to the utility.[48] Unfortunately, present British Columbia Hydro policies for purchasing surplus energy from industries that generate their own electricity have tended to disuade rather than encourage co-generation.[49]

A fourth perceived deficiency relating to strategy selection is the failure of existing planning procedures to ensure that environmental and social impacts are adequately considered. In the case of the Site C Project, for example, opponents of the scheme pointed out major deficiencies in data presented and in the analyses employed.[50] They were especially critical of the small amount of effort devoted to these aspects compared with the engineering dimensions.

Incorporation of Public Views

The final major issue facing British Columbia's energy managers concerns the incorporation of public views in the decision-making process. In contrast with the situation a decade ago, there are now numerous ways in which the opinions of the citizenry can be sought and incorporated into decision-making. Provincial government officials point to the attempts to involve various interests through public hearings, workshops, task forces and advisory committees. They note that millions of dollars have been spent on several major energy-related inquiries, such as the Site C Hearings or the West Coast Oil Ports Inquiry.

Despite all of this, however, there remain substantial criticisms of present mechanisms. Some of the problems are concerned with the provision of information in a form that is readily understandable and available in sufficient time for the public to consider prior to offering an opinion on the merits of a given project. Other difficulties stem from the fact that reports of inquiries seldom indicate the role that particular factors played in the recommendations that were made. Because of this, some of those who have made presentations at hearings feel that the process is little more than window-dressing. In addition, other critics have suggested that the requirement for the proponent to provide evidence of the likely impacts minimizes the likelihood that all the adverse effects will be adequately assessed.[51]

TOWARD THE 1990s

The difficulties posed by the economic recession of the early 1980s have presented British Columbia with a rare opportunity to pause and reflect. While it is probable that the policies of the 1950s and 1960s did much to provide the energy required to build a thriving economy, it is also evident that such policies lack a critical characteristic, that of resilience, or the ability to respond to major change. It is also clear that social values have altered considerably in the past 30 years. There is now much more concern about the environment and the rights of individuals and minority groups than was the case in the 1950s. There are also lingering worries about the power wielded so far by the province's major electric power utility. Perhaps the time has come for new policies and new kinds of administrative structures. Most important of all, there is now an opportunity to reconsider the role of energy in the province's economic and social affairs.

REFERENCES

1. The author wishes to offer his sincere thanks to the following individuals who provided information or who commented upon various drafts of the chapter: Anthony Dorcey, John Chapman, Robert Durie, Chad Day, John Allen, Ray Crook, Harold Foster, and Robert Bell.

2. SCHURR, S.H. and ELIASBURGH, V.F., "Energy and Economic Growth in the United States", in SCHURR, S.H., *et. al.*, *Energy in the American Economy*. Baltimore: Johns Hopkins Press, 1960, pp. 1-18.

3. DAVIS, J., "The Market for Energy", *Proceedings of the British Columbia Natural Resources Conference*. Victoria: British Columbia Natural Resources Conference, 1960, pp. 10-25.

4. "Put Oil Checks on Hold". *The Province*, January 22, 1986.

5. "Petro-Canada Gets Long Term Natural Gas Contract in the U.S.", *Vancouver Sun*, March 14, 1986.

6. BRITISH COLUMBIA, Ministry of Energy, Mines and Petroleum Resources, press releases, January 31, 1985; March 14, 1985; and April 25, 1985.

7. "Some Hurdles Facing Site C", *Times-Colonist*, February 24, 1986.

8. "Kemano Project Shelved", *Vancouver Sun*, October 27, 1984, p. A1.

9. "Tories Shatter Gas Pipe Dream", *Times-Colonist*, October 6, 1984, p. A1.

10. "Home of Ailing Quintette Coal Mine, Tumbler Ridge Uneasy About Future", *Globe and Mail*, February 1, 1986.

11. "Teck Reduces Coal Price for Japanese", *Vancouver Sun*, October 11, 1984.

12. "Interest Charges Up 60%", *Vancouver Sun*, October 31, 1984, p. D18.

13. FOX, I.K., "An Assessment of the Site C Proposal of B.C. Hydro and Power Authority", a presentation of the Review Panel for the Proposed Site C Hydro-Electric Power Project, Vancouver, December 31, 1981.

14. "Sagging Prices May Jeopardize Oil Projects", *Globe and Mail*, January 22, 1986, p. B1.

15. BRITISH COLUMBIA, *British Columbia Coal Development.* Victoria: Ministry of Energy, Mines and Petroleum Resources, 1983.

16. BRITISH COLUMBIA, *Guidelines for Coal Exploration*. Victoria: Ministry of Energy, Mines and Petroleum Resources, 1981.

17. BRITISH COLUMBIA, *Energy: The Canadian Picture*. Victoria: Ministry of Energy, Mines and Petroleum Resources, 1980, p. 11.

18. REID, COLLINS AND ASSOCIATES, *Hog Fuel Availability in British Columbia*. Vancouver: Wood Waste Co-ordinating Committee, 1978.

19. McDANIELS, T.L., *Forest Biomass Energy in British Columbia: Opportunities, Impact, and Constraints*. Victoria: Pacific Forest Research Centre, Environment Canada, 1982.

20. GOVIER, G.W., *Commissioner Inquiry on British Columbia's Requirements, Supply and Surplus of Natural Gas and Natural Gas Liquids: Summary Report*. Victoria: Queen's Printer, May 1982, p. 35.

21. *Ibid.*, p. 66.

22. BRITISH COLUMBIA, *Energy: The Canadian Picture*, *op. cit.*, p. 19.

23. BRITISH COLUMBIA HYDRO, personal communication; B.C. Ministry of E.M.P.R., personal communication. The figure 222,240 gigawatt hours refers to the average energy capability if the rivers in question were completely harnessed.

24. BRITISH COLUMBIA HYDRO, personal communication.

25. BRITISH COLUMBIA HYDRO, *Power Perspectives, '79: A Look at Today's Planning for Tomorrow's Power*. Vancouver: British Columbia Hydro and Power Authority, 1979.

26. SWAINSON, N.A., *Conflict Over the Columbia: The Canadian Background to an Historic Treaty*. Montreal: McGill-Queens University Press, 1979.

27. BRITISH COLUMBIA HYDRO, *Power Perspectives '79: A Look at Today's Planning for Tomorrow's Power, op. cit.*.

28. BRITISH COLUMBIA HYDRO, personal communication.

29. BRITISH COLUMBIA, Ministry of Environment, personal communication.

30. ACRES CONSULTING SERVICES, *Solar Energy Resources: Assessment Study for British Columbia*. Victoria: British Columbia Ministry of Energy, Mines and Petroleum Resources, 1980.

31. BRITISH COLUMBIA HYDRO, *Tidal Power in British Columbia*. Vancouver: British Columbia Hydro and Power Authority, June 1979.

32. BOTT, R., BROOKS, D., and ROBINSON, J., *Life After Oil; A Renewable Energy Policy for Canada*. Edmonton: Hurtig, 1983.

33. CANADA, *National Energy Program, 1980*. Ottawa: Department of Energy, Mines and Resources, 1980.

34. CANADA, *The Western Accord: An Agreement Between the Governments of Canada, Alberta, Saskatchewan and British Columbia on Oil and Gas Pricing and Taxation*. Ottawa: Department of Energy, Mines and Resources, March 1985.

35. BRITISH COLUMBIA, *An Energy Secure British Columbia: The Challenge and the Opportunity*. Victoria: British Columbia Ministry of Energy, Mines and Petroleum Resources, February 1980.

36. BRITISH COLUMBIA, *Site C Report*. Victoria: British Columbia Utilities Commission, May 3, 1983.

37. GOVIER, G.W., *op. cit.*

38. BRITISH COLUMBIA, Ministry of Energy, Mines and Petroleum Resources, press release, August 1, 1985.

39. BRITISH COLUMBIA, Ministry of Energy, Mines and Petroleum Resources, press release, March 21, 1984.

40. "Put Oil Checks on Hold", *The Province*, January 22, 1986, p. 27.

41. BRITISH COLUMBIA, Ministry of Energy, Mines and Petroleum Resources, press release, February 21, 1984.

42. BRITISH COLUMBIA, *Guide to the Energy Project Review Process*. Victoria: British Columbia Ministry of Energy, Mines and Petroleum Resources, 1982.

43. COLUMBIA-PACIFIC RESOURCES GROUP, *Assessment of B.C. Hydro Electric Load Forecasts Including Comparisons with B.C. Government Forecasts*. Vancouver: Columbia-Pacific Resources Group, November 5, 1981.

44. BOHI, D., *Analyzing Demand Behavior: A Study of Energy Elasticities*. Baltimore: Johns Hopkins Press, 1981.

45. "Power System Near Bankruptcy", *Vancouver Sun*, January 12, 1983.

46. SOCIETY PROMOTING ENVIRONMENTAL CONSERVATION and PEACE VALLEY ENVIRONMENTAL ASSOCIATION, *The Case Against Site C*. Vancouver: Society Promoting Environmental Conservation and Peace Valley Environmental Association, July 3, 1983.

47. MUNASINGHE, M. and WARFORD, J.J., *Electricity Pricing: Theory and Case Studies*. Baltimore: Johns Hopkins Press, 1982.

48. SOCIETY PROMOTING ENVIRONMENTAL CONSERVATION and PEACE VALLEY ENVIRONMENTAL ASSOCIATION, *op. cit.*, p. 17.

49. FOX, I.K., *op. cit.*, p. 6.

50. SOCIETY PROMOTING ENVIRONMENTAL CONSERVATION and PEACE VALLEY ENVIRONMENTAL ASSOCIATION, *op. cit.*

51. FOX, I.K., *op. cit.*

PLATE 16 Yachting on the west coast. *B.C. Government Photo*▶

PLATE 17 English Bay, Vancouver. *B.C. Government Photo*▶

3187
4833

12 MARINE-BASED RECREATION

Philip Dearden
University of Victoria

INTRODUCTION

The widely-acclaimed Canadian author Jack Hodgins has termed the coastal zone of British Columbia "the ragged green edge of the world". It is the purpose of this chapter to provide a selective overview of the key dimensions of recreational use of this "edge". First, however, it is necessary to provide a brief biophysical description of the area under question.

Although it is only some 850 kilometres (530 miles) from Prince Rupert in the north to Victoria in the south, the actual length of coastline is in the order of 41,000 kilometres (25,625 miles). This is due to the highly indented nature of the coastline, punctured by fiords extending deep into the heart of the Coast Mountains. Burke Channel, for example, extends 90 kilometres (56 miles) and Dean Channel 100 (62) from the outer coast. The fiords are typical drowned glaciated valleys with steep rock walls rising to over 2,000 metres (6,780 feet) in many cases. The east coast of Vancouver Island and the Fraser delta provide the only major areas of respite from this rugged topography.

The climate is dominated by the onshore flow of Pacific airstreams which release considerable precipitation in their passage over the mountain ranges. Winters are generally mild and very wet, whereas summers are cool, but less humid. The main exception to this pattern is the recreationally-significant climate of the southeastern coastal portion of Vancouver Island and the Gulf Islands. Here, in the rain shadow of the mountains to the west, a Mediterranean-type climate prevails, giving prolonged periods of warm, sunny weather, ideal for boating and other recreational activities in the summer period.

These climatic differences are reflected in prevailing vegetation patterns. The west coast climate engenders a climax forest dominated by species such as sitka spruce, cedar, hemlock and Douglas fir, whereas the drier

Mediterranean climate produces a distinctive association, including Mediterranean species, such as the arbutus and garry oak.

This biophysical tapestry provides the basis for a wide range of recreational activities for residents and tourists alike. Many studies have confirmed that the primary attraction in British Columbia for tourism is the scenery, particularly the mountains and the coast.[1] Between them, the Vancouver Island Tourism Region and the Southwestern Tourist Region (including the Lower Mainland and the Sunshine Coast) account for over 50 percent of the province's tourism revenue, and this proportion is growing. The continuing popularity of activities such as fishing, boating, scuba diving and beachcombing indicates that the coastal zone is a major attraction in these regions. In addition, most of the population of the province lives in the two major urban areas of the Lower Mainland and southeastern Vancouver Island, both within the coastal zone. These two forces — the growth of tourism and the demographic characteristics of the province — have led to an increasing recreational pressure on the fragile resources of the coastal zone. Against the background of this brief introduction, the rest of this chapter will be devoted to an examination of the various recreational activities of the coast and a summary of the major emerging problems.

For the purposes of this paper the coastal zone will be taken to include not only the physical meeting place of land and sea, but also the inshore waters and cruising grounds farther offshore, as well as the visual backdrop to the coast from the water. Along this continuum a wide range of recreational activities occurs, ranging from the cruise ship operations where the recreationist scarcely sets foot on land to those activities, such as beachcombing, which are totally land-based. These two categories, water-based and land-based activities, form the main organizational framework for the chapter, although in reality some activities fit equally well into both categories (Figure 1,12).

WATER-BASED ACTIVITIES

The coastal waters of British Columbia are not warm, meaning that water-based activities tend to be located on rather than in the water. Activities can be classified according to this distinction, with by far the majority of recreational use being on the water in the form of boating, and a much smaller amount being actually in the water, usually with the participants wearing some sort of thermal protective wear, such as in scuba diving and surfing. Exceptions do occur, however, with free swimming being popular in sheltered and warmer waters in Georgia Strait (Figure 2,12).

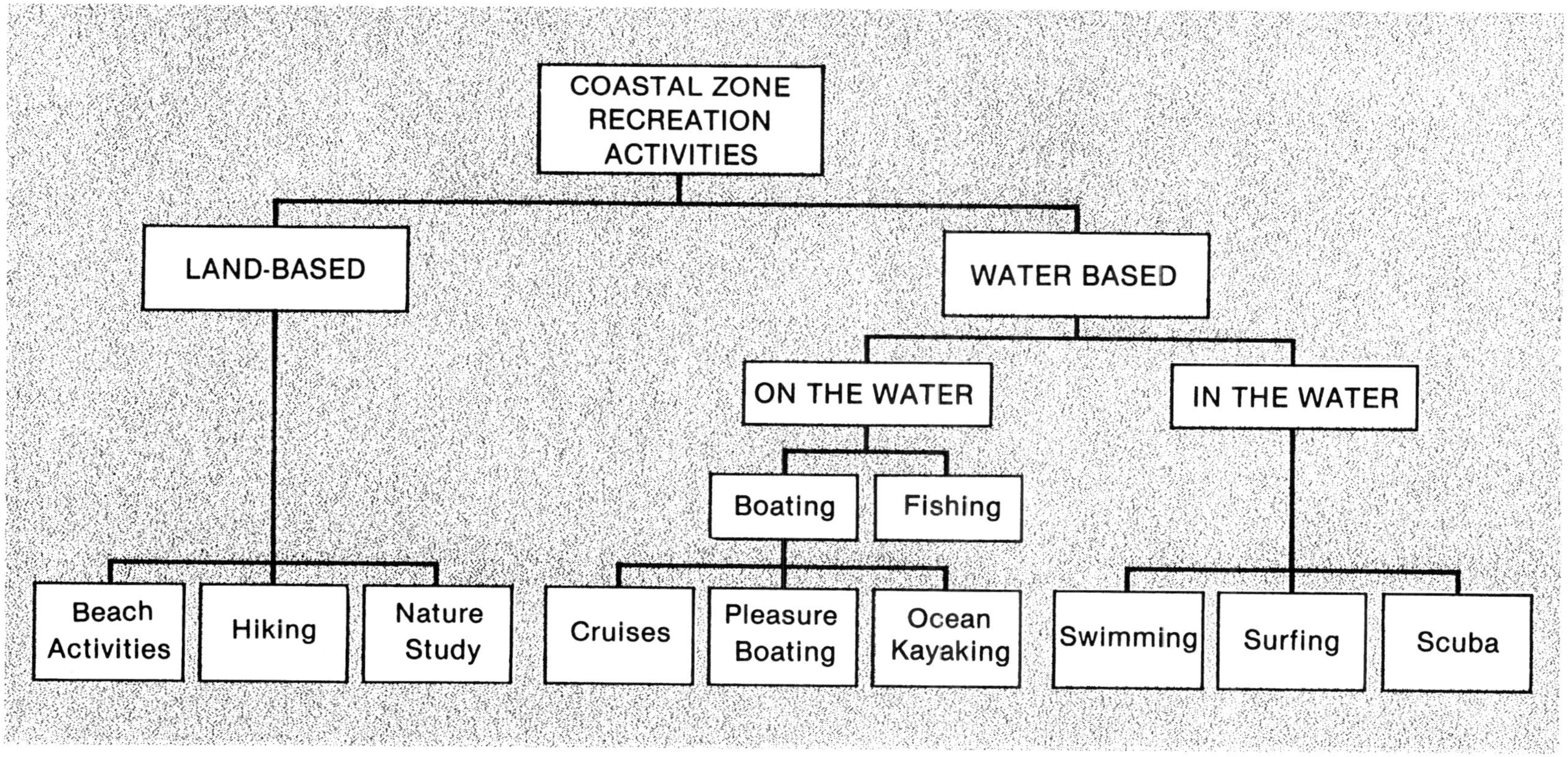

FIGURE 1,12 Major coastal zone recreational activities

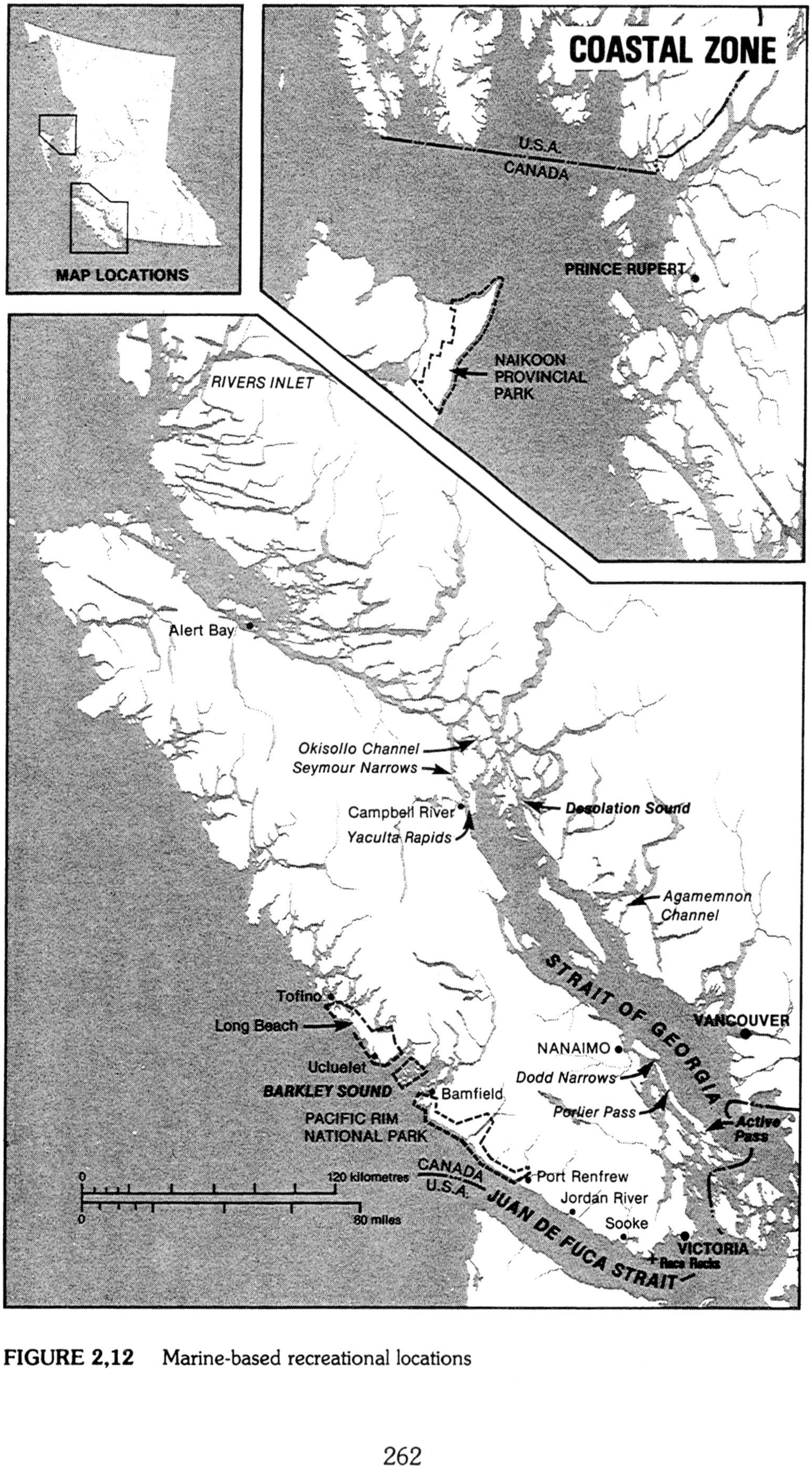

FIGURE 2,12 Marine-based recreational locations

Cruises

Two main kinds of cruises can be identified, depending upon the size of the vessel and goals of the cruise. At one end of the continuum large luxury liners ply the British Columbia coast en route to Alaska. Although the major attractions are scenery and wildlife,[2] there is little physical contact with the environment and the life on board the vessel constitutes a significant part of the experience. At the other end of the spectrum, numerous smaller vessels operated by much smaller businesses provide a limited, but equally high-paying, number of passengers the opportunity for extensive physical contact with the environment. Education is a major goal for such cruises. Both market segments have experienced very rapid growth. A paucity of data makes this difficult to measure for the small operators, although the growth in advertising for such cruises provides indirect evidence, whereas the luxury "Love Boat" cruises have increased at an annual rate of 60 percent since 1970.[3]

"LOVE BOAT" CRUISES

The Inside Passage cruise to Alaska is one of the most lucrative in the world. Sixteen vessels were engaged on the seven-day cruise from Vancouver to Alaska in May to September, 1983, with an average base price of 1,525 dollars per person. As all the major vessels are foreign-owned, most of this money leaves the country. However, in 1979 total revenues generated by ship and passenger spending were estimated at 44 million dollars.[4]

The ships engaged in the trade are among the largest passenger liners afloat today. They include the 150 million dollar, 1,200 passenger "Nieuw Amsterdam", completed in 1983, the 1,111 passenger, 34,000 tonne (38,000 ton) "Rotterdam", and the 1,200 passenger "Tropicale", built in 1981 at a cost of 130 million dollars. A new 150 million dollar, 1,200 passenger vessel for P. and O. Lines of London, which already operates three vessels on the run, has recently commenced operations. Whether the market can support such capacities remains to be seen.

The benefits to the province of this trade largely accrue through ship and passenger expenditures. Vancouver is currently the only embarkation-debarkation port, with other calls being made at Victoria, Nanaimo, Alert Bay and Prince Rupert. It is difficult to ascertain the economic impacts of this trade on the communities, although Victoria estimates that its 30 cruise stops in 1983 generated 1,200,000 dollars in passenger expenditures.[5] Certainly the visitation of 1,200 passengers, largely from southern California, into the small, dominantly native Indian fishing village of Alert Bay provides insight into free market economics! Prices of everything from ice cream cones to totem poles double instantly. With a projected increase of up to 413,000 passengers per year by 1990, the local economic impacts of the trade can only increase.

In terms of its purely recreational impact the luxury liner trade is not significant. Although the passengers are undertaking a recreational experience for which they are willing to pay large amounts of money, the principal recreational environment is the actual vessel, a strictly controlled and artificial environment. However, it should not be forgotten that the coastal scenery is the factor that has made the Alaska run one of the most popular in the world. Furthermore, the 1980 ruling limiting the number of vessels entering Glacier Bay, Alaska, in order to protect whale breeding grounds, serves to remind one of the potential impact of the trade on coastal wildlife habitats.

NATURE CRUISES

Of lower visibility but of increasing importance are the smaller, independently-operated vessels concentrating on immersing the participant in the natural and cultural milieus of the British Columbia coast. Such cruises vary in length from one day trips to observe the grey whale migration off Ucluelet, or to view the sea lions congregated on Race Rocks off Victoria, to much more extensive two week tours to more remote areas of the coast. Most are on small vessels under 16 metres (52 feet) in length where the actual operation of the boat is part of the recreational experience.

These trips are part of the more general recreational phenomena of recreational education that has also benefited membership in conservation organizations and the sale of nature guide books of all descriptions, binoculars, cameras and other equipment. They cater to the educated, middle classes interested in learning more about their environment, as a recreational experience. One operation, Pacific Synergies, advertises:

> If you're the sort of person for whom the very idea of an educational holiday sounds like a contradiction in terms, look into our sail expeditions in the Queen Charlotte Islands. A good many people have had the time of their lives living and travelling aboard our 45-foot sailing cutter while learning about the people, history and ecology of the little-known and largely unspoiled Queen Charlottes, the ancestral home of the Haida people.[6]

The vessel sleeps 10 people, houses a good library on coastal British Columbia, has microscopes, an aquarium, diving equipment and other equipment to allow an in-depth look at the environment. Most cruises also include an anthropologist, biologist, ornithologist or similar expert to provide more detailed insight. One week trips at roughly 1,200 dollars a head are scheduled from May through to October and all are fully booked.

Pleasure Boating

There are more boats per capita in British Columbia than in any other province, and 70 percent of the boat owners are found within the Strait of Georgia region.[7] Table 1,12 gives an indication of fleet composition changes in that area between 1973 and 1978 and displays a trend toward larger boats. In addition to the self-owned, resident category, a further five percent of households rent boats[8] and substantial numbers of boaters from the United States visit British Columbia waters. Accurate data on the latter are not available, but several figures give an indication of the magnitude of American visits. Customs estimates, for example, indicated that approximately 10,800 pleasure boats registered at Sidney, Bedwell Harbour and Victoria in 1979,[9] while Statistics Canada recorded 41,039 American visitors who arrived in British Columbia by private boat for a one or more day's visit in 1979. Cooke's research[10] on transient boat populations cruising in Georgia Strait revealed the numerical importance of non-resident boaters and the significant increases in their numbers that can be expected. Personal observation based on extensive coastal cruising would certainly lend credence to the trend of an overall dominance of American over Canadian boats. A survey by Rhodes[11] of non-resident boating populations on the attraction of British Columbia waters listed 21 factors mentioned by respondents, with scenery, sheltered waters, fishing, less crowding, and birds and wildlife being the top five features. About 75 percent of these non-resident boaters came from Washington, with the balance from Oregon, California and other states in roughly equal proportions.[12] Figure 3,12 shows some trends in boating populations.

This large number of boaters, both resident and non-resident, requires various facilities, such as moorage, fuel and supplies. These are provided both by private enterprise in the form of marinas, launching ramps, fuel barges, stores, restaurants and similar facilities, and by government in the provision of moorage facilities and launching ramps.

Two main kinds of moorage facilities can be recognized — permanent, where the boat is kept, and transient, for cruising purposes. A recent study projects substantial deficits in the provision of both types of facility in the future.[13] It was calculated in 1978 that fixed costs to resident boat-owners, such as moorage, insurance, and repairs, totalled some

Table 1.12 FLEET COMPOSITION TRENDS IN GEORGIA STRAIT, 1973-1978

Category	*Percent of Fleet* 1973	1978
BOAT TYPE		
Sailboat	10	12
Outboard	58	51
Inboard, Inboard/Outboard	14	24
Rowboat, Canoe, other	18	14
LENGTH		
0 - 3 metres (0 - 10 feet)	13	5
3.5 - 4.5 metres (11 - 15 feet)	43	40
5 - 6 metres (16 - 20 feet)	30	35
6.5 - 9 metres (21 - 30 feet)	10	16
9.5 metres + (31 feet +)	4	5

NOTE: The percentages may not total exactly, due to rounding.

SOURCE: Harrison, M.C., *Resident Boating in Georgia Strait – 1979 Update*, Vancouver: Environment Canada, 1979.

61,000,000 dollars, and that operating costs, such as fuel, food, and temporary tie-up fees, totalled some 134,300,000. Non-residents contributed a further 12,000,000 dollars in operating costs.[14] The provision of adequate onshore facilities for boaters is obviously critical in maintaining this high economic return. Harrison's survey indicated that inadequate berthing facilities were a major concern among visiting boaters on many parts of the coast.[15]

Private marinas, of which there are over 200 along the coast, are located mainly in the lower Georgia Strait area. They vary greatly in size from fully serviced docks with attached resorts and restaurants to small floats and launching ramps. Cooke found that 63 percent of the transient boaters interviewed preferred such private facilities compared with public ones, despite the higher cost.[16] Indeed, the most isolated facilities provided by the government rarely require fees of any kind for overnight moorage. Such wharves, administered by the Small Craft Harbors Branch, also vary greatly in degree of service provided, but with less likelihood of having on-dock power and water than the private facilities. This difference, perhaps, accounts for the preference for private facilities noted above. The Department of Transport also provides simple floats with no services, or charges,

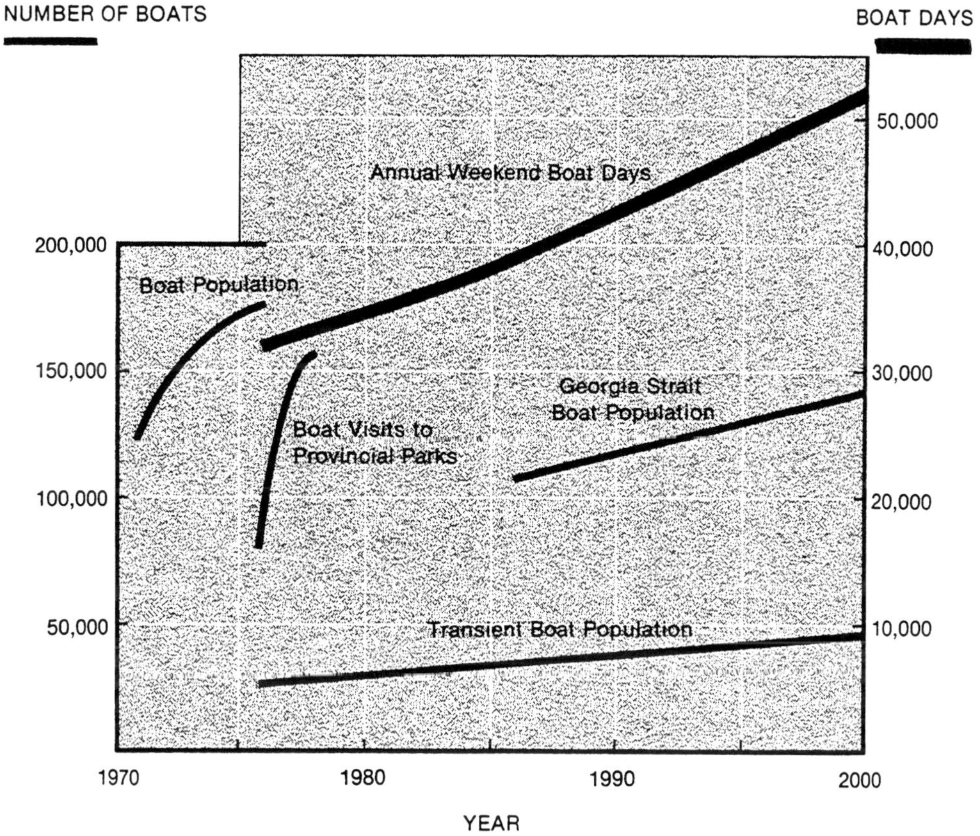

FIGURE 3,12 Trends and projections of boating populations

in strategic locations as ports of refuge in inclement weather conditions. Although these federally provided facilities can be used by pleasure craft, they are principally designed for small commercial boats and, in particular, fishermen.

By way of contrast, the provincial government provides facilities largely for the recreational boater. Primary among these are the marine parks scattered mainly along the lower coast. This system of provincial marine parks has developed largely on an ad hoc basis prompted by boaters' demands, rather than as a preconceived and planned marine park system. The first park, Montague Harbor on Galiano Island, was established in 1957, with 11 more added during the 1960's, 13 in the 1970's and 5 so far in this decade. The primary criteria for establishment are good anchorage combined with available uplands. Most are within 10 to 20 hectares (24 to 48 acres) in size and contain little or no facilities, other than perhaps a small dinghy-mooring float. Desolation Sound at the northern tip of the Strait of Georgia is the largest park, with more than 60 kilometres (37 miles) of shoreline, several offshore islands, an upland area containing lakes and waterfalls, and the whole surrounded by snowcapped peaks some 2,400 metres (7,800 feet) in height. Only five of the parks have any attendance records and charge for moorage. However, personal observation of most of the parks indicates their extreme popularity during the summer months and their growing clientele during the rest of the year.

The federal government is also proposing to create a system of national marine parks. Unlike the recreationally-oriented provincial parks, the national marine parks would be established "to protect and conserve representative examples of the diversity of Canada's coastal zone and oceans for the benefit of present and future generations",[17] and not be primarily recreation-oriented. Little mention is made of recreational boating in the proposed policy for the parks and it is not anticipated that the creation of national marine parks in British Columbia would influence pleasure boating to the same degree as the provincial parks.

Several studies have investigated the major points of origin and destination of recreational boaters in British Columbia.[18] By far the majority of boating activity takes place in the sheltered waters of the Strait of Georgia, with Victoria, Vancouver and the United States being major points of origin and the principal destinations being the Gulf Islands, Sunshine Coast and Desolation Sound. Hence, the major points of origin and destination are in relatively close proximity. Much greater time, effort and experience is required to navigate to more distant cruising grounds along the west coast of Vancouver Island or the mainland coast to the north. The weather, also, is likely to be less favorable in such locations where there are fewer support facilities available. However, these areas are attracting increased numbers of boaters who are perhaps disenchanted with the overcrowded situations on the more popular inland waters.

The final trend in recreational boating to which attention should be drawn is the very rapid growth of ocean kayaking. Empirical data to substantiate this trend are difficult to obtain, but indirect evidence, such as the growth in specialty shops and clubs, public lectures on the topic, articles in boating magazines and personal observation over the last five years, indicates substantial growth. Ocean kayakers were very rarely encountered in the main inland cruising grounds, such as the Gulf Islands, Desolation Sound and Johnstone Strait, five years ago, but now provide a very common sight, particularly in Desolation Sound, Robson Bight, the Broken Group Islands off the west coast of Vancouver Island, in part of Pacific Rim National Park and in the Queen Charlotte Islands. It appears that this form of low impact, low technology boating is going to become quite an important mode of coastal recreation enjoyment in the years to come.

Fishing

Various estimates suggest that 60 percent of pleasure boats on the west coast are also used for sport fishing, dominantly for Pacific salmon.[19] Until recently, little attention has been directed to the sport fishery, either as an important component of the salmon resource allocation, or as a major form of coastal recreation. However, recent studies have indicated the importance of the sport fishery, especially in economic terms.[20] No comprehensive treatment of this complex topic is attempted in this chapter, and readers are referred to Chapter 9 on the fisheries.

The sport fishery is largely based on three of the five species of Pacific salmon — chinook, coho and pink. However, increasingly anglers are beginning to realize the value of other fish, such as snapper and ling cod, both as a fishing challenge and culinary delight. Tidewater fishermen are now required to purchase licences to fish and this has helped provide better data on the numbers participating. For example, 282,247 licences were sold for a total value of 1,730,540 dollars in 1981.[21] It is estimated that the sport fishery takes something like four percent of the total salmon catch and, yet, the sport fishing fleet is worth considerably more than the commercial fleet. This has led many economists to recommend a far greater share of the catch be allotted to the sport fishermen, at the expense of the commercial fishermen.[22]

Geographically most of the sport fishing activity is concentrated within the Strait of Georgia, particularly at the northern end near Campbell River, and also in Juan de Fuca Strait off Sooke. However, salmon fishermen and fishing resorts can be found throughout the area between these two points. North of Campbell River the fishing is generally good, but the remoteness deters many family fishermen. An additional large concentration of fishing activity occurs at Rivers Inlet where anglers come from all over the world to fish for trophy-sized chinooks. A similar fishery is now beginning to develop on the west coast of Vancouver Island in the Barkley Sound area, as a result of the abundant success of enhancement activities in the area.

In-the-Water Activities

Three main in-the-water activities take place in British Columbia waters. Two of them, surfing and swimming, are somewhat localized and likely to remain that way, whereas the third, scuba diving, appears to be undergoing a major expansion all along the coast. Surfing is an important but rather localized activity occurring mainly on the beaches of the west coast of Vancouver Island, chiefly at Jordan River and Long Beach. By way of contrast, ocean swimming occurs mainly in the warmer waters of the Strait of Georgia. Although tolerable conditions for the hardy exist throughout the Strait in summer, and swimming is a popular activity off many beaches, the waters at the northern end are noticeably warmer.

The third activity, scuba diving, takes place all along the coast although, of course, diver numbers are higher in the south. There are some 41,000 active and participating divers in the province, a figure which is growing at the rate of about six percent per year.[23] As well, there are an estimated additional 4,600 diving tourists per year, of which 64 percent are from out-of-province.[24] It is estimated that judicious marketing could boost the latter figure by 800 percent to make a considerable increase on the 1,000,000 dollars currently generated by diving-related accommodation and charter boat sales.[25] Furthermore, in terms of economic importance, it should be noted that diving equipment is not cheap, with an average basic kit costing in the region of 1,500 dollars. In response, there are a large number of diving shops catering to the needs of divers and forming the focal point for diving activity. It is estimated that the total yearly expenditure on sport diving is in the region of four million dollars with, in 1979, some 169 people being directly employed in the industry.[26]

The activity is based upon the excellence of the British Columbia marine environment, with a wide assortment of marine life nurtured by the rich nutrients, and miles of rocky headlands and reefs off the coast. Tidal passages provide some of the most spectacular sites for marine invertebrates and areas such as Race Rocks near Victoria, Active Pass, Porlier Pass and Dodds Channel in the Gulf Islands, Agamemnon Channel on the Sunshine Coast, and Seymour Narrows, Yuculta Rapids and Okisollo Channel farther north are becoming known internationally as spectacular diving sites. In addition, the coast has 16 different marine mammals, over 200 species of fish, the world's largest octopus and the greatest abundance of starfish species found anywhere in the world. Despite such abundance, local area closures, harvest limits and seasonal restrictions have to be enforced to protect the wildlife from over-zealous collectors. Even so, many populations, such as abalone, crab and ling cod, have become severely depleted through overharvesting in some areas. This is seen to be a major problem in the future if recreational diving is to continue to be a major tourist activity, and emphasizes the need for some form of protective status for the most commonly used diving sites.

LAND-BASED RECREATIONAL ACTIVITIES

Land-based recreational activities in the coastal zone are severely limited by access. Most of the coast north of Vancouver Island is virtually inaccessible from the land and recreational activities are limited to those areas where roads provide some degree of access. Consequently, most land-based recreational use of the coastal zone is concentrated around the heavily populated Georgia Strait region and the southwest coast of Vancouver Island. In these areas extensive use is made of beaches and coastal trails for activities, such as beachcombing, nature study and sunbathing. The east coast of Vancouver Island, between Cowichan Bay and Campbell River, in particular, is intensively developed by beach-oriented resorts, whereas the west coast of the Island provides more opportunity for wilderness-type recreational activities.

Pacific Rim National Park is a major focus for land-based, non-urban, coastal zone recreation in the province and is the only national park within the coastal zone of British Columbia. The park, although still not officially gazetted as a national park because of financial disagreements between the federal and provincial governments, was established in 1970. Located on the west coast of Vancouver Island, it is comprised of three geographically separated areas spread along 100 kilometres (60 miles) of the coast (Figure 2,12). The most northerly unit, and that which receives the bulk of the visitors, totalling 523,000 in 1981, is the Long Beach unit, occupying 138 square kilometres (53 square miles) between Tofino to the north and Ucluelet to the south. Most of the park naturalist and interpretation facilities are concentrated in this unit. Broad expanses of beaches make it a favorite for family recreation during the summer, when available campsites within the park are in very short supply. Potential campers vie with each other every morning to secure sites in the park campgrounds. The overflow is to some extent accommodated by minimal facilities, temporary areas within the park, and private campgrounds outside the park.

Off-season camping is also becoming increasingly popular in the area. Due to the relatively mild winter conditions, campers in tents can be found in the campgrounds all year. The total number of visitors is also bolstered by those staying in the comfortable private resorts surrounding the parks. Personal observation has indicated that a large number of these off-season visitors are engaged in various nature-watching exercises. In particular, whale-watching has blossomed over the past few years, with the twice annual migration offshore of the Pacific grey whale (*Eschrictius glaucus*) being a focal point. Up to 50 whales appear to reside along the Vancouver Island-Washington coasts during summer, rather than making the trek from Mexico to the Arctic and back again, and can sometimes be seen feeding in the shallow waters of Wickaninnish and Schooner bays. The park naturalist staff now undertake major interpretation efforts in this area. An unexpected opportunity for this arose in 1984 when one whale spent most of the summer feeding in the confined waters of Grice Bay, to the east of Tofino, and readily accessible by road.

The second unit of Pacific Rim National Park is the Broken Group Islands to the south of the Long Beach unit. It consists of over 100 small islands at the mouth of Barkley Sound separated from the two closest villages of Ucluelet to the north and Bamfield to the south by open expanses of water over 10 kilometres (six miles) wide. The area is becoming very popular with ocean kayakers attracted by the relative isolation and the contrast between the high energy outer coasts and the secluded beaches, as well as the abundant wildlife of the inner coasts. Although camping there is designated as primitive, there are increasing signs that significant ecological changes are taking place, including the wide-scale destruction of vegetation, due to the sheer numbers of people now visiting the islands. The park's administration is also receiving increasing numbers of letters regarding the conflict between kayakers and power boaters in the area. These kinds of conflict between the increasing number of users and environmental preservation, and between different kinds of uses, cannot but intensify in the future, as all trends point to increasing numbers of visitors.

These problems of environmental degradation and inter-use conflict are, perhaps, most evident in the third and most southerly unit of the park, the West Coast Trail, a linear section stretching from Bamfield in the north to Port Renfrew in the south. The roughly 80 kilometre (50 mile) long trail is based mainly on the "lifesaving trail" that was originally built along this "Graveyard of the Pacific" to help shipwrecked sailors. Since its reconstruction by Parks Canada, between 1973 and 1980, the trail has become extremely popular as a coastal rain forest wilderness experience that is unequalled in natural characteristics. It now attracts an estimated 8,000 visitors per year from all over the world. A recent letter to the Minister of Environment from a hiker who had hiked similar long-distance trails in New Zealand reads in part:

> The West Coast Trail has the potential to be one of the finest wilderness tracks in the world . . . . Despite its rugged beauty and ocean wildlife . . . my enthusiasm for British Columbia's West Coast Trail has diminished . . . . The litter on the West Coast Trail is, frankly, an embarrassment. At each and every campsite human excrement and toilet paper is found. Paper,plastic, cans, and broken glass despoil the otherwise beautiful environment. The problem is most acute on the hiking trail itself.[27]

This is not an isolated viewpoint and indicates one of the severe management problems that will have to be met by Parks Canada in the future. It is also symptomatic of the problems of land-based recreation on the British Columbia coast. Vast areas are inaccessible and sustain virtually no land-based recreational use, but, the areas that are made accessible, such as through Pacific Rim National Park, have experienced many problems in accommodating the large numbers of people, with resulting disastrous effects on both the local ecology and the recreational experiences of many of the visitors. One of the effects of this has been to force increasing numbers of the recreationists who are looking for relative solitude on the coast farther north, where the Queen Charlotte Islands, in particular, have taken the brunt of the attack. Presently Naikoon Provincial Park provides for wilderness-type activities on Graham Island, whereas if the current proposal to establish South Moresby Island as a national park comes to fruition this area will also start to accommodate more visitors.

Before leaving the topic of land-based coastal recreation, it should be noted that the provincial parks on the coast of the Strait of Georgia are among the heaviest used in the entire system. Campgrounds are full throughout the summer at parks such as Miracle Beach, Rathtrevor and Bamberton, even though they have the highest per night fees in the province. Major activities are those associated with camping, building fires, for example, and beach and water-oriented activities.

CONCLUSION

In this chapter an attempt has been made to draw together the many diverse threads that comprise recreational usage of the coastal zone in British Columbia. The range is vast. Luxury cruisers, complete floating hotels, leisurely ply the waters of the Inside Passage for those little interested in close encounters of the environmental kind, while others pound the West Coast Trail looking for wildlife, wilderness and physical challenge. All along this spectrum the demand for activities appears to be rapidly increasing, both from residents and non-residents alike. The main technical report of the Canada-British Columbia Travel Industry Development Subsidiary Agreement indicates the highly significant role that marine recreational activities will play in the future as tourist attractions.[28]

Despite the disparate nature of the activities and infrastructure required to support them, all are vulnerable in the future to threats from the same kinds of forces. First, all these activities are, to a greater or lesser extent, dependent upon the maintenance of key elements in the coastal environment for which there is keen competition from other resource-based industries. For example, the sport fishery is dependent upon there being a reasonable opportunity for, and probability of, the sport angler catching a salmon. Without the salmon there would be no associated recreational activity, nor business built to serve it. A corresponding decline would be felt throughout the pleasure boating industry. Yet, as has been mentioned earlier, and discussed more fully in Chapter 9, salmon stocks are seriously depleted, largely because of overfishing by the commercial sector. Following proposals from the Department of Fisheries and Oceans to severely limit sport fishing to conserve stocks, the sport fishermen suddenly became aware of their minimal influence on the way in which the resource was managed and allocated. Now a strong sport fishing lobby has managed to considerably strengthen the voice of the recreational fishermen in managing the resource.

The point to be made here is that all the coastal recreational activities are dependent to some extent for their existence and enjoyment upon resources over which they have no control. Until the recreational opportunity becomes so badly impaired as to be negligible, this resource dependency is not always obvious and, hence, of little concern to the recreationists or those responsible for management. A recent letter to the *Victoria Times-Colonist* provides another ready example:

> On a recent trip up the Inside Passage, I was appalled at the destruction from logging operations, ... the area across from Alert Bay has undergone a huge clear-cutting. It appears the logging companies are given far too much leeway to clearcut, and no visible attempt is made to prevent erosion.
>
> Numerous cruise ships travel the Inside Passage and each year the view deteriorates. We should try to keep this passage as presentable as possible if we want to maintain the tourist industry.[29]

A major component for many of the coastal recreational activities is the aesthetic pleasure derived from the scenery, yet, indiscriminate clear-cutting along the coast is causing vast changes in the landscape. How far can aesthetic satisfactions be compromised before the activity is no longer pleasurable becomes the relevant question, especially, as noted in the letter above, with reference to the tourist industry. This point has been discussed in more detail elsewhere.[30]

It is obvious that there are many competing and conflicting demands for the resources of the coastal zone. However, despite the widely acknowledged significance of the area as a recreational resource, recreation is seldom one of the factors taken into account in managing coastal resources. As a result, the recreational potential of the coast often is impaired as an externality of the operation of other resource users. This will continue to be the case unless recreationists and the recreation industry join together to stake a claim in the management of the coastal resource. It is too late in many areas. However, there are signs, with the formation of the Sport Fishing Advisory Board and the widespread publicity over Meares and South Moresby islands, that this is happening.

The second major cause for alarm in the future relates, not to the external threats, but to the threats from within. The phenomenal rise in popularity of outdoor recreation over the last three decades has been felt very strongly in the coastal zone. As essentially a linear feature, there is a greater concentrating effect of recreationists than is generally felt in purely terrestrial situations. In short, there are more people doing more recreational activities on the coast than ever before. In all likelihood, there will be more next year, and even more the year after that. This creates many resource management problems, leading to degradation of the biophysical habitat and conflict between overcrowded recreationists.

One short example from pleasure boating will suffice. In her study on transient boating populations in Georgia Strait, Cooke received the following response from one local person:

> . . . everyone has a right to use and enjoy our local waters. However, the water now is so polluted with so many boats dumping sewage into our enclosed harbor that swimming is no longer possible. We have not been able to swim on the beach in front of our home in over three years. This abuse I strongly object to as we are not allowed to dump raw sewage into the harbor, so why should our visitors?[31]

Furthermore, local boaters voiced the following complaints about transient boating (in decreasing order of priority): hazardous navigation by other boats, increased lineups for fuel, shortages of supplies and bait, and unable to moor at customary place. These kinds of complaints are symptomatic of the general trend all the way from the West Coast Trail hikers through to the numbers of luxury liners plying the Inside Passage. Only so many can be accommodated without leading to degradation of the environment and overall impairment of the recreational èxperience. To satisfactorily resolve these problems constitutes a major challenge in recreational resource management in the future.

REFERENCES

1. BROOKS, L., *Recreation and Tourism in the British Columbia Environment*. Victoria: Ministry of Environment, 1979; BRITISH COLUMBIA, *Tourism Highlights*. Victoria: Ministry of Tourism, 1979; CANADA AND BRITISH COLUMBIA, *Technical Report of the Province of British Columbia Tourism Development Strategy*. Ottawa and Victoria: Canadian Office of Tourism and British Columbia Ministry of Tourism, 1979; VAR, T., BECK, R.A.D. and LOFTUS, P., "Determination of Touristic Attractiveness of the Touristic Areas in British Columbia", *Journal of Travel Research*, Vol. 15, 1977, pp. 23-29.

2. MONTGOMERY, G., *An Evaluation of the Tourism Potential of the Cruise Ship Industry of British Columbia*. Ottawa and Victoria: Canadian Office of Tourism and British Columbia Ministry of Tourism, 1981.

3. *Ibid.*

4. *Ibid.*

5. *Times-Colonist*, July 9, 1983, p. D-1.

6. Pacific Synergies, promotional brochure, 1983.

7. PFISTER, R.E., "Pleasure Boating in British Columbia" in *Lands, Parks and Housing Review*. Victoria: Ministry of Lands, Parks and Housing, 1979.

8. CANADA AND BRITISH COLUMBIA, *op. cit.*, reference 1.

9. EBY, P., *Recreational Boating Facility Requirement to 1985 in British Columbia*. Vancouver: Small Craft Harbors Branch, Pacific Region, 1979.

10. COOKE, K.A., *Transient Boating in the Strait of Georgia*. Vancouver: M.A. Thesis, University of British Columbia, School of Community and Regional Planning, 1977.

11. RHODES, A., *A General Survey of Non-Resident Boating Populations in the Gulf of Georgia*. Vancouver: Department of Fisheries and Oceans, 1979.

12. CAPITAL REGIONAL DISTRICT, *Boating within the Capital Regional District*. Victoria: Capital Regional District, Regional Planning Department, 1979; COOKE, K.A., *op. cit.*; RHODES, A., *op. cit.*

13. EBY, P., *op. cit.*

14. CANADA AND BRITISH COLUMBIA, *op. cit.*, reference 1.

15. HARRISON, M.C., *Resident Boating in Georgia Strait - 1979 Update*, Vancouver: Environment Canada, 1979.

16. COOKE, K.A., *op. cit.*

17. CANADA, *National Marine Parks, Draft Policy*. Ottawa: Ministry of Environment, Parks Canada, 1983.

18. EBY, P., *op. cit.*; RHODES, A., *op. cit.*; and HARRISON, M.C., *op. cit.*

19. HARRISON, M.C., *op. cit.*

20. PEARSE, P.H., *Turning the Tide: A New Policy for Canada's Pacific Fisheries*. Vancouver: Department of Fisheries and Oceans, 1982.

21. *Ibid.*

22. *Ibid.*

23. ERNST and WHINNEY, *An Evaluation of the Tourism Potential of the Scuba Diving Industry in British Columbia*. Ottawa and Victoria: Canadian Office of Tourism and British Columbia Ministry of Tourism, 1980.

24. *Ibid.*

25. *Ibid.*

26. *Ibid.*

27. ALMACK, R., "West Coast Hiking Trail, Pacific Rim National Park". *Outdoor Reports*, August 1983, p. 11.

28. CANADA AND BRITISH COLUMBIA, *op. cit.*, reference 1.

29. *Times-Colonist*, June 1, 1984, p. A-4.

30. DEARDEN, P., "Tourism and the Resource Base" in MURPHY, P.E. (ed.), *Tourism in Canada: Selected Issues and Options*. Victoria: University of Victoria, Department of Geography, Western Geographical Series, 1983, pp. 75-94; DEARDEN, P., "Forest Harvesting and Landscape Assessment Techniques in British Columbia, Canada", *Landscape Planning*, Vol. 10, 1983, pp. 239-253.

31. COOKE, K.A., *op. cit.*, p. 93.

PLATE 18 Tuchodi Lake in the Rocky Mountains west of Fort Nelson. ▶
B.C. Government Photo

PLATE 19 Cross-country skiing in the Interior. *B.C. Government Photo* ▶

13 LAND-BASED RECREATION

Bruce K. Downie
Parks: Research and Planning Incorporated

INTRODUCTION

If one word could describe the essence of recreational opportunities in British Columbia it would be "diversity". Throughout the province special regional characteristics have spawned important recreational activities: hunting in the Kechika River Valley; canoeing on the Stikine River; heli-skiing in the Bugaboos; trail riding in the Cariboo; boating in the Okanagan; rafting on the Chilko River; and hiking in Spatsizi. In each case these are recreational activities that have an interaction with the natural environment as a major element. It is this aspect of the experience that links the diversity of the natural environment of the province with the breadth of recreational opportunities available.

Certainly there are many recreational activities that are not dependent upon the character of the landscape, but the focus of this chapter is on those that do provide a natural environment experience. Even with such a focus, the scope of the subject is immense and an overview of the current status of outdoor recreation supply and demand is all that is attempted. However, throughout this review, the broader perspective of the complexity of recreational behavior and the fulfilment of recreational needs should be recognized. Supplying facilities or areas for activity is only part of the process of responding to recreational needs. There must be an awareness of the personal and social factors that also influence the quality of the recreational experience. Wall states:

> The state of knowledge of outdoor recreation is such that many students of recreation recognize that they are investigating opportunities for relaxation, exercise, socialization, challenge and many other motives for participation, but they still design parks, camping sites, and ski areas for particular activities and hope that somehow they will satisfy their motivations . . . . Recreation behavior is a function of a continuous process of multidirectional interaction . . . . For example,

> mountaineering involves interaction with the environment, with other people, and with oneself (in the form of a personal challenge or test). Each of these interactions influence the other components of the system. The resulting experience, through a feedback mechanism will influence subsequent decisions to participate and subsequent emphases placed on interaction with self, others and the environment.[1]

In reviewing the status of outdoor recreation supply and demand in British Columbia, it is impractical to address the topic as comprehensively as implied by Wall's perspective, as important as this may be to understanding the true nature of recreational demand. Rather, this chapter focusses on the resources and facilities for outdoor recreation and on the mechanisms by which they are made available to the public. Describing the range, nature and distribution of the recreational activities to be addressed will be the first task. This will be followed by an overview of the providers of recreational opportunities: who are they, and what kinds and how much recreational opportunity do they provide? The final segment of the chapter will discuss current problems in the outdoor recreation system of British Columbia and, in summary, a few observations and recommendations for the system's improvement will be provided.

RECREATIONAL ACTIVITIES

For the purpose of this analysis, two categories of recreational activities can be identified: front country and back country.[2] Front country recreational activities are easily accessible. They are available to the widest range of users; accessible by private vehicle; supported frequently by extensive facilities and equipment; and do not require prolonged time commitments. These activities can be water-based, such as swimming, boating, fishing and scuba diving, or land-based, such as picnicking, day-hiking, cycling and camping. A special grouping of activities is found in the recreational resort. Although much of what attracts visitors to such facilities is not a natural environment experience, for example, dining, accommodation and entertainment, many resorts have a natural environment experience focus, such as a ski resort or a hot springs resort, and thus will be discussed from that perspective as an element of front country recreation.

Back country recreational activities are not easily accessible. They are less available to the full range of recreationists, due to factors such as time, money, physical ability and personal preferences. They tend to be much less accessible by private vehicle and are generally less supported by facilities and equipment. Examples include backpacking, canoe tripping, hunting, mountain climbing and horse packing.

Inevitably, there is not a distinct separation between the two categories and, indeed, many activities can be classed in both categories. One can pitch a pup tent next to a car in a K.O.A. campground adjacent to the Trans-Canada Highway or in a Forest Service campsite at the end of a logging road. Although the activity, vehicle camping, is the same, the experience is not. The category of the activity is affected by a variety of factors and their interaction is described by Wall earlier. Thus, although the categories may be useful generalizations for an overview such as this, it is more accurate to visualize recreational activities as being part of a continuum. Some activities clearly fall into one category or the other, such as mountain climbing or driving for pleasure, while others are not as easily classified, such as swimming, boating, and camping.

Front Country

Since the late 1940s and early 1950s when road transportation came of age in Canada, driving for pleasure or sightseeing has consistently been a primary recreational activity.[3] This is, perhaps, especially true in British Columbia where scenery is a major attraction for visitors to the province.[4] As a passive form of recreation, a very large segment of the population may participate and, indeed, virtually all recreational travel, as well as destination activity, requires that some degree of importance be placed on the scenic qualities of the environment.[5] The growth of a network of support facilities has encouraged and added to the viability of road-based recreational travel. Accommodation of widely varying types, food services and transportation services provide the fundamental support for the traveller. Add to this recreation opportunities provided by both natural resources, such as beaches, and facility development, such as picnic sites, and the image of front country recreation becomes more complete. Picnicking, swimming, camping, boating and day hiking are very common activities throughout front country areas. Other activities are site-specific and, although not widely popular, contribute very significantly to the mix of recreational opportunities in a particular local area. An example is the thrill and views of a sky ride, such as at Grouse Mountain in North Vancouver, or Hell's Gate along the Fraser Canyon. Also site-specific, but more widespread in British Columbia, are downhill skiing sites. The quality of downhill skiing opportunities in the province is excellent and the activity has continued to be a very popular one, as well as one which generates significant investment and economic return to British Columbia.[6] More recently, other activities have also been gaining popularity due to changes in recreational preferences. An example is the growing popularity of river rafting. Commercial operators on the Fraser and Thompson rivers offer everything from three-hour to eight-day trips, May through September.

Of special note in a discussion of front country recreational opportunities is the destination resort. Usually focussed on a particular recreational activity, such as downhill skiing, the tendency has been to supplement this focus with an increasing range of other leisure pursuits, such as golfing, tennis, hiking, boating, biking and swimming, in the hope of adding to the resort's attractiveness and expanding its appeal to all seasons of the year. But these pursuits are not necessarily outdoor recreational activities and, thus, a trend has been to create mini-urban environments in a superb natural setting. The setting does not contribute to the discos and specialty shopping opportunities, but does serve as an appealing backdrop to the full range of outdoor recreational activities, as well as the necessary services of accommodation and restaurants. Harrison Hot Springs is a well-established example, however, Whistler, only 119 kilometres (74 miles) north of Vancouver, is probably the province's leading destination resort.

> Whistler is a mega-mountain, a 7,118-foot giant that attracts skiers from every part of the world, a ski area vast in scope which includes Blackcomb Mountain. Here you'll find everything from bunny slopes to wicked downhill racecourses, on the longest vertical drops in North America . . . .
>
> Outdoors activity in the valley is by no means restricted to downhill skiing. Crosscountry trails abound, there's ice fishing and miles and miles of trails for the snowshoe brigade. Ice skating, hills for sleds and toboggans and Ice Stock Sliding (Australian curling, gaining rapid popularity in the Whistler area), are other diversions provided by the valley.
>
> In both winter and summer, the resort village of Whistler really swings. Clubs, pubs, good restaurants and apres-ski facilities fill the town centre, and the nightlife in this alpine outpost is getting to be as good as the skiing.
>
> In the summer, Whistler/Blackcomb is an alpine wonderland — the heart of the B.C. Coast Mountains and as scenic an area as you'll find anywhere on the planet. Bring extra film for your camera — you'll use it all.
>
> In the town itself, much money, planning and emphasis has been placed in recent years on turning the area into a year round visitor destination. Hotels, condomin-

iums, a recreation centre, shops, swimming pools, golf course, restaurants and convention facilities have either been or are currently being added to the blossoming little town.

The area is full of clear alpine lakes, perfect for swimming, fishing and boating. Radiating from the town centre itself are many hiking trails, ranging in difficulty from Sunday afternoon strolls to fairly strenuous hauls up the mountain, to actual rock-climbing conditions.

Also near the town centre is a new 18-hole golf course designed by Arnold Palmer, no less. A newly completed recreation centre boasts a swimming pool, squash and racquet ball courts, tennis courts, and an olympic-sized skating rink. It converts to a theatre/convention centre.[7]

Back Country

Back country activities are probably best characterized as the antithesis of the easy-to-reach passive forms of recreation. In back country activities the character and quality of the natural environment becomes more significant to the experience. Backpacking is probably the most readily identifiable and popular back country activity.[8] This activity and canoe tripping are comparable in the back country to sightseeing and recreational travel by vehicle in the front country: only the power source has changed from the motor to the individual. In so doing, the portion of the population able to participate is significantly reduced. Nevertheless, with the environmental movement consciousness of the 1960s and the fitness consciousness more recently, back country activities have become very popular. Other influencing factors, such as age of the population and cultural origin, may negatively influence participation rates and absolute numbers of participants in the future.[9] Where there is a diversity of high quality associated recreational activities, participation will be greater, just as is the case with recreational travel in the front country. Scenery, the potential for viewing wildlife, opportunities for nature study, highly desirable routes (whether they are especially challenging or especially enjoyable), and high quality camping sites all contribute to the attractiveness of one area compared with another. Fuelling participation has been a plethora of popular magazine accounts of some of British Columbia's multitude of opportunities: Monkman Provincial Park,[10] Cathedral Provincial Park,[11] the Warner Pass area of the southern Chilcotin Mountains,[12] and the Akamina Valley.[13]

More significant, in terms of numbers of participants in the activity, are hunting and fishing. An indicator of the level of participation can be found in the licencing system, showing about 150,000 resident hunters and 300,000 resident fishermen in 1983.[14] Not all of these participants pursue the activity in what could be considered back country areas. Fishing from a major Trans-Canada Highway bridge or duck hunting along the roadside edge of farm fields in the Peace River country does not reflect the concept of back country recreation. However, a large portion of the participation in these activities takes place on and, in fact, depends upon, the lands and waters of the province that are removed from the more frequented landscapes of settlement, industry and agriculture. Similar to other back country activities, there is a high degree of interdependence between the recreational experience and the resource, although access assistance is inevitably present whether by aircraft, truck, all-terrain vehicle, snowmobile, boat or horse. Participation in hunting and fishing is encouraged by the formation of clubs or groups and by the popular magazines which cater to these activities. But perhaps the single most important factor in continued participation is the experience of participation as a youth.[15]

Another back country activity, truly exemplary of the category, is mountain climbing. It is a significant activity in British Columbia, due to the extensive natural resources virtually at the doorstep of residents throughout the province. A very active Federation of Mountain Clubs of British Columbia co-ordinates the activities and interests of local clubs around the province. Club members total over 6,000.[16] A national organization, the Alpine Club of Canada, is also active.

Two other highly specialized activities are worthy of note which are characteristic of back country recreation activity types, but which are not necessarily restricted to back country locations. They are diving and spelunking. In each case the activity itself takes place in an environment totally removed from that which we frequent, yet the location of these sites may be readily accessible and by the classification standards used here may be found in front country locations.

Diving is a very popular sport in British Columbia and the province has many exceptional areas of national and international quality. As many people recognize, the coastal areas of the Lower Mainland, Vancouver Island, the Gulf Islands and the Sunshine Coast provide the majority of high quality diving sites, but a much greater proportion of diving activity takes place in inland freshwater lakes than is generally acknowledged. Prime lakes within the province are Shushwap and Kootenay.[17] British Columbia currently has in excess of 60,000 registered divers resident in the province, of which 75 percent might be considered active.[18] Although not organized in the same way as the mountain clubs, provincial and national associations are presently in the process of being formed.

Spelunking is also a very important element in the recreational character of the province because of the excellent natural features providing opportunities for the activity. The geology is such that caves are common in the Kootenays and on Vancouver Island. Some of the best known are Nakimu Caves in Glacier National Park, Cody Caves in the Kootenays, and Horne Lake Caves Provincial Park on Vancouver Island. Caving clubs are also present in British Columbia, giving support to members and providing opportunies for shared experiences.[19]

The specialized nature of these types of recreational activities is a significant factor influencing the degree to which club membership reflects the proportion of the actively participating public. Other forms of recreational activity, such as hunting, fishing, hiking, camping or trailering, all have clubs associated with their particular areas of interest. However, the proportion of the general public participating in these activities tends to be much greater than club membership would indicate.

Future

In British Columbia outdoor activities are very popular. A recent survey of residents conducted for the Parks Branch summarizes some of the observations noted above and shows a high level of participation (Table 1,13).[20] Trends suggest that participation rates will remain high. Total numbers of users and, therefore, demands on the resource base, are likely to increase as the provincial population increases and as increased emphasis is placed on the development of tourism as a stimulus to the provincial economy.

DISTRIBUTION OF RECREATIONAL OPPORTUNITIES

The resident population of the province is widely dispersed beyond the major focus in Vancouver and the Lower Mainland and the secondary regions of southern Vancouver Island and the Okanagan Valley. Demands for recreational opportunities and open space are similarly distributed. This pattern of demand is accentuated by the tourism patterns within the province. Vancouver and the Lower Mainland, the Okanagan and Vancouver Island absorb the most significant proportions of tourist activity, each having its own distinctive mix of activities to offer.[21]

Much of this tourism is urban oriented, focussing on the commercial attractions of shopping and entertainment, as well as activities such as sightseeing and visiting museums. Even within or adjacent to the urban environment, however, a wide range of outdoor recreational activities is also important — beaches for swimming and strolling, and parks for walking,

Table 1.13 PARTICIPATION IN OUTDOOR RECREATIONAL ACTIVITIES BY BRITISH COLUMBIA RESIDENTS

Rank/Order	*Activity*	*Percent of Respondents*
1	Picnicking (away from home)	79
2	Swimming/Sunbathing (at beaches)	78
3	Sightseeing	67
4	Historic Parks/Museums	55
5	Hiking	49
6	Fishing	46
7	Boating/Sailing	43
8	Downhill Skiing	22
9	Backpacking	21
10	Canoeing	20
11	Trailbike/Snowmobile	18
12	Cross Country Skiing	17

SOURCE: Jeroski, S.F. and Conry, R.F., *Provincial Parks Survey: Summary Report*. Victoria: Prepared by Horizon Research and Evaluation Affiliates for the Parks and Outdoor Recreation Division, Ministry of Lands, Parks and Housing, 1983.

cycling, nature study, picnicking and playing. The demand for such space is exceptionally high and, in spite of many excellent resources in the greater Vancouver area, the continuing supply of quality recreation areas is still a concern. Major areas, such as Stanley Park, Spanish Banks, and the University Endowment Lands, are urban lands that contribute significantly to the outdoor recreation demands of both resident and tourist alike. Other adjacent areas that expand the range of recreational opportunities are: Delta Marsh, Mount Seymour Provincial Park and Campbell River Valley Regional Park.

Beyond the urban oriented recreational activity in the Okanagan, the recreational focus is clearly on the lakeshore for swimming, boating, sailing and other beach activities. The hot summer weather in the valley, combined with warm water temperatures, makes beaches and beach activities there more sought after by residents and tourists than those of any other area in the province. In the winter, the predominant attraction is skiing, with three major ski areas, Big White, Silver Star and Apex Alpine, accounting for most of the activity.

Southern Vancouver Island also has a water focus, but because of its coastal location and colder water, beachcombing and other beach activities are more prevalent than swimming. Boating is also very popular, especially associated with fishing. Very popular beach resources are Rathtrevor Beach Provincial Park, French Beach Provincial Park, Island View Beach Regional Park and a number of intensively used portions of the City of Victoria shoreline. A wealth of upland recreational opportunities complements the shoreline focus of the area. Day hiking, picnicking and nature study are popular activities in Goldstream and Sooke Potholes provincial parks, Mount Work and East Sooke regional parks and along the Cowichan River Valley.

Beyond the primary activity regions just described, overall visitor numbers and recreational participation may be less, but the quality of special activities well suited to the region remains high. Shuswap Lake has long been noted for its extensive and excellent boating and fishing potential. As a result, shoreline development in support of recreation has been considerable, including summer homes, marinas, service facilities and lakeshore parklands. Boating use data have never been collected with accuracy, but a mid-1970s estimate suggested a resident and non-resident boating population of over 13,000 and a boat rental supply of approximately 600.[22] Activity on the lake has continued to increase and the Parks Branch has actively sought to maintain a balance between private lakeshore developments and opportunities for public access to the lake's recreational opportunities.[23] The gold rush of the Cariboo County has created an image that has become the focus of attention for many visitors to the region. Barkerville, as well as restoration work in the town of Wells, attracts many visitors each summer. However, the rugged outdoors character of the region is perhaps still its biggest drawing card, with hunting and fishing being the most popular outdoor activities. More recently, the region has also become increasingly known for guest ranch vacations, including winter season cross-country skiing opportunities. This type of recreational opportunity is expected to be of growing importance in the future.

SUPPLIERS OF RECREATIONAL OPPORTUNITIES

Within British Columbia a variety of government agencies, private groups and individuals contribute to the supply of outdoor recreational facilities and opportunities. Consistent with the previous discussion, these suppliers have been grouped according to their involvement in either front country or back country recreation. Only the major providers have been identified. As was the case with individual recreational activities, it is not possible to state that a supplier only provides one type of recreational activity. Rather, the suppliers described are categorized according to their predominant involvement in certain forms of recreational activity. Due to the diverse nature of involvement of the two largest agencies, British Columbia Parks and Outdoor Recreation Division and Parks Canada, categorization was not possible and they are described separately.

Front Country

Private Lands

The lands held in private ownership are together the largest suppliers of recreational opportunities and facilities. This is true largely because of two factors: first, privately-owned lands include a diversity of lands and activities, including private cottage lots, marinas, lodges, motels, and campgrounds; and, second, the private sector is able to provide facilities when use levels are high and lands are intensively used. Because much front country recreation reflects these factors the majority of activity takes place on private lands. It should be noted that this includes commercial opportunities, as well as the personal recreation taking place on privately-owned lands.

It is impossible to generalize regarding the purpose, activities and facilities that characterize private recreational lands. However, it is possible to categorize certain types of private recreational lands and gain a perspective on the breadth of characteristics they possess. A first and very important division is between lands held privately for personal or group use as opposed to lands held for commercial use. A third category, lands held for or even used for other purposes than recreation which incidentally provide recreational opportunities, is also important.

Personal use lands are made up largely of cottage lots or, in the case of organized group ownership, lands used for Y.M.C.A. and church camps. The purpose of acquiring such properties is as a retreat from the urban environment where leisure time can be spent in a natural setting. People enjoy a more relaxing pace, a different routine and a different set of activities. Often cottage lots and camps are located alongside or near water which provides the change in setting and ready opportunity for new activities. Shawnigan Lake, near Victoria, is an excellent example of the demand for such recreational land. The lake is ringed with cottages, many of which have become permanent residences, due to the easy access to the urban centres of Victoria and Duncan.

The United Church Camp on the lake is also typical of organized group ownership and use. A series of cabins with central service and meeting buildings provides accommodation for a total of about 100 people. The lakeshore has also been developed for recreational use, with the undertaking of beach improvements and the construction of docks and boat launch ramps. Children's camps, family camps, and congregational retreats are planned to provide much of the function of the private cottage: a change of routine, a change of place and new activities, but also, in this case, a chance to meet and interact with other members of the church. The church camp, and also the private cottage, but to a lesser degree, is partially a commercial recreational opportunity, since other groups, perhaps connected with a school or organization, may rent the facilities when church activities have not been planned. Camps of this type are common throughout the province, although their collective impact on the recreational activity patterns of residents is not well known.

Lands held for commercial recreational use serve a seemingly endless variety of functions. In general they attempt to provide facilities and services for the relaxation and enjoyment of visitors. Because of the necessity to profit from the fees for such facilities and services commercial recreational lands tend to be smaller and/or more intensively used than either individually-owned recreational lands or those that are publicly owned. A diversity of activities is common, in order to appeal to a wide variety of potential users. This activity diversity is not only developed and promoted as characteristic of an individual owner's facility, but also of its particular location relative to other recreational and service facilities. Quotations from the Okanagan Travel Guide serve to illustrate the importance of this diversity:

> Open year round. Excellent fishing, swimming and golfing. (p. 41)
>
> Close to beautiful Okanagan beach, golf courses and community convention centre. (p. 30)
>
> Barbeques - Picnic Tables - Spacious Grass Play Areas and Large Trampoline. Situated 350 ft. from Okanagan Lake Beach across from Riverside Park. Par Three Golf and Pitch and Putt. (p. 26)
>
> Large shaded campground with full and partial hookups, clean washrooms and showers with hot water. Safe, private sandy beach near two golf courses, theatre and restaurants.[24] (p. 8)

Increasing use of commercial facilities in British Columbia has prompted greater awareness of the potential benefits of expanded and improved commercial recreational opportunities, not only on the part of the entrepreneurs who develop them, but also on the part of the visitors that they serve.

Regional Parks:

A regional parks function is within the mandate of the 28 regional districts in British Columbia. Each district has circumstances that affect its ability to, or need to, pursue this parks function. Naturally, in the Greater Vancouver Regional District and the Capital Regional District of Victoria where population pressures on adjacent lands are very great there has been considerable effort to protect natural areas and plan for their recreational use. Not all regional park systems have developed similarily, yet investigation of one, the Capital Regional District park system, will offer some perspective on the functions and opportunies such recreational lands provide (Figure 1,13).

Of approximately 2,506 hectares (6,192 acres) of regional-scale parkland in the regional district about 46 percent (1,153 hectares or 2,849 acres) is within the regional park system.[25] These lands are extremely popular and heavily used for a variety of activities (Table 2,13). Major parks within the system that receive highest use include: Elk-Beaver Lake, Witty's Lagoon, East Sooke and Island View Beach. In the continued evolution and development of the regional park system a classification scheme has been developed to assist in identifying and providing a balance of recreational opportunities. Seven categories of lands have been identified, although not all are yet represented within the system.[26]

(1) *Regional Shoreline* (for example, Island View Beach)

The purpose is to conserve river, lake and ocean shorelines for public use where significant recreational, educational, natural and scenic values exist. Predominant outdoor recreation opportunities include: beach activities, swimming, boating, nature study, picnicking, photography, scuba diving, and sightseeing.

(2) *Regional Preserve* (for example, Francis King, Witty's Lagoon)

The purpose is to protect the natural, historical or scenic features for the benefit of public enjoyment and education. Less intensive or manipulative recreational activities, such as park interpretation, nature study, walking and photography, are encouraged. Motorized vehicles are prohibited and horseback riding is limited.

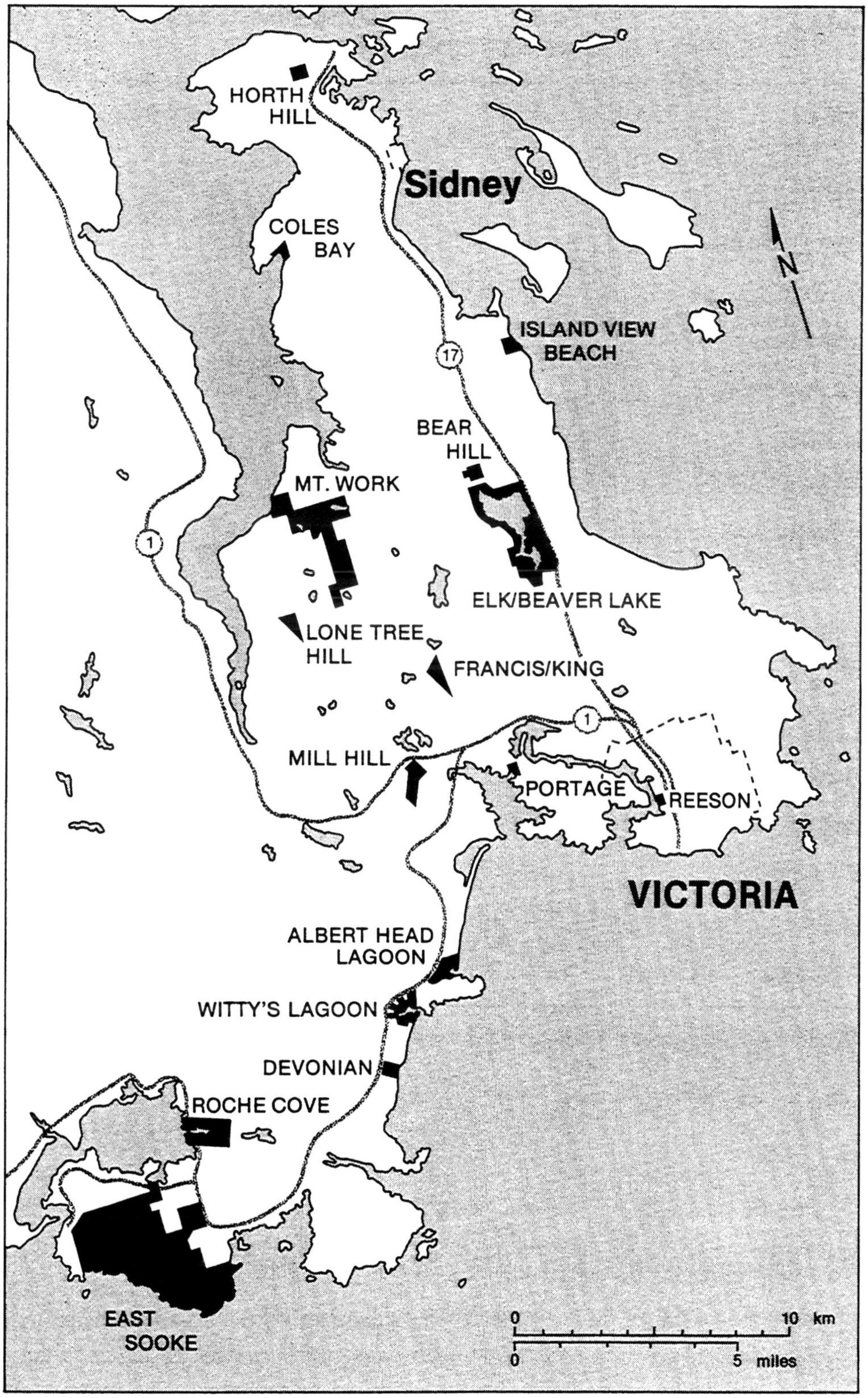

FIGURE 1,13 Capital Regional District parks system

Table 2.13 ESTIMATED PARTICIPATION FOR OUTDOOR RECREATIONAL ACTIVITIES IN THE CAPITAL REGIONAL DISTRICT

Activity		*Percent of Respondents*
BEACH ACTIVITIES (include swimming, sunbathing)		75
PLEASURE DRIVING		60
PICNICKING		50
WALKING/HIKING		50
PLEASURE CYCLING		40
FISHING		35
PLEASURE BOATING		30
Motorboating	25	
Canoeing	10	
Sailing	10	
Kayaking	1	
Windsurfing	1	
NATURE STUDY		30
GOLFING		15
GROUP CAMPING		5
HORSEBACK RIDING		5
TRAIL BIKING		3

SOURCE: Capital Regional District, *Review of Potential Parkland.* Victoria: Capital Regional District, Parks Division, 1983.

(3) *Regional Hilltop* (for example, Bear Hill, Lone Tree Hill)

Because hilltops have natural values associated with their floral, faunal or scenic values, the purpose of these parks is to conserve such values for public enjoyment and education. Similar to regional preserves, activities such as hiking, photography, sightseeing and nature study are most suitable. Off-road vehicles are prohibited.

(4) *Regional Recreation Area* (for example, Elk-Beaver Lake)

The purpose is to provide a variety of outdoor recreation experiences. These activities will be prominent in the region and the park must be capable of providing for intensive use, as well as preserving a high quality of experience. Predominant activities include: a variety of water-based activities, group and family picnicking, sightseeing, lawn and field sports, walking, jogging and horseback riding.

(5) *Regional Wilderness* (for example, East Sooke)

The purpose is "to conserve an area with dominant natural qualities where park visitors can be isolated from the urban area while enjoying hiking, horseback riding, nature study and other wilderness activities".[27] Such areas require minimal facility development and support only non-motorized activities.

(6) *Regional Trail* (for example, Gowlland Ridge Trail - impending)

The purpose is to provide natural, recreational links among existing parkland and other regional points of interest or activity. Non-motorized forms of travel are encouraged, including horseback riding, hiking, walking, jogging and cycling.

(7) *Regional Vehicular Recreation Area*

Although no areas presently exist in the system which fulfil this role, certain regional district holdings are presently being considered for such designation. The purpose is to provide a variety of motorized recreational experiences, including opportunities for trail biking, four-wheel driving and model plane flying, but not excluding other recreational activities where compatible.

Back Country

The most extensive system of lands in the province is held by the Crown. Particularly beyond major urban areas and transportation corridors, Crown land predominates because it is not economical for the private sector to own large tracts of land that is inaccessible and undeveloped. Crown land, however, is not always simply vacant, unencumbered land. It is often given over to temporary uses (timber licence) or held as a reserve for future use (British Columbia Hydro flooding reserve). Nevertheless, such lands provide many opportunities for dispersed recreational activities that are dependent upon large tracts of land for the activity itself, such as hiking or mountain climbing, or for maintaining the resources upon which the activity depends, such as hunting and fishing.

Access to such lands is the key to their enjoyment and use by individuals. Since recreation is not the focus of the use of these lands, provisions for recreational access and facilities for support are generally absent unless private initiatives have been taken. This is entirely consistent with the character of back country recreation and is one of the reasons Crown land is valued so highly by some who seek the isolation of untouched landscapes. Pristine lakes, fishing camps, hunting lodges, wild rivers and mountain ridges exemplify the character of significant portions of the present Crown lands of British Columbia. Another major segment of back country use takes place on lands utilized for forestry. The rotational nature of the resource extraction activity and the prevalence of road access to logged areas are two reasons for the significant use of forest lands for back country forms of recreation. The area of the province so utilized is extensive and accounts for a significant segment of back country use. Such activities are encouraged and have become an increasingly important focus of the mandate of the Ministry of Forests.[28]

National Parks

National parks in Canada are established:

> . . . to protect for all time representative natural areas of Canadian significance in a system of national parks, and to encourage public understanding, appreciation and enjoyment of this natural heritage so as to leave it unimpaired for future generations.[29]

In British Columbia five national parks currently exist covering a total of approximately 5,000 square kilometres (3,125 square miles) (Figure 2,13). Pacific Rim National Park on the west coast of Vancouver Island is a relatively recent addition to the national parks system and more adequately reflects Parks Canada's present attitudes to the balance between use and preservation. Although highway access is provided through the Long Beach unit of the park, recreational opportunities are not focussed on facilities but, rather, on the natural features of the park, such as beaches, rain forest trails and seascape which are the hallmarks of the park. Besides road access, camping, picnicking and interpretive facilities, support services for park users have remained outside the park, in order to protect park resources and to stimulate local, private sector initiatives in the tourist industry. In the other units of Pacific Rim National Park, the Broken Group Islands and the West Coast Trail, an even more back country recreational orientation is prevalent. There road access does not exist. Minimal facilities are

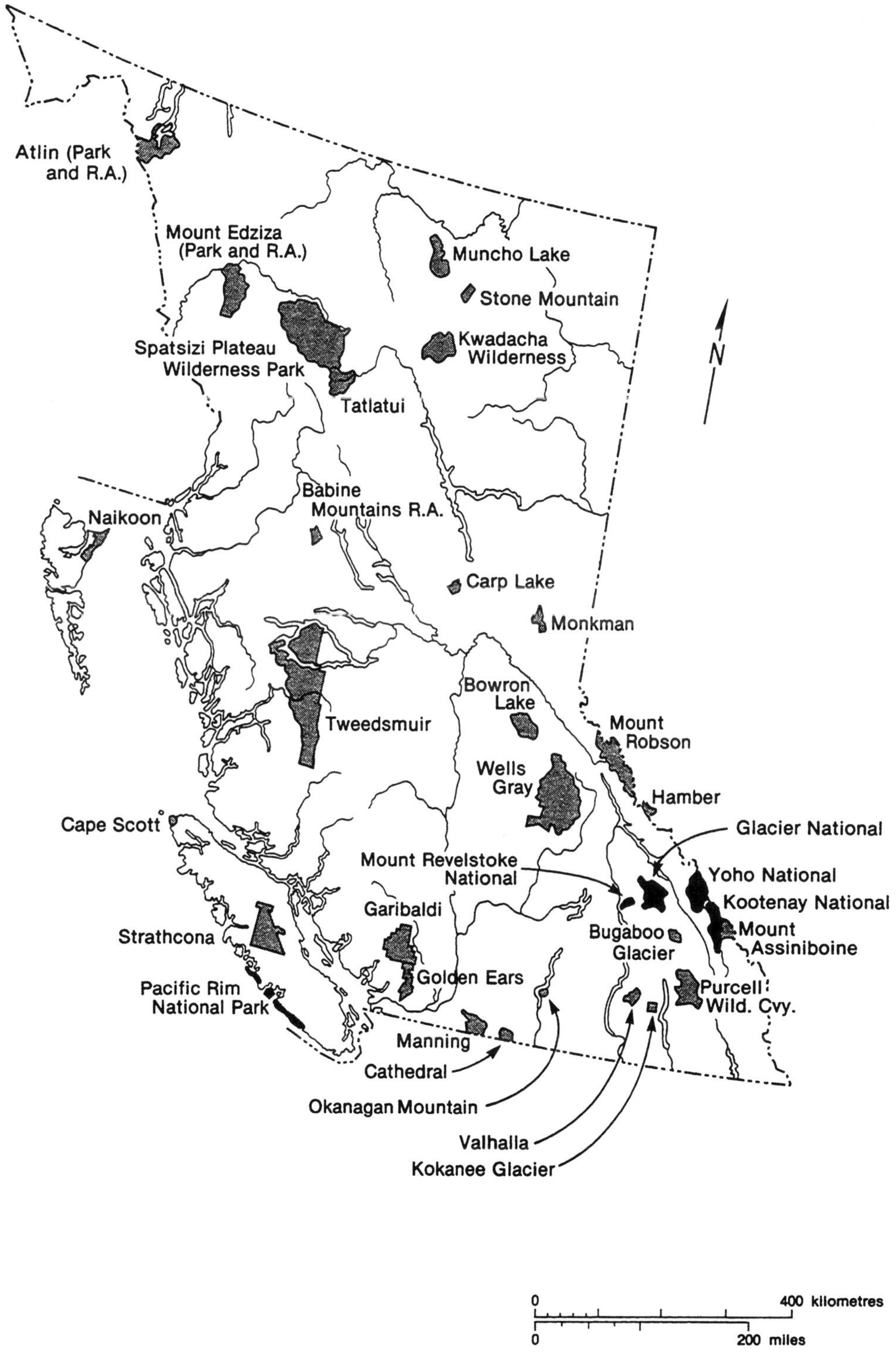

FIGURE 2,13 National and provincial parks

provided to assist visitors in coping with environmental conditions and to assist park managers in protecting the park's resources. Trails, bridges and campsites are examples of the facility support Parks Canada has provided.

The other four national parks in British Columbia have a much longer history and have evolved under conditions and policies that are very different from those which guide Parks Canada today. Glacier, Revelstoke, Yoho and Kootenay were all established prior to 1920 as generators of tourist travel, initially along the Canadian Pacific Railway line and later along the Trans-Canada and Banff-Windermere highways. These major travel corridors are central to the parks and provide a heavy concentration of visitors seeking recreational opportunities in what have been described as front country areas of the parks. Beyond these travel corridors that provide camping, picnicking, day hiking, commercial accommodation and other services, are large areas of scenic back country dotted with primitive facilities, such as trails and campsites, and occasionally more elaborate facilities, such as back country lodges, for example, Lake O'Hara Lodge in Yoho National Park, or back country shelters, such as the Elizabeth Parker Hut operated by the Alpine Club of Canada near Lake O'Hara.

Seeking a balance between the preservation of natural environments, with their associated back country recreational activities, and the facilities and use levels typical of front country recreation along the travel corridors is not an easy task. It is especially difficult with the increasing pressures of visitor numbers, commercial road and rail traffic, and the resource utilization activities in areas adjacent to the parks. In attempting to reach that balance Parks Canada has become intimately involved in the provision of both front country and back country recreational experiences in each of the parks it administers.

Provincial Parks

The provincial parks system of British Columbia is a large, highly diverse system of protected areas (Figure 2,13). Conservation of the lands encompassed by the system for the recreational benefit of the residents of the province is its primary goal.[30] In achieving this goal the managing agency, the Parks and Outdoor Recreation Division of the Ministry of Lands, Parks and Housing, attempts to provide an equitable distribution of year round recreational opportunities which will complement those of other park systems and those provided by the private sector. The division is also committed to an extensive program of public information regarding the attributes and opportunities provided by the system. Consequently, awareness of the system both within and beyond British Columbia has served to enhance the tourism industry in the province.

Presently, there are approximately 370 provincial parks, recreation areas and conservancies, covering about 4,600,000 hectares (11,000,000 acres) (4.8 percent of British Columbia).[31] Park size varies dramatically, with the vast majority of the parks (over 300) being less than 500 hectares (1,235 acres) and some as small as half a hectare (1.25 acres). At the other end of the spectrum, a dozen parks contain 85 percent of the land base, with only 30 exceeding a size of 5,000 hectares (12,350 acres).

The informal classification of parks within the system clearly indicates the diversity of recreational opportunities offered.

PICNICKING AND DAY USE

Such parks are provided for rest and recreation, focussing on day use activities, such as swimming, boating, walking and picnicking. Spectacle Lake, just north of Victoria, is a good example of such an area where trail development around the lake complements the fishing and swimming opportunities along the shore. Day use parks also may have special features as a focus for activity, such as in the case of MacMillan Park. The giant Douglas fir trees of Cathedral Grove in this roadside park provide an awe inspiring experience, as visitors can walk the self-guiding trail.

HERITAGE

A number of parks commemorating the cultural heritage of British Columbia are also part of the system. They provide opportunities for visitors to see and appreciate some of the characteristics of life in circumstances and times gone by. Two of the best known examples are Barkerville and Fort Steele, early settlements of the gold rush days of British Columbia's Interior.

CAMPING

Many of the provincial parks serve an accommodation function, both as destination areas associated with other primary activities, such as swimming or boating, and as roadside stops fulfilling the needs of the travelling public. Of the destination variety, Cultus Lake and Scotch Creek parks are good examples. Large campgrounds near excellent beaches provide the major foci of activity there. Roadside campgrounds, such as Emory Creek and Johnstone Creek, are also common along many of the province's highways.

WILDERNESS

Large areas within the park system are without developed access and facilities, providing exceptional wilderness opportunities in highly scenic and challenging country. Hiking, mountaineering, ski touring, nature study and photography are predominant activities. Examples of parks fulfilling this role in the system are Garibaldi, Spatzizi and the newly established Valhallas.

WINTER RECREATION

The focus of some of the established parks is on the opportunities they provide for winter recreation, primarily downhill skiing, cross country skiing and snowshoeing. Mount Seymour Park and Silver Star Recreation Area are two good examples.

BOATING

One of British Columbia's most noteworthy features is its scenic and protected coastline. Opportunities for cruising have led to the development of a whole system of coastal marine parks focussed on the waters of Johnstone Strait and the Strait of Georgia. Examples of the parks in this system are Sidney Spit, Princess Louisa Sound and Desolation Sound provincial parks where anchorages and a few basic on-shore facilities provide the basis for an enjoyable stay. In addition to the coastal parks, a similar concept has been developed on major lakes in the interior of the province, most notably on Shuswap Lake, with parks such as Shuswap Lake Marina which is accessible only by water and provides picnicking and walk-in campsites, as well as opportunities for swimming and fishing.

The provincial parks system obviously is a broad mixture of back country and front country opportunities for recreation, from its vast wilderness areas to the over 10,000 campsites, the almost 24,000 vehicle day use parking capacity, and the five downhill ski operations.[32] It is a well used system, recording almost 20,000,000 visits in 1981, and is highly valued by the people of British Columbia.[33]

CONCLUSION

There are some important general observations that can be made regarding outdoor recreation in British Columbia. The first, which has been demonstrated throughout this chapter, is that the recreational opportunities available to the residents and visitors are extremely diverse. The second is that, to a large extent, the quality of these recreational experiences is dependent upon the maintenance of a high quality resource base. It is inevitable that concerns for the future of outdoor recreation in the province will, in some way, relate to these primary characteristics.

To maintain and improve the diversity and quality of outdoor recreational opportunities in the future there is a need to address three important aspects:

(1) co-ordinated planning of recreational opportunities;

(2) the relationship between tourism and the natural environment; and

(3) the competition for land between recreational and other resource users.

Co-ordinated Planning

The recognition of the need for co-ordinated planning of recreational opportunities is by no means new. With so many agencies, groups and commercial enterprises involved, identifying public needs and their distribution, and efficiently responding is indeed a challenge. But certain present circumstances have made co-ordinated planning more than just an idle dream or academic construct. The growth of recreational participation is putting tremendous pressure on the existing facilities and opportunities — so much so that the demand simply cannot be met in some cases, without declines in the quality of the experiences offered. At the same time, increased pressure from the competing non-recreational land uses is resulting in similar pressures on existing recreational areas. Added to these difficulties is the impact of the recent economic recession. The sagging economy has encouraged the slogan "do more with less" and a "luxury" item like recreation becomes a prime target for austerity.

In spite of these difficulties some progress is being made. The initiative of sub-regional system planning by the Parks and Outdoor Recreation Division is an attempt to identify a particular role for the division among the variety of suppliers of recreational opportunities in an area. Recognition is made of the role of other agencies and the private sector, and co-operation and support to seek such a balanced approach to recreational development are encouraged. More such planning which attempts to integrate the efforts of a variety of recreational suppliers is needed.

Tourism

Budowski states, "For the majority of cases at present the relationship between tourism and conservation is usually one of coexistence moving toward conflict, mainly because of an increase in tourism and the shrinking of natural areas".[34] He suggests that symbiosis, in which both tourism and conservation benefit from being related, should be the goal. This is difficult to achieve. The economic pressure of our present situation makes it increasingly essential, and at the same time increasingly difficult. The tendency is for each agency or entrepreneur to become more independent — hang on to and even further exploit those resources and economic opportunities that will yield a return, without regard for the larger system or long term future. Yet, an integrated and efficient system of resource utilization to provide recreational opportunities is now an even more critical need than in the past when resources were abundant and the demands on our resources were less.

This relationship between tourism and conservation, although present throughout the spectrum of front country and back country recreation providers, has been specifically identified as a major issue in protection and use of parklands.

> The simultaneous preservation and use of park areas has long been recognized as a fundamental problem facing park establishment and management. An industry such as tourism promotes the recreational use of our parks in order to generate economic benefits. This use contributes significantly to the imbalance of the preservation/use spectrum if planning and management have not been adequate.
>
> Tourism is a growing industry and a particularly significant one in British Columbia. The Parks and Outdoor Recreation Division of the Ministry of Lands, Parks and Housing is increasingly emphasizing economic arguments for new park establishment. There seems to be little doubt that the relationship between parks and tourism is both critical and current.[35]

Accordingly, the National and Provincial Parks Association of Canada has developed a draft position paper on parks and tourism. The association "believes that a clearly defined balance is required to ensure preservation of ecological integrity of the resources, while allowing sensitive use of park resources for both recreation and economic return".[36] In the paper key issues are identified, such as the need for planning and design of tourism activities and facilities that are compatible with natural integrity; the need for synthesis of park and tourism systems plans; the need for strategies to achieve common goals; and the need for strategies to minimize conflict in achieving incompatible goals.[37] These are clearly major issues which are particularly important in the British Columbia context and deserve concerted effort to improve prospects for the future symbiosis of parks and tourism.

Other Resource Uses

It is no secret that in British Columbia strongly opposing forces have clashed consistently on the issue of land control and use. The province is rich in natural resources and continues to depend upon their extraction and utilization as the mainstay of the economy. As the impact of these activities spreads and the significance of the natural environment increases, both for the enjoyment of residents and the attraction of visitors, so too will the frequency and intensity of conflict. Consider some past examples of the conflict between resource use and recreational opportunities: the Peace River dam, the Cascade wilderness proposal, the northeast coal development, and the Valhalla park and Slocan Valley plan. Consider too some of the unresolved issues: the Stikine (in a state of quiescence, but still a potentially explosive issue);[38] the wildlife/old growth timber controversy on Vancouver Island;[39] the mining activity adjacent to Spatzizi Provincial Park;[40] the stalled negotiations on the establishment of Pacific Rim National Park;[41] and the extended indecision in the South Moresby area of the Queen Charlotte Islands.[42] It is inevitable that the quality and diversity of the recreational opportunities of the province will be significantly affected by the resolution of such issues. However, such issues need not be the confrontational obstacles they have so often become. Co-operation and co-ordinated planning could alleviate much of the intensity of these issues and bring a more acceptable decision-making process to bear.

In all three important aspects of the future of outdoor recreation discussed above under separate headings one further common element is worthy of note. Progress in resolving issues and improving the system of outdoor recreational opportunities will be significantly improved if there is public interest and support for co-operative and concrete actions. The citizens of this province need to show their concern for the quality and diversity of recreational opportunities they have come to take for granted. They need to be prepared, along with industry and government, to work together to plan and implement a growing system of high quality recreational opportunities that is in harmony with the other resource interests of the provincial economy and will continue to be recognized and sought after internationally as among the best.

REFERENCES

1. WALL, G., "Recreation and Canadian Identity", in WALL, G. and MARSH, J. (eds.), *Recreational Land Use*. Ottawa: Carleton University Press, 1982.

2. The terms back country and front country are used in this chapter to differentiate between intensively used, readily accessible areas, and more remote areas where access is limited and use is dispersed. The terms have been developed and used in the Four Mountain Parks Planning Program. See CANADA, *Four Mountain Parks Planning Program: Introduction*. Calgary: Parks Canada, Western Region, 1982.

3. CHUBB, M. and CHUBB, H.R., *One Third of Our Time*. New York: John Wiley, 1981.

4. BROOKS, L., *Recreation and Tourism in the British Columbia Environment*. Victoria: Ministry of Environment, 1979.

5. DEARDEN, P., "Tourism and the Resource Base", in MURPHY, P.E. (ed.), *Tourism in Canada: Selected Issues and Options*. Victoria: University of Victoria, Department of Geography, Western Geographical Series, Vol. 21, 1983, pp. 75-93.

6. MARSHALL, MACKLIN, MONAGHAN LIMITED, *Executive Summary of the Province of British Columbia Tourism Development Strategy*. Victoria: prepared by Marshall, Macklin, Monaghan Limited for the Government of Canada and the Government of British Columbia under the Travel Industry Development Subsidiary Agreement, 1979.

7. SOUTHWESTERN BRITISH COLUMBIA TOURIST ASSOCIATION, *Southwestern British Columbia Visitor's Guide*. Vancouver: Key Publishers, Vol. 11, No. 13, 1982.

8. JEROSKI, S.F. and CONRY, R.F., *Provincial Parks Survey: Technical Report*. Victoria: prepared by Horizon Research and Evaluation Affiliates for the Parks and Outdoor Recreation Division of the Ministry of Lands, Parks and Housing, 1983.

9. *Ibid.*

10. HANCOCK, L., "The Trail to Monkman Pass", *Explore*, No. 6, July, 1982.

11. FAIRHOLM, M. and DUNCAN, R., "Titanic Secrets", *Beautiful British Columbia*, Vol. 25, No. 4, 1984.

12. CAMPBELL, R., "Wandering Through Warner", *Explore* , No. 17, July-August, 1984.

13. SHAW, K., "Akamina Valley", *Park News* , Vol. 20, No. 1, 1984.

14. BRITISH COLUMBIA, *Summary of Revenue, 1983-84*. Victoria: Ministry of Environment, 1984.

15. BOND, R.S. and WHITTAKER, J.C., "Hunter-Fisherman Characteristics: Factors in Wildlife Management and Policy Decisions", in *Recreation Symposium Proceedings*. Upper Darby, Pennsylvania: United States Department of Agriculture, 1971.

16. FEDERATION OF MOUNTAIN CLUBS OF BRITISH COLUMBIA, *Cloudburst*. Vancouver: Federation of Mountain Clubs of British Columbia, Newsletter, October, 1984.

17. ALLEN, D., personal communication.

18. ALLEN, D., personal communication.

19. WHITFIELD, P., personal communication.

20. JEROSKI, S.F. and CONRY, R.F., *Provincial Parks Survey: Summary Report*. Victoria: Prepared by Horizon Research and Evaluation Affiliates for the Parks and Outdoor Recreation Division, Ministry of Lands, Parks and Housing, 1983.

21. MARSHALL, MACKLIN, MONAGHAN LIMITED, *op. cit.*

22. THOMPSON, D., *The Shuswap Lakes Recreational System*. Victoria: Department of Recreation and Conservation, Parks Branch, Planning Division, 1975.

23. *Ibid.*

24. OKANAGAN-SIMILKAMEEN TOURIST ASSOCIATION, *Okanagan Travel Guide 1984*. Vernon: Skyword Marketing, 1984.

25. CAPITAL REGIONAL DISTRICT, *Review of Potential Regional Parkland*. Victoria: Capital Regional District, Parks Division, 1983.

26. *Ibid.*

27. *Ibid.*

28. OUTDOOR RECREATION COUNCIL OF BRITISH COLUMBIA, *Recreation Access in British Columbia: A User's Handbook*. Vancouver: Outdoor Recreation Council of British Columbia, 1984.

29. CANADA, *Parks Canada Policy*. Ottawa: Parks Canada, 1979.

30. BRITISH COLUMBIA, *Provincial Parks System Plan: Draft*. Victoria: Ministry of Lands, Parks and Housing, Parks and Outdoor Recreation Division, 1983.

31. BRITISH COLUMBIA, *Park Data Handbook 1983*. Victoria: Ministry of Lands, Parks and Housing, Parks and Outdoor Recreation Division, 1984.

32. *Ibid.*

33. JEROSKI, S.F. and CONRY, R.F., *Provincial Parks Survey: Summary Report*, *op. cit.*, reference 20.

34. BUDOWSKI, G., "Tourism and Conservation: Conflict, Co-existence or Symbiosis", *Parks*, Vol. 1, No. 4, 1977.

35. DOWNIE, B. and PEART, B. (eds.), *Parks and Tourism: Progress or Prostitution?*. Victoria: National and Provincial Parks Association of Canada, Conference Proceedings, 1982.

36. DOWNIE, B., "Parks and Tourism: A Position Paper of the National and Provincial Parks Association of Canada", in DOOLING, P., *Symposium on Parks in British Columbia: Conference Proceedings*, in press.

37. *Ibid.*

38. FRIENDS OF THE STIKINE, VANCOUVER, *Newsletter*, No. 9, October, 1983.

39. BRITISH COLUMBIA, *Reservation of Old Growth Timber for the Protection of Wildlife Habitat on Northern Vancouver Island.* Victoria: Ministry of Environment and Ministry of Forests, 1983.

40. FOX, F.J., "Provincial Park Management Issues: Spatsizi Plateau Wilderness Park", in DOOLING, P., *Symposium on Parks in British Columbia: Conference Proceedings*, in press.

41. NATIONAL AND PROVINCIAL PARKS ASSOCIATION OF CANADA: BRITISH COLUMBIA CHAPTER, VICTORIA, *Newsletter*, June, 1982 and March, 1984.

42. SOUTH MORESBY RESOURCE PLANNING TEAM, *South Moresby Land Use Alternatives*. Victoria: Ministry of Forests, 1983.

PLATE 20 A pole raising ceremony at Alert Bay. *B.C. Government Photo* ▶

14 POPULATION AND ETHNIC GROUPS

Colin J.B. Wood
University of Victoria

INTRODUCTION

The population of British Columbia in 1981 was officially enumerated at 2,744,467, an increase of 11.3 percent over 1976.[1] The rate of growth was almost double the national average of 5.9 percent during the same period. Since entering Confederation, the trend has been one of steady expansion, so that today the province accounts for 11.3 percent of the national population. From the initial European colonization in the early nineteenth century onwards, much of the population has been concentrated in the southwest portion of the province, associated with the settlement and expansion of Vancouver, Victoria, the capital, and Nanaimo: together these three cities comprise what is often termed the Strait of Georgia urban region (Figure 1,14).

In contrast, the remainder of the province is either uninhabited or thinly populated with hamlets, small towns and occasional small cities strung in linear fashion along the main transportation arteries like occasional beads on long necklaces. The mountainous terrain and indented, fiord coastline have forced an uneven settlement pattern to develop, with abrupt changes in density from suburbs to uninhabited hills and mountains.

In this chapter the general evolution of this unusual pattern of population geography will be described first and then discussed in relation to quality of life indicators. The second part will focus on the ethnic composition of British Columbian society and evaluate briefly its contribution to the variety in human landscape.

POPULATION GROWTH AND DEMOGRAPHY

British Columbia, the province with the third largest population, and Alberta are the only ones to have increased continuously their share of the national population: for British Columbia this has been at the rate of

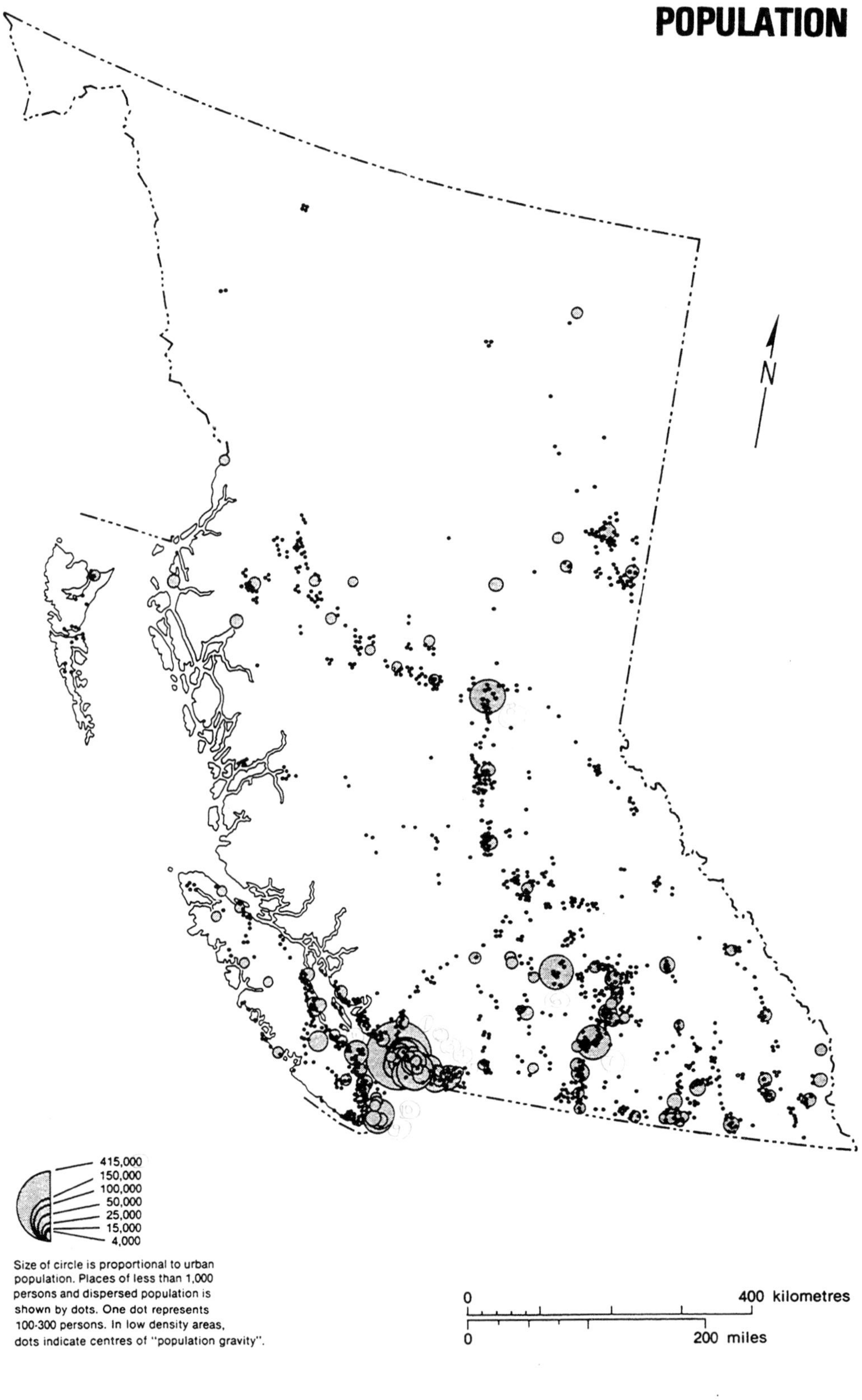

FIGURE 1,14 Population and settlement

approximately one to two percent per decade, reflecting the general westward shift in the axis of North America's population.

European settlement in the early nineteenth century spread inland mainly from the southwest, along the river valleys.[2] Waves of natural resource exploitation were the primary impetus, composed of fur traders, prospectors, miners, loggers, farmers, ranchers, entrepreneurs and speculators. These pioneers rolled across the fledgling colony eager to prosper by unlocking the riches of the region.[3] The gold rush of the 1860s created the first dramatic surge in population as prospectors, mainly American, poured into the Cariboo region. The British hastily expanded the form of government to prevent American annexation by default. At the same time, the Indian population, hard hit by diseases introduced by the Europeans, began its long decline which would not be reversed until the 1940s.

The people of British Columbia decided to join the Confederation of Canada in 1871 on the condition that a rail link was built to the East. The construction and completion of the Canadian Pacific Railway in the 1880s ushered in a new era of growth, as the province became integrated with the Canadian economic system. Railroads were built, mines were sunk, forests cleared, swamps drained and crops planted. However, the optimism of this period was replaced by the doubts and difficulties of World War I and the inter-war period. The late 1940s saw a resurgence as the great housing boom fuelled world demand for construction lumber and minerals, and a period of prosperity ensued. The investment from venture capital of the earlier period was joined by large public expenditure in transportation infrastructure and hydro-electric power. It was an era of road, bridge, railway, port and dam construction, a period inextricably linked with W.A.C. Bennett, the colorful premier. The opportunities for employment attracted many migrants and immigrants. At the same time, however, the influx of young and ambitious people was accompanied by a steady stream of seniors from the Prairies and Eastern Canada, retiring to the pleasant environment of the southern coastal region and the small interior towns. The relative contribution to population increase by natural increase and migration is shown in Table 1,14. Clearly, the province has benefited from the westward drift in population. Although the rate of natural increase has declined, as fertility has declined, migration and immigration have sustained the overall growth pattern.

DEMOGRAPHIC STRUCTURE AND POPULATION PROJECTIONS

Population pyramids based on the 1981 census (Figure 2,14) show that the provincial demographic pattern is similar to the national, with a characteristic indentation at the base related to the continuing decline in fertility

Table 1,14 NATURAL INCREASE AND MIGRATION BY REPRESENTATIVE PROVINCES IN THOUSANDS

	1881-1931		*1931-1961*		*1961-1979*	
	Natural Increase	*Net Migration*	*Natural Increase*	*Net Migration*	*Natural Increase*	*Net Migration*
NOVA SCOTIA	293	-218	293	- 70	152	- 42
QUEBEC	1589	-391	2186	198	1283	-244
ONTARIO	1342	21	1721	1083	1292	972
SASKATCHEWAN	284	356	370	- 368	198	-167
BRITISH COLUMBIA	127	418	389	546	425	513

SOURCE: Beaujot, R. and McQuillan, K., *Growth and Dualism*. Toronto: Gage, 1982, p. 150.

and the progression of the postwar baby boom generation to the middle tier of the graph. Some differences are apparent in that the percentage composition of cohorts up to 20 years is lower, while in the middle tiers it is the same or slightly higher and at the retirement levels it is significantly higher. The below average representation of young people is also reinforced by family size statistics of 3.1 persons compared with 3.3 nationally. Thus, the general demographic picture for the 1980s is one of a slightly older population, 30.9 years average age compared with the Canadian population of 29.6 years, with more retireds, singles and smaller families.[4]

A major objective of demographic analysis is to provide population projections which, in turn, are a basis for economic and social planning. Projections of the provincial population for 1981 that were made during the period 1963 to 1974 ranged between 2.4 and 3.1 million, incurring errors of between 1.5 and 3.4 percent.[5] With hindsight it can be seen that the forecasts were overly optimistic and did not anticipate the decline in fertility and reduced immigration. Unfortunately, the projections were used to justify significant public investment in infrastructure in the 1970s, which subsequently turned out to be surplus to need. The recent trends in demographic characteristics have led to noticeable reassessments of population projections for the next two decades.[6]

The general pattern for the wealthier provinces is one of below average birth rates, thus, British Columbia currently has the lowest natural increase rate.[7] From the demographic viewpoint, the problem is whether

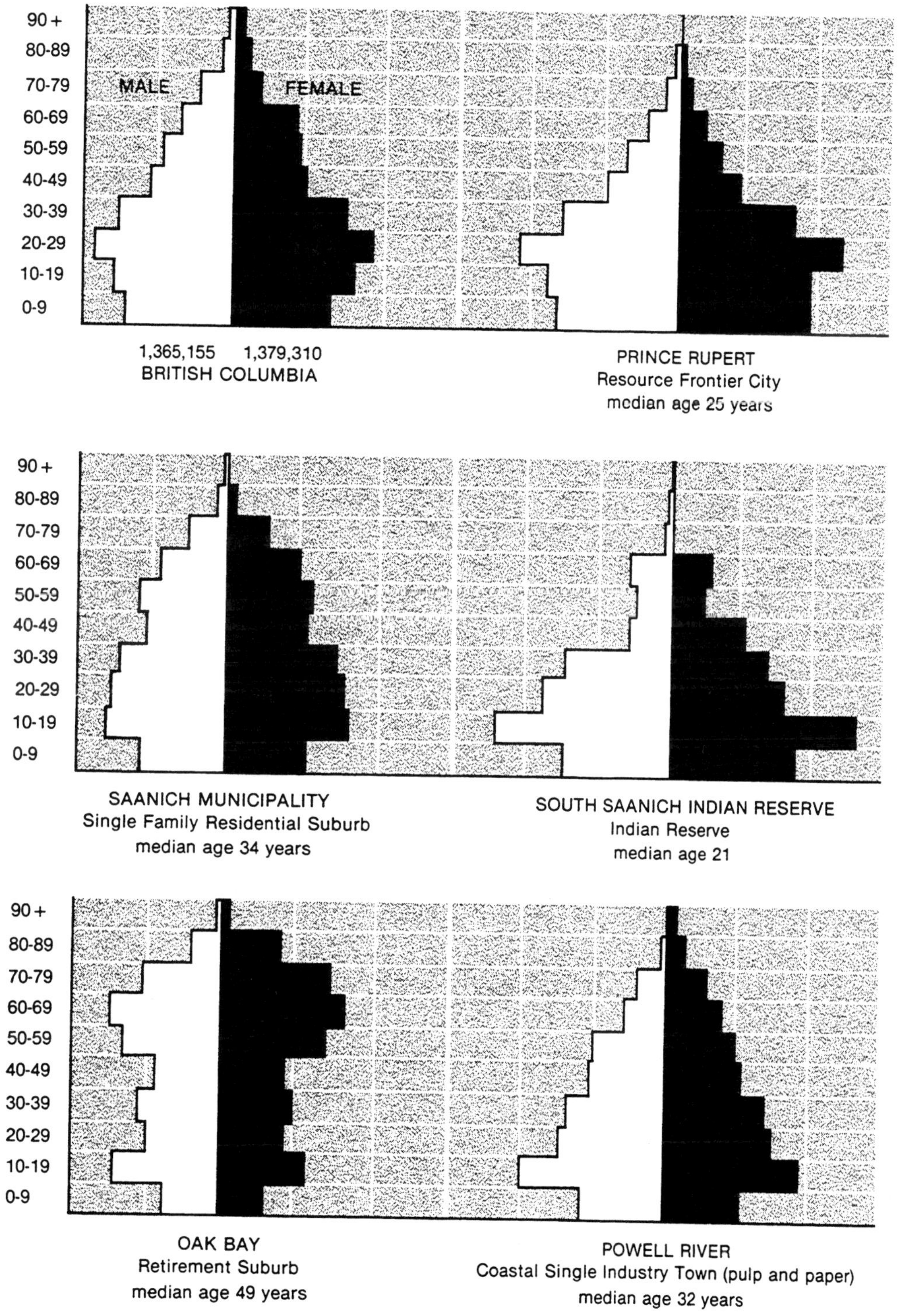

FIGURE 2,14 Selected population pyramids, 1981

the experience of the 1970s represented a long-term trend or a short-term aberration.[8] By varying the basic components of fertility, migration, and immigration, several projections are possible. Foot's 1982 forecasts for 2001 are between 3.31 (low) and 4.02 (high) million people for the province.[9] The crucial inference to draw from the figures is that the dependency ration generally will tend to decline well into the next century. Care has to be taken in anticipating the consequences of such a development. For example, the demand for housing may continue to increase, due to changing family structure and greater longevity, even though fertility levels decline.

REGIONAL GROWTH

With 66 percent of the population concentrated in the southwest, the province has a distinct core and periphery pattern. Since 1971 this region has been responsible for 60 percent of the population growth. The two main cities of Vancouver and Victoria have reached a sufficient size for the "ratchet effect" to occur and are unlikely to decline. Greater Vancouver's expansion has engulfed the city of New Westminster, although the latter strives to maintain its separate identity. Private and public sector initiatives have prevented the inner city areas of Vancouver and Victoria from acquiring "doughnut-ring" features of a decayed core and booming freeway ring zone. The rapid development of suburban shopping centres has affected downtown Nanaimo significantly. Elsewhere in the remainder of the province expansion has been uneven. Prince George, Kelowna and Kamloops have continued to grow, while some resource oriented towns which boomed in the 1950s and 1960s have remained stable or even declined, such as Port Alberni, Powell River, and in the southeast, Nelson and Trail. At the local neighborhood level very marked differences in demographic structure exist between the youthful, dormitory suburbs of Richmond and Delta (Lower Mainland) and Saanich and Colwood (Capital Region) and the retirement neighborhoods of Inner Vancouver and Oak Bay (Figure 2,14).

QUALITY OF LIFE

Automobile licence plates proudly proclaim the province as "Beautiful British Columbia", an assertion that few would dispute, given the breathtaking views and pristine landscape. Indeed, the notion of a "lotus land beyond the Rockies" may not just be Fotheringham hyperbole. If the environment can be couched in superlatives, is the way of life of the inhabitants also sybaritic? As a net receiver of migrants there would seem to be truth to these perceptions if the adage that people vote with their feet is to be believed. However, quality of life is, like beauty, in the eye of the beholder and is open to subjective interpretation.[10] Nevertheless, despite the difficulties of definition and evaluation, there is a general consensus on the broad, objective indicators of quality of life. The Organization for Economic Co-operation and Development, using a method of analysis based on nine broad categories of indicators, specifies the basic requirements of modern life as: health; individual development; employment and conditions; personal economic condition; regional environment and leisure; safety and justice; social environment; and social opportunity.[11] Each of these broad categories is comprised of more detailed subsets, such as, for health: lifespan, healthfulness, and mortality. While it is not possible to explore the whole range of detailed indicators, some insight can be derived from examining six of the main categories. The information shown in Table 2,14 attempts to summarize some of the more salient features of British Columbia's way of life in relation to the rest of Canada. The figures, and the following interpretation, must be viewed as primarily descriptive and without statistical inference. Furthermore, averages must be interpreted with caution, since they can mask extremes and bimodality, such as rich and poor, with no middle group.

Health[12]

Life expectancy is close to the national average for both sexes and is generally higher than in Eastern Canada, however some caution in interpretation must be exercised, since the 60-plus cohorts are bolstered by retiring migrants. The major causes of death and their spatial variation have been documented in the Mortality Atlas of Canada which shows that British Columbia generally is rated favorably compared with other areas of the country. The exceptions are those ailments related to over-indulgence and stress, notably liver complaints related to high alcohol consumption, with its reciprocal effect on traffic accident rates. In terms of hospital care, bed provision is high and the ratio of doctors to patients is close to the highest among the provinces, but care facilities tend to be oriented to the heavily populated southwest. The mental health of the population is characterized by an admission rate to mental hospitals that is below the national average rate for 1971-1979, but the suicide rate is high and close to an epidemic for the young adult male. Infant mortality figures are marginally above the national average, although not the worst and well below the United States figures. Native Indian infant mortality rates are three times higher than those of the non-native population. Explanatory variables are complex: the population at risk, that is pregnant women, may be revealing through their lifestyles the effects of a higher proportion of native people in isolated areas and some environmental influences. Many native people live in isolated reserves where confinement and both pre and postnatal care are well below average.

Development and Learning[13]

Along with the rest of Canada, British Columbia is experiencing a gradual decline in student enrollment, a demographic "transition" which has significant repercussions throughout the educational system. The retirement rate for teachers, many of whom were recruited for the boom years, is slower, so that pupil-teacher ratios have declined and costs per student have increased. Comparison of "investments" in this human resource over the period 1970-1980 generally shows the province to be at the lower end of the scale. While significant post-secondary facility expansion has occurred, recruitment to the major professions still draws heavily on out-of-province migrants. The 1970s saw the expansion of the regional college system, while the 1980s is witnessing its retrenchment and the development of distance, university-level course provision via television. The provincial government is currently following a policy of increasing student to teacher ratios.

Employment and Conditions of Employment[14]

The index of employment for the 1970s and early 1980s has generally been higher than in most other provinces and unemployment, while higher than in the Prairies, did not reach the severe levels of maritime Canada until the mid-1980s. With an economy closely geared to resource extraction, and particularly oriented to the demands of the housing market in the United States and the steel industry in Japan, the economy, and hence labor, quickly feels the effects of market downswings. Therefore, although the economy has the highest weekly earnings rate in the country, this advantage is offset by the effect of periodic booms and recessions. Furthermore, employment conditions in the resource extraction industries are often hazardous, even though unions and governments strive to achieve safe working environments. Accident rates are well above the national average, reflecting the hazardousness of the logging, fishing and mining industries.

Personal Economic Condition[15]

Historically, the province has had above average disposable incomes, reflecting the strong economy, high weekly wage rates and union organization in key industries. This is offset, however, by the high cost of living, especially due to the impact of house prices which in the Strait of Georgia urban region are the highest in the country. The housing and urban property markets generally have also been fuelled by speculation and foreign investment. The distribution of family incomes by region shows that the province has average incomes over 50 percent above the national and has below average frequencies in the low income group (Table 2,14). However, averaging masks the pockets of low income groups who often lead a precarious existence. British Columbia has three such groups — the retired, native people and single parent families. During the period 1978-1983, the proportion of families classed as low income actually increased from 7.8 to 11.2 percent.

Regional Environment (Man-Made and Natural) and Leisure[16]

Housing conditions are generally above average, reflecting the recent construction of much of the accommodation. The percentage of rental housing is slightly below both the national average and those of the prosperous provinces. Security of tenure has again become an issue affecting lifestyles, as economic and political pressures on the provincial government have led to drastic reductions in tenant protection by government agencies. Furthermore, the "restructuring through development" of urban infrastructure has significantly affected the stock of accommodation available to tenants. One response has been the development of co-operative housing schemes. Housing conditions on some Indian reservations are deplorable.

The natural environment is one of the great attractions of the province, by virtue of its spectacular scenery and often minimal impact by man. The easy access to outdoor recreation of many kinds for even the urban dweller of Vancouver is probably unequalled elsewhere. Problems do exist during peak recreation periods, particularly as a result of the influx and pressure of out-of-province visitors in certain park areas. Maintaining the quality of the natural environment is often a vexatious issue in the way of life and affairs of the people. The government has the difficult dual role of encouraging resource development for revenues and employment, such as the forests for lumber and pulp and the coal and ore bodies for minerals, but at the same time protecting and conserving the environment. Every

Table 2,14 QUALITY OF LIFE INDICATORS – ONE EQUALS "BEST" AND TEN EQUALS "WORST"

	HEALTH					*DEVELOPMENT & LEARNING*			*ECONOMIC CONDITION*					*LIFESTYLE*			*SAFETY & JUSTICE*			
	Infant Mortality	Pop./ M.D. Ratio	VD Rate	Reported Suicide M+F / 2	Admit. Rate 1971-81	Educ. Expend.	Elem. Expend.	No. of Teachers to Labor Force	Weekly Earnings	Unemployment Rate	Percent Rented Housing	Percent Pop. in Lowest Quintile	Injuries to Work-force	Divorce	Beer / Wine	Traffic Accidents	Violent Crimes	B & E	Reported Rape	Drug Convictions
BC	4	2	8	8	5	(10)	(10)	8	1	5	7	1	9	(10)	2/(10)	(10)	(10)	(10)	(10)	(10)
ALTA	5	6	9	9	6	9	9	(9)	2	2	8	3*	8	9	1/7	9	9	7	9	8
SASK	9	5	(10)	6	9	7	7	7	8	3	6	3*	7	5	6/6	6	7	8	8	2
MAN	8	7	7	7	1	6	3	4	3	1	3	3*	4	4	4/4	8	6	6	7	5
ONT	2	1	6	5	(10)	8	4	(9)	4	4	9	1	4	8	8/8	5	8	5	4	7
QUE	1	3	2	(10)	4	2	1	5	5	6	(10)	4	(10)	7	9/9	7	5	9	5	9
NB	6	(10)	1	3	1	5	6	3	6	8	5	(5*)	4	3	7/2	4	3	3	2	6
NS	7	4	5	2	7	4	5	2	9	6	4	(5*)	2	6	3/5	3	4	4	6	3
PEI	(10)	9	3	4	8	3	8	6	(10)	9	2	(5*)	1	2	4/3	2	1	2	1	4
NFLD	3	8	4	1	3	1	2	1	7	(10)	1	(5*)	2	1	(10)/1	1	2	1	2	1

* Prairie and Atlantic regions are grouped together.

development proposal generates opposition from environmental groups. This does not necessarily mean that people want their cake and want to eat it too, but that there is often conflict over methods or lack of evaluation of alternatives in resource development.[17]

Personal Safety and Justice[18]

The decline in personal safety in cities is one of the more unpleasant features of late twentieth century life. However, placed in perspective, Canadian crime rates tend to be substantially lower than those in the United States, but slightly higher than those in Europe. Rates tend to increase westward, being low in the Maritimes and high in British Columbia. Vancouver has been dubbed the crime capital of Canada, with a violent crime rate per 100,000 that is double the average for large cities. However, extreme caution has to be exercised in using such statistics for, while it is clear that rates are higher in British Columbia, the areal unit for collecting and calculating urban crime statistics can have a marked effect on the rate.[19] For example, the city of Victoria has a violent crime rate of 1,194 per 100,000, compared with the national average of cities of that size of 618. The adjacent and contiguous suburb of Saanich has a very low rate of 348. If the areal unit is reshaped according to a metropolitan context (including the law-abiding suburbs of Oak Bay and Esquimalt) the overall rate is below average for cities of 100,000 to 249,000. Significant variations do occur in the category of small cities, with Prince Rupert, Kamloops and Campbell River experiencing high rates and Powell River, Trail, Langley and Matsqui being close to or below the national average. Explanations for British Columbia's high rates can be found in the highly urbanized population, the mobile lifestyles, the drug trade of the Pacific Rim, the environment which is conducive to crime all year, thereby attracting a seasonal migration of criminals, the higher incomes, along with the greater range in incomes, and a higher than national average native Indian population.[20] The incarceration rate also provides a clue to understanding the nature of this social turmoil, particularly in the smaller cities and towns with high violent and property crime rates. The prison population of native Indians, over the past 30 years, has climbed to 16 percent (almost 20 percent in some years), despite the fact that the proportion of native Indians is only about five percent of the total population.[21] Some commentators have wryly suggested that this represents not a crime problem, but a race war in some parts of the province.

ETHNICITY

The peculiar mix of history, geography and politics has resulted in the broad brush of settlement being predominantly English speaking, with Anglo-Canadian norms regulating social order and economic behavior. At the local and occasionally regional scale, certain ethnic groups have left distinguishable individual marks on the landscape which show through the broad pattern of the majority.

An ethnic group can be defined objectively as a minority of people having traits distinguishable from the majority, such as language, dress, nationality and race, in sum, those elements which denote a common cultural tradition.[22] The majority are representative of nationality or mainstream Canadian ethnicity. These broad objective definitions must be used carefully, since not all members of an ethnic group may participate in the group. Notwithstanding this caveat, aspects of ethnicity can be understood as landscape phenomena.[23] This section will outline the ethnic composition of the province, and the place association, distinguishable landscape structures and landscape impacts characteristic of the ethnic groups.

Ethnic Composition

The present ethnic composition, using "mother tongue" as the definition, is shown in Table 3,14. This illustrates well the Anglo-Canadian flavor of society. Compared with the remainder of Canada there is a higher percentage of native Indians and people of Asian origin. French-Canadian representation is low. One anomaly created by this definition is that Americans are not distinguished as an ethnic group because obviously they have the same mother tongue as Anglo-Canadians. On the basis of other variables, such as level of income, religious affiliation, and occupational grouping, some differentiation would be possible between Canadians and Americans.

The number and sizes of ethnic groups are the outcome of the interplay of many complex elements which include "push" factors (population and economic pressure, oppression), pull factors (awareness, government policies, relatives) and the effect of intervening distance. The main events in the evolution of the ethnic "mix" of British Columbia are shown in Table 4,14. The complexities of immigration, government policy and race relations are beyond the scope of this chapter, however, there are several detailed publications which deal with the topic.[24]

The result of a century of European settlement is a mainstream Anglo-Canadian culture based on natural resources extraction and processing tied into the national and world economies. The lifestyle is predominantly urban, salary or wage earning, institutionalized, consumer oriented and

Table 3,14 MAJOR ETHNIC GROUPS BY MOTHER TONGUE IN PERCENTAGE

	1931	*1981*
BRITISH	71.0	80.0
OTHER EUROPEAN	16.0	10.0
NATIVE INDIAN	4.0	4.0
CHINESE	3.9	2.7
FRENCH CANADIAN	2.0	1.6
INDO-PAKISTANI	0.2	1.5
JAPANESE	3.2	<1.0

SOURCE: Census of Canada, 1931, 1981.

heavily influenced by the popular culture of the United States. The impact of an ethnic group within the mainstream is related to its size, growth rate, power, length of residence, assimilation and distance from the majority in cultural terms. The remainder of the chapter will examine the regional and urban place associations, as well as activity associations of various ethnic groups. Table 4,14 summarizes the ethnicity-place associations at the provincial scale, illustrating where particular groups have contributed, and in some instances still are contributing to the meaning of the landscape.

Regional Ethnicity

NATIVE INDIANS

At the time of European contact in the nineteenth century the original inhabitants of the region had a relatively stable village society, with an economy based on fishing, hunting, collecting and some horticulture. Technologically it was a stone age society, rich in artistic and oral traditions.[25] Significant ethnic variety existed within the broad European term of 'native Indians' (Figure 3,14) which partly explains why a united opposition to the Europeans was not possible.

The region, one of the last in North America to be colonized by Europeans, saw few conflict situations between whites and native Indians

Table 4,14 ETHNICITY TIMELINES

	NATIVE INDIAN	*EUROPEAN*	*CHINESE*	*JAPANESE*	*EAST INDIAN*	*FEDERAL POLICIES*
		Spanish Exploration	Chin. Exploration(?)			
		English Exploration				
1850	First Treaties	British Colonization Begins				
			First Immigrants Gold Rush			
1860		Gold Rush: American & Europeans				
	Missionary Contact			First Immigrants		
1870	Denial of Title (Trutch)	Joins Confederation				
	Committee on Reserves					
1880			Railway Laborers			
	Potlatch Outlawed		Head Tax $50			
1890		Rail link to East & Eastern Canadians	Vancouver Riot			
		Italians Doukhobors Scandinavian	Federal Head Tax $100			
1900		Utopian Settlements	Head Tax $500 Reduces Entry	Informal Restriction	First Immigrants	Report on Immigration
			Attacks on Chinatown	Vancouver Riot	Denied Vote	
				Issei Wave		Immigration Act Continuous Passage Clause
1910	Royal Commission			Picture Brides		
		Anti-German Riots	Riot v Chinese in Mines		Komagaa Maru	
				Nisei Wave		
1920	Veteran Indians Vote		Immigration Act: No Entry	Licences for Entry		Duff Commission
		Greeks		Fishing Licences Reduced	Illegal Entry Only	
		Mennonites				
1930	Native Indian Brotherhood			No Picture Brides	Married Males' Wives & Children Enter	
		Doukhobors Lose Vote		Veterans Franchise		
		Sudetan Germans				
1940	Native Voice Published Franchise			Forced Evacuation	Immigrant Status	Citizenship Act
		Dutch	Act Repealed Franchise	Repatriation Franchise (Fed) Franchise (Prov)		
1950	Anti Potlatch Dropped					
	Relaxed Liquor Law	Doukhobors Franchise Restored		Sansei Wave		
		Hungarian Refugees	Relatives Eligible to Enter			
1960	Indian Act Amended	Doukhobor Violence	Internal Conflicts (Mainland v Taiwan)		Fijian East Indians	No Discrimination
		American Draft Resisters				
1970	BCANSI Union of Chiefs Alliance	Czech Refugees				Multiculturalism
1980	Land Title Claims Renewed	Central & South American Refugees				
1985		Polish Refugees		Discussions on Wartime Compensation	Internal Conflicts	

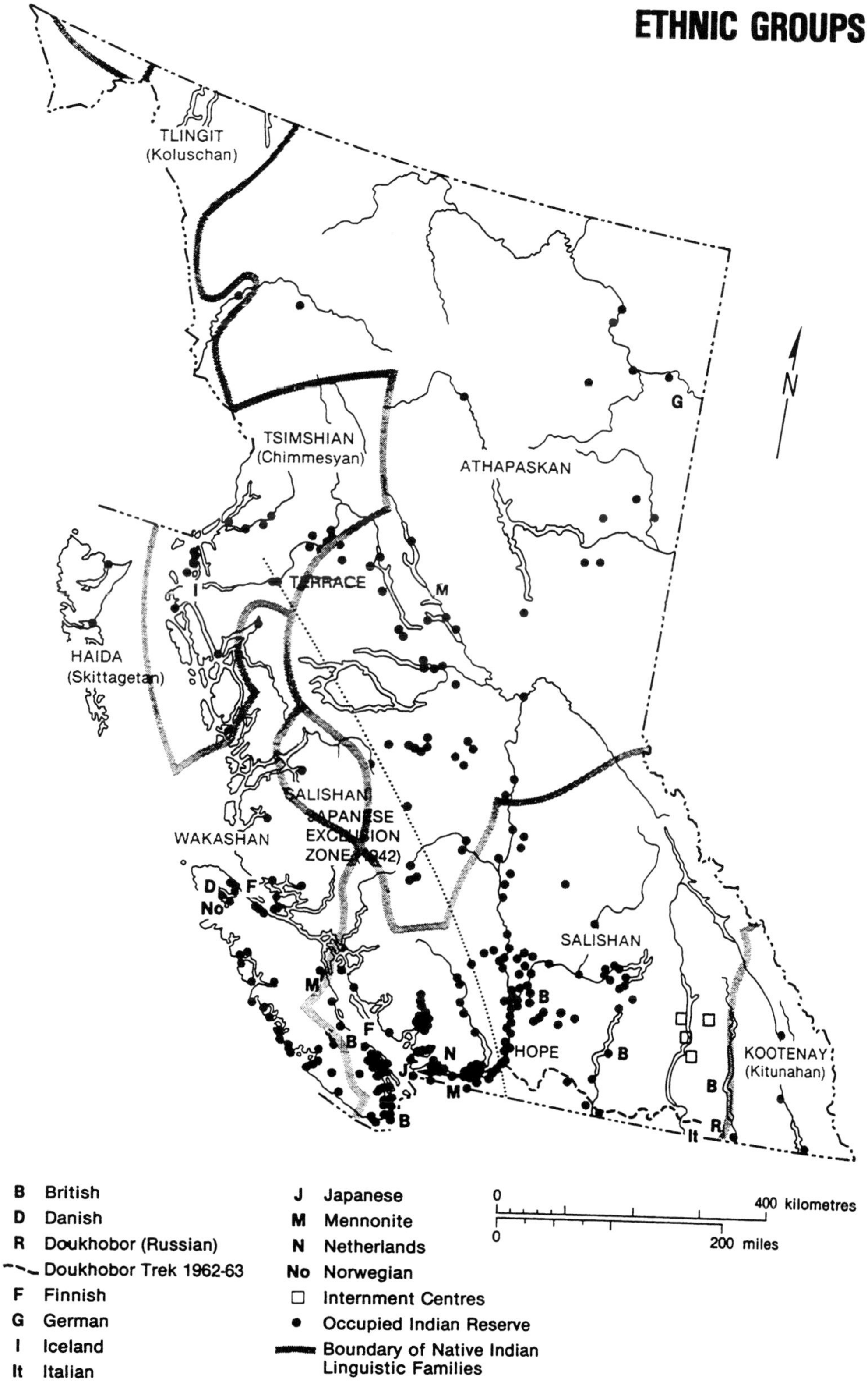

FIGURE 3,14 A summary of ethnicity-place associations

occur, mainly because the Indians, whose numbers were relatively small, were located in riverine and often isolated coastal communities, while the Europeans, whose population increased quickly, sought minerals, lumber and grazing land. Weakened by introduced diseases, the native Indians were no match for the settlers or the colonial government who assumed the title to land and denied Indian land title rights. The only treaties signed were on Vancouver Island (1850-1854) and in the northeast (Treaty Number 8, 1900-1914). A government committee designated Reserve Lands (1876-1906) which basically meant specifying the land in the vicinity of Indian villages as native land[26] and vesting the remainder with the Crown.

The place association of the native people today is the archipelago of Indian reserves which are scattered across the province. The west coast north of the Georgia Strait urban region, the offshore islands and the northwest region are generally areas where native society predominates and white settlement has been minimal. The contemporary place association is therefore a much shrunken version of the former tribal lands. Reserve residents continue many traditional activities associated with their original locations, but in a modernized, modified and spatially truncated form. The toponymy of the province gives a good indication of the palimpsest of Indian place name association which shows through the broad Anglo-Canadian text. One interesting Euro-Indian landscape hybrid phenomenon which has developed is the Indian Rodeo.[27]

EUROPEAN MINORITIES

It was seen earlier how European settlement evolved as the resources of the province were developed. The Europeans introduced a resource export economy with company towns, coastal settlements which evolved port functions, and a dispersed rural farm pattern as the region became integrated with the market economy-based world of Anglo-American capitalism. The map of ethnic British Columbia indicates both the mainstream pattern and exceptions to the general pattern in the place association of European minorities (Figure 3,14).[28]

Scandinavians

Danish (Holberg), Norwegian (Bella Coola) and Finnish (Sointula) coastal communities were established in the late nineteenth and early twentieth centuries, often with a strong utopian and co-operative spirit. Most of these communities, in the face of isolation, failed and subsequently have disappeared, with the descendants of the pioneers migrating and integrating with the mainstream society.

Doukhobors (Russian)

Early in the present century Doukhobors migrated to southeastern British Columbia, mainly to the Kootenay Valley, and established a very distinctive form of communal settlement. Internal feuds and conflict with the government have led to violence and protest. In 1962-63 events came to a climax and some leaders were imprisoned at Agassiz which led to a trek of followers to the prison. In recent years agitation has diminished. While the traditional settlements have been abandoned, many of the remaining group traits have been maintained.[29]

German Speaking Mennonites and Sudetans

German speaking Mennonites who immigrated from the Soviet Union in the 1920s began establishing communities in British Columbia in the late 1920s, particularly in the Yarrow and Abbotsford areas of the Fraser Valley. Settling on small farms, they gradually established a prosperous farming community.[30] While the group maintains many of the cultural traits of Mennonite society, they have assimilated in landscape terms because, unlike the old order Mennonites of the East, they have adopted many of the cultural practices and technology of the majority culture.

A second German speaking group with a distinct place association must also be mentioned. In 1938 about 600 refugee Sudetan Germans settled in the Peace River region. The remarkable feature of these pioneers was that they were primarily from urban professional and industrial occupational backgrounds, yet they were pioneering one of the most physically demanding frontiers in North America.[31]

Dutch

In the postwar period Canada received many immigrants from the Netherlands. Dutch families settled especially in the Lower Fraser Valley in the Pitt Polder area. Many revived the formerly depressed dairy farming in the region and also invigorated the horticultural sector.[32] There has been a noticeable landscape impact, both in terms of the quality of the farming and the neatness and care given to the farms' appearance.

ASIAN MINORITIES

It was shown in Table 4,14 that Asian immigration into the province has a very tangled history. For many years the immigrants were single males. The Chinese (see Chapter 15) provided labor for railway construction and also had an historical association with gold prospecting. Generally, there was a small community of Chinese in each evolving mining town throughout the province. Although the Chinese have been active in peri-urban market gardens, their landscape association tends to be more oriented to the urban region, both in Chinatowns and corner stores.

Figure 3,14 illustrates the place associations of the Japanese ethnic group who initially had a coastal location and were involved in the fishing industry and, to a lesser extent, market gardening. A distinct Japanese community evolved at Steveston and smaller Japanese communities existed in other coastal towns. Fears of a Japanese invasion in 1942 led to a forced evacuation of all Japanese to interior camps and to eastern Canada: in one stroke the community was smashed and its capital sequestrated.[33]

East Indian immigration has had a similarly checkered history, as seen in the infamous Komagata Maru incident in 1914 when a boatload were denied entry. A significant percentage of East Indians have been Sikhs from the Punjab who provided labor for the forest products industry and agriculture. Many settled close to the sawmills of False Creek and later, as the locus of this industry shifted to the southern part of Vancouver along the Fraser River, they moved there also. Recently, this group has been troubled with internal conflict over the Sikh struggle for independence in India.

Urban Ethnicity

Just as ethnic minorities have given a colorful diversity to the broad, regional cultural landscape, so too their impact and local place association can be appreciated at the neighborhood urban scale. Generally, these groups have impacted upon the texture and detailed architectural form of the landscape and its individual buildings. This is most noticeable in ethnic shops and restaurants, religious and community centres and in the outward use, maintenance and decorative detail of individual buildings.

As an example of this ethnic variety, the main concentrations of ethnic place associations for Vancouver are shown in Figure 4,14. Other ethnic groups have also located in the metropolitan area, but are either dispersed

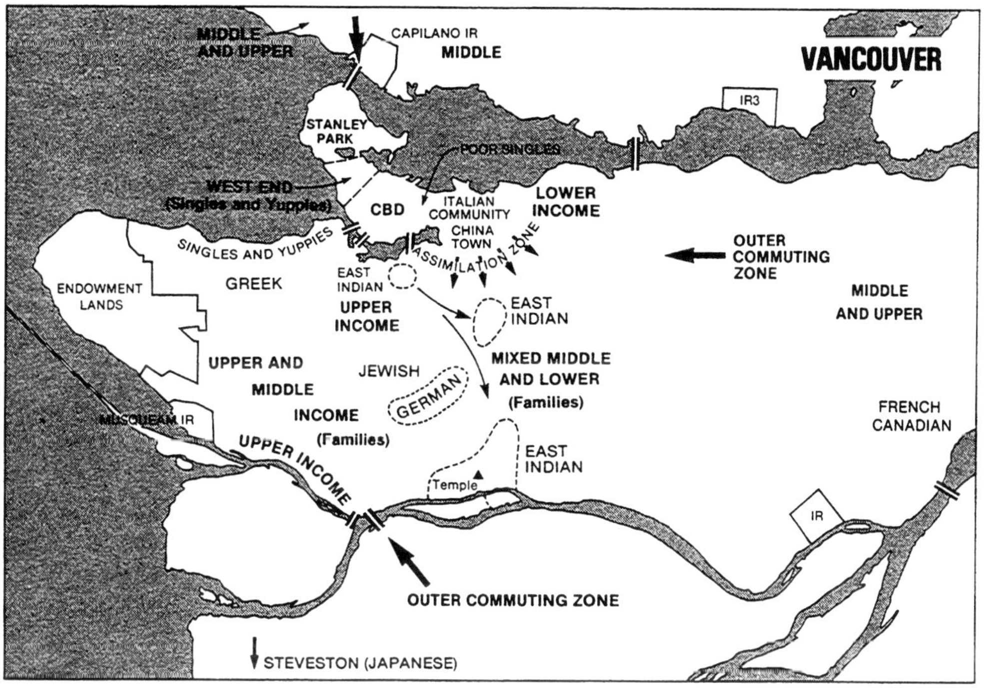

FIGURE 4,14 Ethnic place associations in Vancouver

or insufficient in numbers to have a landscape impact.[34] The contribution of each group to the detailed texture of the urban landscape is summarized in Figure 4,14.

SUMMARY

British Columbia is a growth province with a continuous stream of migrants and immigrants eager to improve their personal wealth or, conversely, to live out their sunset years in a pleasantly mild and attractive environment. The settlement geography consists of an urban core in the southwest, a periphery of scattered interior and coastal communities, and an empty northern region. The quality of life enjoyed by the majority of its inhabitants is relatively high, although there are distinct anomalies in several of the ratings for specific indicators. Parallelling the majority is a minority, mostly native Indian, but also single elderly who are at the other end of the quality of life spectrum.

Immigration to the province has not been without turmoil, as minority groups, in addition to the native Indians, sought access and equality in the face of British and later Anglo-Canadian dominance. The racism of the earlier years has been giving way to a measure of equality and tolerance. Many individuals from specific minority groups have risen to economically and even politically powerful positions. The cultural landscape has acquired some cosmopolitan features, giving contrast to the more sober Anglo-Canadian norm.

REFERENCES

1. CANADA, *Census of Canada*. Ottawa: Statistics Canada, 1976 and 1981.

2. FARLEY, A.J. (ed.), *Atlas of British Columbia*. Vancouver: University of British Columbia Press, 1979, pp. 2-20; SIEMENS, A.H., "Settlement" in ROBINSON, J.L. (ed.), *British Columbia*. Toronto: University of Toronto Press, Studies in Canadian Geography, 1972, pp. 9-31.

3. For a detailed history of the province see ORMSBY, M.A., *British Columbia: A History*. Toronto: MacMillan, 1958.

4. CANADA, *Census of Canada*. Ottawa: Statistics Canada, 1981.

5. For example, B.C. RESEARCH COUNCIL, *B.C. Population Projections 1974-1996*. Vancouver: B.C. Research Council, 1974; LOWER MAINLAND REGIONAL PLANNING BOARD, *Population Trends 1921-1986*. New Westminster: Lower Mainland Regional Planning Board, 1968.

6. For example, see FOOT, D.K., *Canada's Population Outlook*. Ottawa: Canadian Institute for Economic Policy, 1982; BEAUJOT, R. and McQUILLAN, K., *Growth and Dualism*. Toronto: Gage, 1982, pp. 111-136.

7. CANADA, *Population*. Ottawa: Statistics Canada, annual.

8. FOOT, D.K., *op. cit.*, p. 41.

9. *Ibid.*, p. 94.

10. Compare the message of SCHUMACHER, E.F., *Small is Beautiful*. London: Sphere Books, 1974, with the reality of FUCHS, V.R., *How We Live: An Economic Perspective on Americans from Birth to Death*. Cambridge: Harvard University Press, 1983.

11. "Progress on Social Indicators", *O.E.C.D. Observer*, No. 85, March 1977, pp. 24-27.

12. CANADA, *Mortality*, vol. III, annual; *Hospital Statistics*, annual; *Mental Health Statistics*, annual. Ottawa: Statistics Canada; CANADA, *Mortality Atlas of Canada*. Ottawa: Health and Welfare, 2 vols., 1980.

13. CANADA, *Education*, annual; *Expenditures on Education*, annual. Ottawa: Statistics Canada; SIGGNER, A. and LOCATELLI, C., *An Overview of Demographic, Social and Economic Conditions Among British Columbia's Registered Indian Population*. Victoria: Province of British Columbia, 1980.

14. CANADA, *Labor Force Annual Averages*, occasional publication, 1983; *Mortality*, annual. Ottawa: Statistics Canada.

15. CANADA, *Family Incomes*, annual; *Income Distribution*, annual. Ottawa: Statistics Canada. Note that family income statistics are based on a sample which excludes Indian reservations. Consequently, the average incomes are too high, especially for British Columbia.

16. CANADA, *Canada's Special Resource Lands*. Ottawa: Environment Canada, Map Folio No. 4, 1979, pp. 68-104. Note that British Columbia is rated high scenically, but has lower than average cottage ownership.

17. See, for example, JONES, T., *Wilderness or Logging?* Vancouver: Federation of Mountain Clubs of British Columbia, 1983.

18. CANADA, *Crime and Traffic Enforcement Statistics*, annual. Ottawa: Statistics Canada.

19. *Vancouver Sun*, July 29, 1983, p. 1.

20. GIFFEN, P., "Official Rates of Crime and Delinquency" in McGRATH, W.T. (ed.), *Crime and Its Treatment in Canada*. Toronto: MacMillan, second edition, 1976. This may be reflected in the urban-rural ratio of 489:1,624 in British Columbia compared with the national ratio of 564:553 for juvenile court appearances. Note also that average incarceration rates in Canada are 111 per 100,000; in United States 270; and in United Kingdom 87.

21. BRITISH COLUMBIA, *Annual Report*. Victoria: Ministry of the Attorney General, Corrections Branch, 1951-1983.

22. See, for example, ANDERSON, A.E., *Ethnicity in Canada*. Toronto: Butterworths, 1981.

23. See, for example, SIEMENS, A.H., *op. cit.*, reference 2; MINGHI, J. (ed.), *Peoples of the Living Land; a Geography of Cultural Diversity in British Columbia*. Vancouver: Tantalus Press, B.C. Geographical Series, No. 15, 1972; WAGNER, P. (ed.), *Ethnic Vancouver*. Vancouver: Simon Fraser University, Department of Geography, Dissertation Monograph, No. 16, 1981.

24. MUNRO, I., *Immigration*. Toronto: Wiley, 1978; WARD, W.P., *White Canada Forever*. Montreal: McGill-Queens University Press, 1978.

25. See DUFF, W., *The Indian History of British Columbia*. Victoria: Provincial Museum, Vol. I, 1964; FARLEY, A., *op. cit.*, pp. 6-11.

26. See LAVIOLETTE, F.E., *The Struggle for Survival*. Toronto: University of Toronto Press, 1961; BERGER, T., "Native History, Native Claims and Self Determination", *B.C. Studies*, Vol. 57, 1983, pp. 10-23.

27. Ethnographers have been attempting to piece together a more complete Indian place name geography of the province. See, for example, McDONALD, J.A., "An Historical Event in the Political Economy of the Tsimshian: Information on the Ownership of the Zimacord District", *B.C. Studies*, Vol. 57, 1983, pp. 24-37.

28. NORRIS, J., *Strangers Entertained: a History of Ethnic Groups in British Columbia*. Victoria: Centennial Commission, 1971, pp. 126-138.

29. TARASOFF, K.J., *Pictorial History of the Doukhobors*. Saskatoon: Modern Press, 1969.

30. KRAHN, J.J., *A History of the Mennonites in B.C.*. Vancouver: University of British Columbia, Department of Medicine, M.A. Thesis, unpublished, 1955.

31. AMSTATTER, A., *Tomslake: History of the Sudetan Germans*. Seattle: Hancock, 1978.

32. VANDERMAY, A., *To All Our Children*. Jordan, Ontario: Paidea Press, 1983.

33. WARD, P., *op. cit.*, reference 24, pp. 97-166; MARLATT, D., *Steveston Recollected*. Victoria: Provincial Archives, 1975.

34. Based on WAGNER, P., *op. cit.*, reference 23; PATTERSON, G.J., *The Greeks of Vancouver*. Ottawa: National Museum, Canadian Centre for Folk Culture Studies, Paper No. 18, 1976; CHADNEY, J.G., *The Sikhs of Vancouver*. New York: AMS Press, 1984; NORRIS, J., *op. cit.*

PLATE 21 A diversity of ethnic groups represented. *J. Ryan Photo*▶

PLATE 22 Chinatown in Vancouver. *B.C. Government Photo*▶

yuen
fong
PORK
FRESH MEAT

15 CHINESE COMMUNITIES

David Chuen-yan Lai
University of Victoria

It is still unknown whether or not the Chinese had come to explore the Pacific coast of North America as early as the fifth century A.D. It is recorded in the *Liang Shu*, (history of the Liang Dynasty) (502 - 556 A.D.) that in his exploration of the eastern sea a Chinese monk called Huishen discovered three countries east of Japan, namely, Wenshen, Dahan, and Fusang, the last one being about 32,000 li away (equivalent to 18,400 kilometres or 11,435 miles).[1] The people of Fusang had developed a written language and made paper from the bark of the Fusang trees, from which the nation derived its name.[2] It is still unknown where Fusang was, but a French sinologist, Stephen De Guignes, propounded an hypothesis in 1761 that the Chinese monk might have crossed the northern part of the Pacific Ocean and eventually reached the southern Californian or Mexican coast where Fusang was situated. His hypothesis triggered off debates in the past 200 years on whether or not the Chinese monk had reached the Americas 1,000 years before Columbus. The 1975 discovery of "stone anchors" in shallow waters off the Palos Verdes Peninsula in California kindled interest in the old debate again, as several American scholars believed that the anchors were Chinese in origin.[3] As early as 1882, it was reported in the *Colonist*, a local newspaper in Victoria, that about 30 Chinese brass coins were dug out of a creek in the Cassiar gold mining district by a party of miners.[4] The coins were brought to Victoria and examined by James Deans, an archaeologist. He found markings of a Chinese chronological cycle on the coins, but he did not date them.[5] The local Chinese people had also examined the coins and said that they were about 3,000 years old (1200 B.C.). This discovery might constitute evidence of ancient migrations between Asia and British Columbia long before Captain James Cook explored the Pacific coast of Canada in the 1770s. On the other hand, it is also possible that these ancient coins were left behind by the Chinese miners who panned for gold in the Cassiar district during the 1870s. The earliest written evidence of the appearance of Chinese in British Columbia is found in Captain John Meares' memoirs of his voyages from China to the northwest coast of America in 1788 and 1789. These early Chinese arrivals were contracted laborers and not immigrants. The real migration did not occur until the Fraser River gold rushes in the 1850s and 1860s. The Chinese immigration and settlement patterns in British Columbia can be studied under five periods, namely, pre-immigration (1788-1857); free entry (1858-1885); restricted entry (1886-1923); exclusion (1924-1947); and selective entry (1948-present).

PRE-IMMIGRATION, 1788-1857

In January, 1788, Captain John Meares recruited 50 Chinese sailors, smiths and carpenters at Macao and Canton and shipped them to an Indian village at Friendly Cove in Nootka Sound on Vancouver Island where they helped build a small fortress, a dockyard and a 40-ton schooner, namely the *North West America*, which was the first ship to be launched in British Columbia.[6] In the following year Captain Meares recruited at Canton another 70 Chinese laborers and shipped them to Nootka Sound as well. The Chinese workers were praised for their obedience, diligence and willingness to work hard. Their performance had impressed the British captain so much that he wrote in his memoirs that "if hereafter trading posts should be established on the American coast, a colony of these men should be a very important acquisition".[7] His remarks were prophetic; the supply of Chinese labor was almost indispensable in the opening up of the virgin land of British Columbia.

The duration of stay of the Chinese laborers in Nootka Sound was very brief. In the summer of 1789 the Spaniards under the command of Captain Martinez arrived at Nootka Sound, took possession of it, seized Meares' vessels, and took his Chinese crew as prisoners. They recorded their names, the list including such names as Jinfo, Allon, Ammey, Ahop-ha, Accog, Accan, and Accah. This was probably the first "census" of Chinese in Canada!

What happened to the 120 Chinese laborers afterward is still a mystery. It was believed that some of them might have been shipped to San Blas in Mexico and forced "to enter into the service of Spain".[8] Some might have returned to China because a large number of sailors captured off Meares' vessels were sent back to China in the *Columbia*, an American vessel, which left Nootka Sound about the end of July, 1789. It was also mentioned in Captain Meares' memoirs that a party of people, including the captured Chinese crew, was heavily guarded by the Spaniards on their way to Gold River, east of Nootka Sound, to search for gold. The log of an American vessel, *Jefferson*, recorded that its captain picked up one of the Chinese when it called at Nootka Sound in 1793.[9] A report of the Hudson's Bay Company at Fort Nisqually also mentioned that two Chinese were held prisoners by the native Indians, and that some Chinese came to Fort Nisqually from Fort Vancouver in 1834.[10] These records revealed that it is possible that some Chinese had escaped from their Spanish captors. For nearly 70 years after the so-called Nootka Sound Controversy between the British and Spanish there was no written evidence of further Chinese arrivals in British Columbia.

FREE ENTRY, 1858-1885

Chinese immigration started in earnest in 1858. The year before, gold had been discovered in the lower Fraser valley. The gold rushes that followed were as important to the future of British Columbia as the discovery of gold in 1849 was to California. Thousands of miners, including a few hundred Chinese, joined in the rush by boat from San Francisco through Victoria, or by overland routes from the Washington territory. Soon more Chinese gold-seekers came directly by ship from Hong Kong. The colonial government estimated that in 1862 the total population of the colonies of Vancouver Island and British Columbia, excluding native Indians, stood at about 11,000, of which 7,700 were whites, 2,300 colored (blacks), and 1,000 Chinese.[11] About 300 Chinese lived in Victoria's Chinatown and the rest were concentrated in the lower Fraser valley between Hope and Lillooet. In that year, hundreds of white gold-seekers rushed to Williams, Lightning and other Cariboo creeks. Virtually overnight Barkerville on Williams Creek emerged as the hub city of the Cariboo district. By 1863, the population of the district had soared from a few hundred to over 5,000. The lower Fraser River was deserted by most of the white miners, and only the Chinese remained. The heyday of the Cariboo Gold Rush was soon over after its shallow diggings were exhausted toward the end of 1864. In the following year, the Cariboo population dropped to less than 1,400 because most of the miners had rushed wildly to the Big Bend of the Columbia, again leaving the Chinese behind. The Chinese miners did not, as the white miners did, desert a place immediately at the next rumor of better diggings. They would remain on the abandoned bars where diggings might pay only two dollars a day because they were satisfied with such a small return which would be scorned by the white miners. As a result, Chinese miners were to be found in every mining area throughout British Columbia.

During the gold rushes, most of the white workers were so impatient to become rich quickly in gold mining that, unless driven by sheer necessity, they would hesitate to take up jobs in the public service. Accordingly, the colonial government had to rely on Chinese laborers for clearing forests, building trails and wagon roads, digging ditches and other back-breaking jobs. Hundreds of Chinese laborers were employed to open the Douglas-Lillooet trail and build the Lillooet-Alexandria wagon road through the dense forests. Chinese quarters were found in the road houses, such as 70 Mile House, 100 Mile House and 150 Mile House, which were designated according to the distances from Lillooet. Chinese laborers were also employed to construct the famous Cariboo Road from Yale to Barkerville, the Dewdney Trail from Hope to Wild Horse Creek in the Kootenay district, and the Collins Overland Telegraph Line from New Westminster to Fort Stager, a supply depot near Hazelton in the Skeena region. The Chinese people seemed to be ubiquitous, as their makeshift camps were found all along these routes. In fact, they were the largest visible minority in the united colony of British Columbia (the colonies of Vancouver Island and British Columbia were united in 1866). In 1868, for example, the colony's population, excluding native Indians, stood at 8,410, of which 6,408 were whites, 1,759 Chinese and 243 colored (blacks).[12] About 40 percent of the Chinese were found in the Cariboo and the main concentrations were around Quesnel and Barkerville (Table 1,15). Victoria's Chinatown functioned as a distribution centre for Chinese merchandise destined to the miners on the mainland.

Generally speaking, the migration of Chinese during the gold rushes followed two main directions. One stream of migration ascended the Fraser River and dispersed northward as far as the northern border of the colony. The other followed the Dewdney Trail from Hope and spread out eastward to the valleys of the Similkameen, Rock Creek and Kootenay. Each migratory wave of Chinese followed closely on the heels of the white miners. Pockets of Chinese settlement were found along these two migration routes. Although it is impossible to map the exact locations, traces of their existence can be detected from the place names of British Columbia (Table 2,15). For example, at least six streams (near Princeton, Lillooet, Robson, Osoyoos, Quesnel and Alberni) are called "China Creek" because the early Chinese miners had concentrated there for gold panning.[13] Many other streams, lakes and mountains in the mining districts also have names which are reminiscent of the early Chinese miners (Figures 1,15 and 2,15).

After the gold rushes of the 1860's were over the colony was in financial difficulties and many miners and laborers were unemployed. Although a large number of Chinese miners still remained in the gold fields on the mainland, many drifted back to Victoria to seek jobs, and many others tried to find employment on farms and ranches or in coal mines. When British Columbia entered Confederation in 1871 its total population was estimated at about 54,480, of which 45,200 were native Indians, 7,600 whites, 1,240 Chinese and 440 colored (blacks).[14]

Table 1,15 ESTIMATED CHINESE POPULATION IN THE UNITED COLONY OF BRITISH COLUMBIA, 1868

Location	*Number of Persons*
CARIBOO DISTRICT	736
LOWER FRASER BETWEEN HOPE AND LYTTON	465
OSOYOOS	197
KOOTENAY DISTRICT	107
MIDDLE FRASER BETWEEN LILLOOET AND SODA CREEK	101
CITY OF VICTORIA	100
NANAIMO	31
NEW WESTMINSTER	32
TOTAL	1,759

SOURCE: United Kingdom, *Blue Books of Statistics, Colony of British Columbia, 1868*. London: Colonial Office, 1868, p. 210.

The second wave of Chinese immigration was associated with the building of the Canadian Pacific Railway in British Columbia during the early 1880s. A railway contractor named Andrew Onderdonk could not find enough white workers locally and had to bring in men from San Francisco. To his dismay, most of the workers who arrived were unskilled and ill-prepared for the hard work they had to face. On the other hand, there was a public outcry that Chinese laborers should not be employed to build the railway. It was simply a question of alternatives, as Prime Minister John A. Macdonald told Parliament in 1882: "either you must have this labor or you cannot have the railway."[15] Accordingly, Onderdonk decided to recruit from the United States the Chinese workers who were formerly engaged in the construction of the Northern Pacific Railway in Oregon and the Southern Pacific Railway in California. In addition, he chartered two ships in 1881 to bring about 2,000 Chinese coolies from Hong Kong.[16] During the long, rough passage across the Pacific they were kept below decks with the hatches closed and given poor food. Lacking in fresh vegetables and fruit, a large number developed scurvy and nearly 10 percent of them died within a short

Table 2,15 SOME CHINESE PLACE NAMES OF GOLD MINING DISTRICTS

District	*Place Names*
LOWER AND MIDDLE FRASER	CHINA BAR, CHINA BAR BLUFF, CHINA FLAT, CHINA GULCH, CHINA HEAD MOUNTAIN
UPPER FRASER AND CARIBOO	AHBAU, AHBAU CREEK, AHBAU LAKE, AHBAU LANDING, AHBAU STATION, CHINA BLUFF, CHINAMAN CREEK, CHINESE CREEK
SIMILKAMEEN	CHINA BUTTE, CHINA RIDGE
KOOTENAY	CHINESE WALL
PEACE RIVER	CHINAMAN LAKE
VANCOUVER ISLAND	CHINA BEACH
CASSIAR	AH CLEM CREEK, AH CLEM LAKE, AHLUK CREEK AHLUK LAKE

SOURCE: Lai, C-Y.D., "Chinese Imprints in British Columbia", *B.C. Studies*, 39, 1978, p. 27.

time after their arrival. More coolies had to be recruited in order to meet the labor shortage. This continuing influx of Chinese workers led to an increase of the Chinese population in British Columbia from 1,240 in 1871 to 4,350 in 1881.[17] As the railway was progressing more Chinese workers were recruited. Between 1881 and 1884, a total of 15,701 Chinese entered the Port of Victoria from China, San Francisco and other places.[18] Many of them remained in British Columbia and helped build the railway. By 1884, the Chinese population in the province had soared to about 10,500.[19]

A study of the distribution of the Chinese population in 1884 revealed that the route of the railway from the Fraser Canyon to Craigellachie was lined with tents where nearly 4,000 Chinese workers were housed.[20] Elsewhere there were still hundreds of Chinese patiently working on the gold fields in the Cariboo around Quesnel Forks, Chist Creek and Stanley; in the Fraser valley from Hope to Quesnelmouth; and in the Thompson

FIGURE 1,15 Cariboo Road, China Bar Bluff. ▶

FIGURE 2,15 Ahbau Creek, Ahbau Lake and Ahbau Station on the British Columbia Railway north of Cottonwood Provincial Forest were named after the Chinese miner named Ahbau (centre).

Valley around Cache Creek, Savonna's Ferry and Kamloops. A small number had reached the northern boundary of British Columbia in the Cassiar gold field. About 300 Chinese laborers were employed as fish hands in the Skeena valley. In Wellington and Nanaimo, where most of the Chinese worked in the coal mines, Chinatowns were established on the properties of the mining companies. In Burrard Inlet, about 100 Chinese settled around Port Moody, Hastings and Granville. The city of Vancouver had not yet been incorporated, and its predecessor, the small village of

Granville, had only one Chinese grocery store and two laundrymen. The largest concentrations of Chinese were in the Chinatowns of Victoria and New Westminster which had communities of over 1,000 residents each.

During the free entry period nearly all of the Chinese who came to Canada settled in British Columbia. In 1881, for example, out of the 4,383 Chinese in Canada 4,350 lived in British Columbia, 22 in Ontario, 7 in Quebec and 4 in Manitoba.[21] In other words, over 99 percent of the Chinese in the country were concentrated in British Columbia. However, the Chinese remained a small minority within the province, accounting for only 8.8 percent of the population of 49,459. In spite of this, they were still the largest and most visible ethnic minority and were made the scapegoats when jobs were scarce, unemployment increased and the economy slackened.

RESTRICTED ENTRY, 1886-1923

When the heyday of the gold rush was over after the mid 1860s, economic recession set in and jobs were scarce. Many Chinese laborers were unemployed. With starvation staring them in the face, they entered almost every type of employment, as laborers in logging camps and coal mines; on ranches and farms; and in the fish canning industry. As they were willing to work for long hours and take any jobs at low wages, they were viewed by the white laborers as a menace to their livelihoods. Throughout the 1870s there was agitation for restricting the flow of Chinese immigrants into Canada. The simmering anti-Chinese feeling started to boil when the large number of Chinese laborers was recruited to build the railway. The outcry against further Chinese immigration became more clamorous after the American government passed the Chinese Exclusion Act on May 6, 1882, by which Chinese laborers who did not possess certificates of residence were denied entry to the United States for 10 years.[22] As a result, the Chinese railway laborers who had gone to British Columbia from the United States without certificates were barred from returning when the railway was completed. To appease the anti-Chinese movement in British Columbia, particularly, the federal government eventually decided to end its traditional free and non-restrictive policy towards Chinese immigration.

On July 20, 1885, the Chinese Immigration Act was passed by which a head tax of 50 dollars was imposed on every person of Chinese origin entering Canada.[23] Those who were exempted from this head tax were consular officers, tourists, merchants, students and returning Chinese residents who had obtained certificates of leave before they left Canada. Under

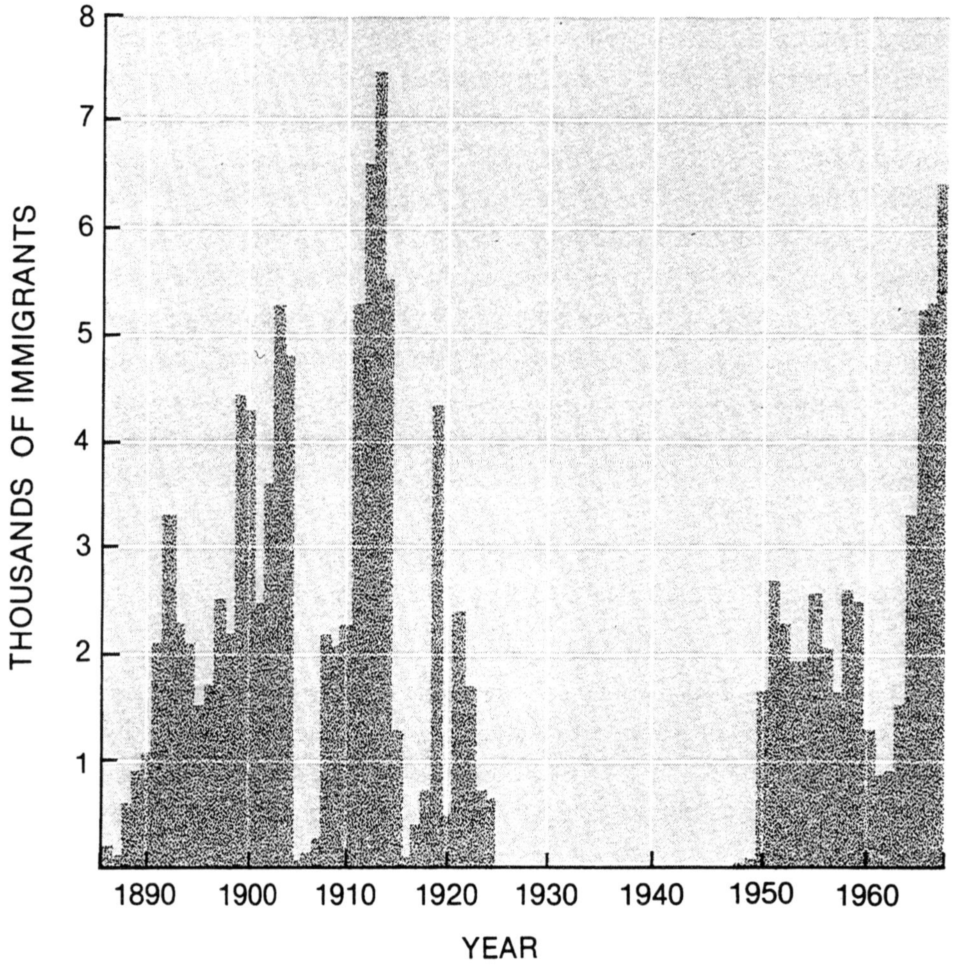

FIGURE 3,15 Annual Number of Chinese Immigrants to Canada, 1886-1967

this Act, any vessel which carried Chinese immigrants was permitted to carry only one immigrant for every 50 tons of its registered tonnage. This measure was used to prohibit companies from chartering vessels to bring large numbers of Chinese immigrants to Canada, a practice by Andrew Onderdonk at the beginning of the railway construction that alarmed the local white laborers. In those days a head tax of 50 dollars was very high, as most of the Chinese laborers could earn only 20 to 30 dollars per month. The Act had an immediate effect upon Chinese immigration which dropped in 1886 to 212 from the thousands who were entering during the early 1880s[24] (Figure 3,15). In the following year it dropped to 124, but the effect was short-lived.[25] In the late 1880's the number of immigrants rose continuously and by 1892 it reached more than 3,000. This increase was partly due to the passing of the Scott Act in the United States in 1888.[26] Under this Act, Chinese laborers, whether they had certificates of residence or not, were prohibited from entering the United States. As a result, many Chinese laborers paid the 50 dollar head tax to enter Canada from which they tried to reenter the United States illegally.

Even after the head tax was imposed on Chinese immigrants the demand for their exclusion was not abated. From time to time, petitions were sent to both the provincial and federal governments requesting the prohibition of Chinese immigration because the Chinese laborers had driven "working men of British race and blood out of many of the fields of labor" and their entry to Canada was not in the best interests of the country.[27] The opposition of the host community to Chinese immigration is well documented in the literature, but it is less widely known that local Chinese merchants in Victoria also tried very hard to check the continuous flow of their fellow countrymen into Canada. The correspondence, circulars, minutes and other documentary records of the Chinese Consolidated Benevolent Association of Victoria indicate that the association tried at least three times between 1899 and 1914 to discourage people in China from coming to Canada and urged the Chinese government to restrict emigration.[28] Without the support of the Chinese government, the association's efforts to reduce immigration resulted in frustration and disappointment. Eventually, the head tax was increased to 100 dollars by the Chinese Immigration Act of 1900.[29] As this new head tax was still ineffective in curbing the migration, it was further increased to a near prohibitive amount of 500 dollars by the Chinese Immigration Act of 1903.[30] The effects were immediately felt. The number of tax-paying immigrants dropped sharply from 4,719 in 1904 to 8 in 1905, 22 in 1906 and 99 in 1907.[31] Having recovered from the initial shock, the poor villagers in China still considered it advantageous to come to Canada where a laborer could earn between 20 and 30 dollars a month, instead of only about two dollars a month in China. If he could spend most of his lifetime in Canada, he would eventually save enough to pay the debts for his passage and head tax, and be able to support his family in the village by remittance. Accordingly, the Chinese continued to arrive. The number of tax-paying immigrants rose from 1,482 in 1908 to 7,078 in 1913.[32] During the years of World War I, immigration dropped to less than 1,000 per year, partly because the "Empress" ships which plied between Asia and Canada were requisitioned as troop ships, and trans-Pacific travel was temporarily reduced. After the war immigration soared again into the thousands. As the heavy head tax failed to solve the problem of the Chinese influx, and the Canadian and Chinese governments could not arrive at a satisfactory solution during their negotiations, the federal government decided unilaterally to exclude the Chinese from Canada in 1923.

During the period of restricted entry, the spatial distribution of Chinese settlement had undergone significant changes. First of all, the City of Vancouver, incorporated in 1886, became the Pacific terminus of the Canadian Pacific Railway and soon replaced Victoria as the gateway to Canada from Asia.[33] Its Chinatown, which was established along Dupont (now Pender Street) and around its intersections with Shanghai Alley, Canton Alley and Carral Street, grew rapidly and eventually outstripped Victoria's Chinatown in population and importance after the turn of the century. In 1921, for example, Vancouver's Chinatown had a Chinese population of over 6,000.[34] Victoria's Chinatown, which was established on Cormorant (now Pandora Avenue) and Fisgard streets between Government and Store streets, had a population of about 3,000 in 1921, although it retained the headquarters of many voluntary Chinese associations and societies.[35] The third largest Chinatown was found in Cumberland, a coal mining town on Vancouver Island, and the fourth in New Westminster where many Chinese were employed in fish canneries and sawmills and on nearby farms. The last major Chinatown was in Nanaimo where most of the Chinese community worked in the coal mines. These five Chinatowns accounted for half of the 24,000 Chinese in the province in 1921. Throughout the 1920's there were still many Chinese panning patiently for the precious flakes of gold on abandoned river bars in isolated places in the Cariboo district.

EXCLUSION, 1924-1947

After World War I, the economy of the country was dislocated. There were factory layoffs, jobless war veterans, and other problems during the general economic slump.[36] The Chinese people, the largest visible ethnic minority in the province, were again made the scapegoats. They were blamed for having taken jobs from the white laborers. In the summer of 1918, a survey of Chinese employment in the province revealed that out of a total of 5,928, 67 percent were employed in sawmills, shingle and planing mills, logging operations, and coal mines.[37] The survey also revealed that the Chinese employees accounted for over 40 percent of the total employees in shingle and planing mills, lumber yards and logging operations, and over 20 percent in sawmills, stone works, fuel dealers, cereal mills and chemical plants.[38] As the number of Chinese arrivals soared from 560 in 1918 to 4,066 in the following year, this quickened the tempo of the demand for exclusion.[39]

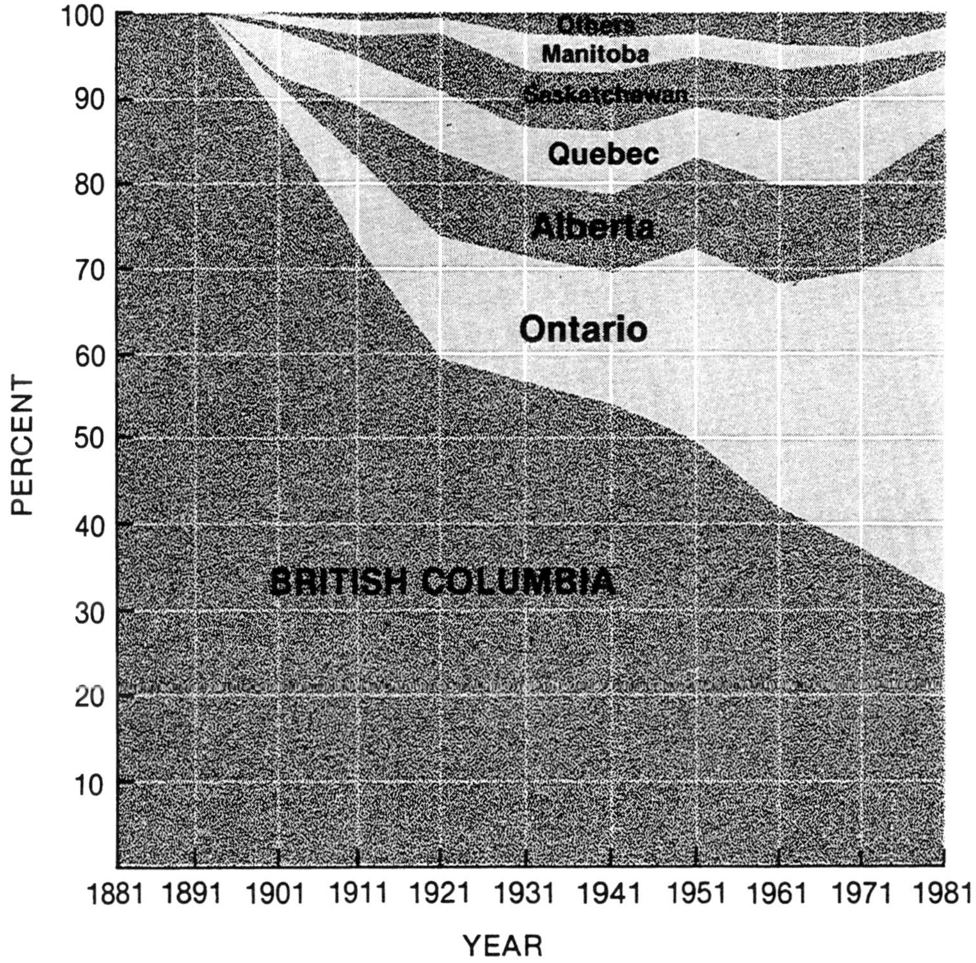

FIGURE 4,15 Provincial Distribution of Chinese Population in Canada, 1881-1981

In the federal election of 1921 Oriental immigration became an issue. During the election campaign, almost every candidate in British Columbia — be he Conservative, Liberal or Progressive — referred to the harmful effects of Oriental competition and promised, if elected, to restrict Oriental immigration.[40] After the election, every one of British Columbia's thirteen members spoke with one voice to warn other provinces that if Canada as a whole did not act quickly the Oriental "menace" would soon spread eastward across the Rockies. It is natural that the outcry against Chinese immigration was noisiest in British Columbia because it had 60 percent of Canada's Chinese population in 1921 (Figure 4,15). MacKenzie King, then Prime Minister, also felt the pressure for excluding the Chinese from Canada. During the debate in Parliament, he stated that "it is not in the interest of Canada to admit to this country large numbers of persons from the Orient", mainly because they had come into competition with white labor.[41] Eventually the Chinese Exclusion Act was passed on June 30, 1923.[42] It prohibited the entry of any person of Chinese origin or descent, except Chinese consular officials, children born in Canada, merchants and students. Even these exempted classes were harassed to the point of exclusion. For example, the term "merchant" was to be defined by whatever regulations the Minister of Immigration and Colonization might prescribe, and "students" referred to those who were going to attend a Canadian university. The Exclusion Act also prohibited Chinese residents from leaving Canada for more than two years. If a Chinese resident failed to return to Canada within two years of his departure, he would not be allowed to enter unless he belonged to one of the exempted classes. After the Act was passed, the Chinese residents whose wives and children were still living in China were not permitted to bring their families to Canada.

After the last group of Chinese arrivals who paid the head tax of 500 dollars in 1924, Chinese immigration to Canada virtually came to a halt. For 23 years between 1925 and 1947 only 12 Chinese entered Canada: 10 belonged to the exempted class and only two were immigrants who were still required to pay the head tax.[43] For more than two decades, many married Chinese men in Canada had been effectively separated from their wives and children and forced to lead a life of celibacy and bachelorhood.

During the period of exclusion, decline in various types of mining operations on the mainland led to depopulation of a large number of small interior towns, from which many Chinese left for employment in big cities such as Vancouver. The decline of the coal mining industry on Vancouver Island after the 1920s also resulted in depopulation and disappearance of Chinatowns. For example, Cumberland had a Chinese community of nearly 900 in 1921, but by 1941 its Chinatown was virtually non-existent, as only 10 elderly men remained there. After the completion of the Canadian Pacific Railway, the Chinese people began to migrate to the East, especially to Ontario. The increase of eastward migration and the lack of new immigrants were major factors accounting for the slow growth of the Chinese population in British Columbia. In 1941, the Chinese population which numbered 34,627 was still heavily concentrated in Vancouver's Chinatown (about 7,000) and Victoria's Chinatown (about 3,000).[44] New Westminster had the third largest Chinese community, with a population of about 400. The Chinatowns in Nanaimo and Kamloops had populations of about 300 each. In the lower Fraser Valley, several hundred Chinese were centred in Coquitlam, Chilliwack, Matsqui and other municipalities where many of them were engaged in agriculture or agriculture-related occupations.

SELECTIVE ENTRY, 1948 - PRESENT

The Canadian government finally changed its Chinese immigration policy in response to changes in the American policy and the growing demands of the Chinese Canadians for equal treatment. The United States government repealed its Chinese Exclusion Act on December 17, 1943, and permitted an annual quota of 105 Chinese immigrants.[45] Less than four years later, on May 14, 1947, the Canadian government repealed its Chinese Exclusion Act.[46] Under the Order-in-Council, P.C. 2115, those Chinese who were Canadian citizens could apply to bring their wives and unmarried children under 18 years of age into Canada to join them. This regulation was still discriminatory against the Chinese. For example, their parents and adult children were excluded from entering Canada under this regulation, which did not apply to Canadians of European extraction. Discrimination against Chinese immigration was still enforced because, as MacKenzie King stated, "the people of Canada do not wish, as a result of mass immigration, to make fundamental alteration in the character of our population. Large scale immigration from the Orient would change that fundamental composition of the Canadian population."[47] Until 1962, the Canadian immigration policy, as implied by the Prime Minister, was still to restrict Oriental immigration.

Throughout the 1950s there were discriminatory restrictions on Chinese immigration. For example, children of Chinese Canadians born of second wives were denied admission.[48] Chinese children seeking admission had to be accompanied by one of their parents. Many Chinese immigrants posed as members of Chinese Canadian families or formed fictitious families in order to meet the requirements. In 1960, the Canadian government introduced the Chinese Adjustment Statement Program, under which all illegal Chinese immigrants who came forward to have their cases reviewed by the Minister of Citizenship and Immigration would be permitted to remain in Canada if they were considered to be of good character. As a result, over 12,000 Chinese had their status adjusted and were allowed to stay in Canada.[49]

In 1962, the Canadian government began a more relaxed policy on Chinese immigration, by which the emphasis on country of origin as a major criterion for admission to Canada was removed. However, there was still some discrimination between Chinese and European immigrants, for example, in terms of the types of relatives who could be sponsored. As a result, the number of Chinese immigrants rose from 876 in 1962 to 1,571 in 1963, 3,210 in 1964, and 5,234 in 1965.[50] In October, 1966, the Citizenship and Immigration Department was replaced by the Department of Immigration and Manpower. Jean Marchand, the minister, presented a White Paper on Canadian immigration policy to Parliament, stating that since the Canadian economy was growing and required more manpower, new immigrants should be absorbed from various countries, and nationality and race should not be used as the major criterion for selecting immigrants.[51] On October 1, 1967, a new set of immigration regulations was introduced which for the first time in the history of Canada treated prospective Chinese immigrants exactly the same as those of all other nationalities and applied a uniform set of criteria for admission, such as level of education, training and skills.

After that year, the immigration statistics no longer identified the ethnic origins of immigrants. Instead, the country of last permanent residence was recorded. In 1967, for example, a total of 6,409 immigrants came from "China", which meant Hong Kong and Taiwan, since emigration from the People's Republic of China was not permitted. It was not until 1973 that Prime Minister Pierre Trudeau and Premier Chou En-lai reached an agreement whereby some Chinese were permitted to leave China to join their families in Canada. Towards the end of the 1970s, China adopted a more liberal policy of emigration which resulted in an increasing number of immigrants coming to Canada. Between 1974 and 1978, for example, the annual average number was about 700, but the number soared to 2,058 in 1979 and 4,936 in 1980, mainly because of the change of Chinese emigration policy.[52]

The postwar immigration has also altered the percentage of Canada's Chinese population in British Columbia, which has been decreasing because a large proportion of the postwar Chinese immigrants went to other provinces. After 1968, the number of Chinese immigrants settling in Ontario exceeded that of British Columbia (Figure 5,15). The 1981 census revealed that Ontario had 118,640 of the total 289,245 Chinese in Canada, versus British Columbia's 96,915, thereby becoming the province with the largest Chinese population (Figure 4,15).[53] The total number of people in British Columbia whose mother tongue was Chinese stood at about 76,000.[54] Nearly 87 percent of them were concentrated in metropolitan Vancouver, and only five percent in metropolitan Victoria. Most of the remaining eight percent were found in other major cities of the province, notably Prince George, Nanaimo, Prince Rupert, Kamloops and Port Alberni.

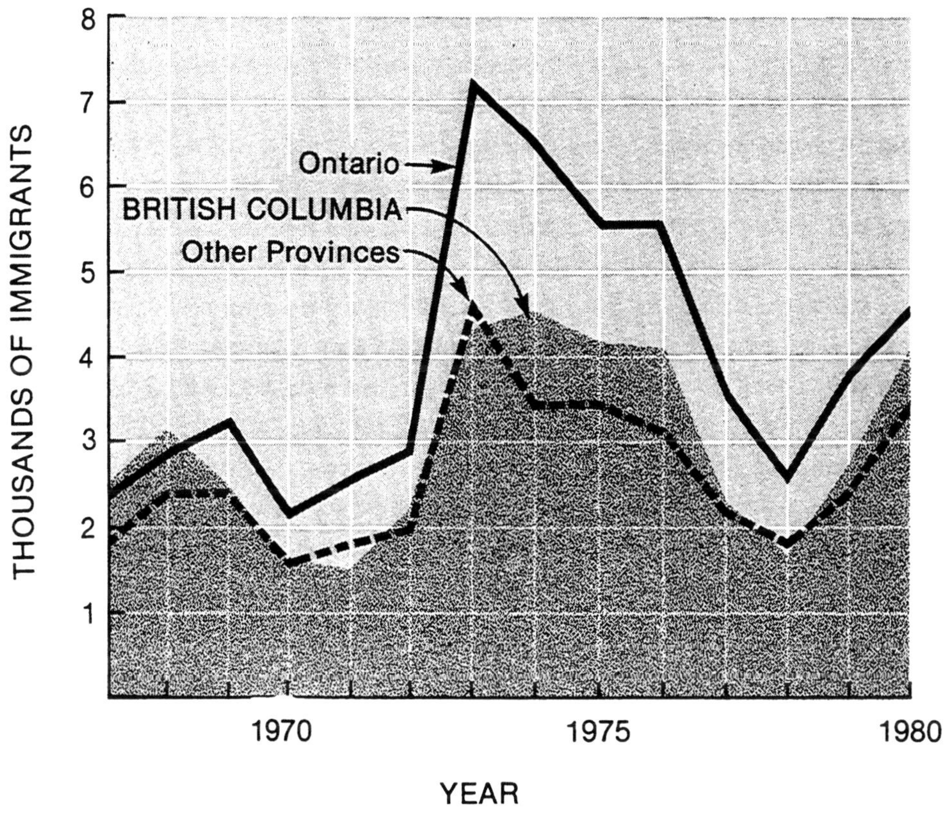

FIGURE 5,15 Provincial Distribution of Chinese Population, 1967-1980

Chinese settlement within the province has undergone significant changes since the end of World War II. A salient phenomenon was the disappearance of Chinatowns in many small towns. The Canadian born Chinese were largely assimilated into the host society and no longer looked to Chinatowns as their refuge. Young families left Chinatowns for better neighborhoods. Throughout the 1950s and early 1960s, the dominant residents of Chinatowns were elderly, single Chinese men. When they passed away their dilapidated tenement buildings were left empty and eventually were wiped out by fire or urban renewal programs. In September 1960, for example, Nanaimo's Chinatown was destroyed by a fire.[55] The few buildings that survived gradually fell into disuse and were eventually demolished. In the case of Cumberland's Chinatown only four elderly Chinese men lived there in 1971 and when they passed away in the late 1970s the already desolate Chinatown reverted to wilderness. Urban renewal also was an important factor in the obliteration of many Chinatowns in the province. In the early 1970s Duncan's

Chinatown was levelled to make way for the construction of government office buildings. Kamloops' once busy Chinatown disappeared gradually during the construction of the Overlander Bridge in the 1960s and the expansion of Victoria Street in the 1970s. Today, there are only two Chinatowns in the province, one in Vancouver and the other in Victoria. They are basically commercial districts and function as tourist attractions, rather than Chinese residential enclaves. According to the 1981 census, there were about 84,000 people of Chinese ethnic origin in metropolitan Vancouver, and about 6,000 in metropolitan Victoria. Most of them were not living in their Chinatowns: in 1981 only about 100 lived in Victoria's Chinatown. About one-third of metropolitan Victoria's Chinese population lived on the periphery of downtown Victoria and in its immediate neighborhoods, such as North Park, Fernwood, Hillside-Quadra and Oaklands. Another one-third was concentrated in the local areas of North Quadra, Gordon Head, Glanford, Blenkinsop and Cedar Hill in the Municipality of Saanich. The remaining one-third was scattered in other parts of metropolitan Victoria.

CONCLUSION

In the past the main concentration of Chinese in Canada has been in British Columbia, although this has changed recently. For nearly 40 years since the Chinese began to immigrate to Canada in the 1850's British Columbia had all of the Chinese residents in the country. Before the completion of the Canadian Pacific Railway there was virtually no eastward migration of Chinese beyond the Rockies. After the 1880s the Chinese immigrants gradually moved eastward along the railway route to the Prairie Provinces and Eastern Canada. Since then, the national percentage of Chinese population in British Columbia has been declining.

Unlike other provinces, British Columbia relied heavily on Chinese laborers because there was a great shortage of labor during its early development period. Chinese laborers were recruited to clear forests, dig ditches, build wagon roads and railways, and engage in other public works connected with the development of the province. For nearly 40 years thousands of these laborers had paid their head taxes, amounting to millions of dollars which were shared by the federal and provincial governments, the province receiving the greater portion of the revenue. Therefore, no other province has benefited so much from the contributions of the Chinese immigrants in Canada.

The Chinese settlements in British Columbia were at one time widely scattered. During the 1880s, for example, nearly every town and city had a Chinatown or Chinese quarter, and nearly every gold mining area had Chinese miners and laborers. Although most of the Chinatowns in the province have now been obliterated, their imprints are nevertheless preserved in old photographs and other artifacts. The two surviving Chinatowns are in Vancouver and Victoria. They are Canada's oldest and still contain cohesive groupings of nineteenth century buildings. They remain intact and are being developed as tourist attractions. Vancouver and its adjacent cities and municipalities attracted most of the postwar Chinese immigrants. This trend toward concentration of Chinese population in the Lower Mainland likely will continue in the foreseeable future.

REFERENCES

1. LUO, R., "Legends, Stone Anchors and The 'Chinese Columbus' Theory", *China Reconstructs*, 32, 1983, p. 8.

2. SHEN, J., "The Theory of Chinese Discovery of America", *Mingbao Monthly*, 3, 1968 (Chinese text), p. 2.

3. FANG, Z., "Did Chinese Buddhists Reach America 1,000 Years Before Columbus?", *China Reconstructs*, 29, 8, 1980, p. 65; PEI, G., "New Research on Chinese Origin of the First Americans", *China Reconstructs*, 32, 1, 1983, p. 38.

4. *Colonist*, October 21, 1882, p. 5.

5. KEDDIE, G.R., "Early Chinese Coins in British Columbia?", *The Midden*, 14, 1972, p. 3.

6. MEARES, J., *Voyages Made in the Years 1788 and 1789, From China to the North West Coast of America*. London: unpublished manuscript, 1790, p. 2.

7. *Ibid.*

8. MEARES, J., *Authentic Copy of the Memorial to the Right Honorable William Wyndham Grenville, Dated 30th April, 1790, and Presented to the House of Commons, 13th May, 1790*. London: unpublished manuscript, 1790, p. 11.

9. HOWAY, F.W., "A Yankee Trader on the Northwest Coast, 1791-1795", *Washington Historical Quarterly*, 21, 1930, p. 93.

10. LAI, M.H. and CHOY, P.P., *Outlines of History of the Chinese in America*. San Francisco: Chinese-American Studies Planning Group, 1971, p. 25.

11. UNITED KINGDOM, *The Colony of Vancouver Island and the Colony of British Columbia Blue Books*. London: Colonial Office, 1862, pp. 225-226.

12. UNITED KINGDOM, *Blue Books of Statistics, Colony of British Columbia, 1868*. London: Colonial Office, 1868.

13. LAI, C-Y.D., "Chinese Imprints in British Columbia", *B.C. Studies*, 39, 1978, p. 27.

14. *Mallandaine's Victoria Directory*. Victoria: Mallandaine, 1871, pp. 94-95.

15. CANADA, *House of Commons Debates*. Ottawa: Queen's Printer, 1882, p. 1477.

16. ORMSBY, M.A., *British Columbia: A History*. Vancouver: Macmillan, student edition, 1971, p. 281.

17. CANADA, *Census of Canada, 1880-81*.

18. CANADA, *Royal Commission on Chinese Immigration: Report and Evidence, 1885*. Ottawa: Royal Commission on Chinese Immigration, 1885, p. 398.

19. *Ibid.*, pp. 363-365.

20. *Ibid.*

21. CANADA, *Census of Canada, 1880-81*.

22. HUANG, T-M., *The Legal Status of Chinese Abroad*. Taipei: Chinese Cultural Service, 1954, p. 82.

23. CANADA, "An Act to Restrict and Regulate Chinese Immigration into Canada", *Statutes of Canada, 1885*. Ottawa: 48-49 Victoria, chap. 71, pp. 207-212.

24. WOODSWORTH, C.J., *Canada and the Orient*. Toronto: Macmillan, 1941, p. 287.

25. *Ibid.*

26. HUANG, T-M., *op. cit.*, p. 87.

27. BRITISH COLUMBIA, *Sessional Papers of British Columbia*. Victoria: Legislative Assembly, 1888, Petition, p. 367.

28. LAI, C-Y.D., "Chinese Attempts to Discourage Emigration to Canada", *B.C. Studies*, 18, 1973, p. 33.

29. CANADA, "An Act Respecting and Restricting Chinese Immigration", *Statutes of Canada, 1900*. Ottawa: 63-64 Victoria, Chapter 32, pp. 215-221.

30. CANADA, "An Act Respecting and Restricting Chinese Immigration", *Statutes of Canada, 1903*. Ottawa: 3 Edward VII, Chapter 8, pp. 105-111.

31. WOODSWORTH, *op. cit.*, p. 287.

32. *Ibid.*

33. NICOL, E., *Vancouver*. Toronto: Doubleday, 1970, p. 56.

34. CANADA, *Census of Canada, 1921*.

35. *Ibid.*

36. WICKBERG, E. (ed.), *From China to Canada*. Toronto: McClelland and Stewart, 1982, p. 137.

37. BRITISH COLUMBIA, *Sessional Papers of British Columbia*. Victoria: Legislative Assembly, 1919, tables of nationality of employees, 31 July, 1918, pp. H12-H35.

38. *Ibid.*

39. WOODSWORTH, *op. cit.*, p. 287.

40. ROY, P.E., "Educating the East: British Columbia and the Oriental Question in the Interwar Years", *B.C. Studies*, 18, 1973, p. 52.

41. CANADA, *Debates of the House Commons*. Ottawa: Vol. III, 1923, pp. 2312-2313.

42. CANADA, "An Act Respecting Chinese Immigration", *Statutes of Canada, 1923*. Ottawa: 13-14 George V, Chapter 38, pp. 301-315.

43. CANADA, *Report of the Department of Mines and Resources*. Ottawa: Department of Mines and Resources, 1947, p. 245.

44. CANADA, *Census of Canada, 1941*.

45. SUNG, B.L., *The Story of the Chinese in America*. New York: Collier, 1967, pp. 80-81.

46. CANADA, "An Act to Amend the Immigration Act and to Repeal the Chinese Immigration Act", *Statutes of Canada, 1947*. Ottawa: 11 George VI, Chapter 19, pp. 107-109.

47. CANADA, *Debates of House of Commons*. Ottawa: Vol. III, May 1947, pp. 2644-2646.

48. SIEN, F., "Yesterday, Today, Tomorrow", *Chinatown News*, November 3, 1956, p. 16.

49. WICKBERG, E., *op. cit.*, p. 217.

50. CANADA, *Quarterly Immigration Bulletin*. Ottawa: Citizenship and Immigration Department, 1962-1965.

51. MA, C., *Chinese Pioneers*. Vancouver: Versatile Publishing, 1979, p. 44.

52. CANADA, *Immigration Statistics Canada*. Ottawa: Department of Immigration and Manpower, 1974-1980.

53. CANADA, *Census of Canada, 1981*.

54. *Ibid.*

55. BREWSTER, E.J., "For the Chinese Quarter of Nanaimo — Day of Disaster", *Daily Colonist*, Victoria, January 20, 1963, p. 3.

PLATE 23 A variety of urban structures, Vancouver. *B.C. Government Photo* ▶

SALE
OPEN

16 URBAN SYSTEM[1]

Charles N. Forward
University of Victoria

The present system of cities is the product of approximately a century of development in British Columbia. Beginning with a seat of government, Victoria, and a couple of other nearby smaller nodes, Nanaimo and New Westminster, cities have arisen throughout the southern two-thirds of the province, arrayed in linear fashion along major valley and transportation corridors. This linear pattern of urban development that is largely topographically controlled contrasts strongly with the more extensive areal distribution of cities in the relatively flat Prairie Provinces to the east. The development of the urban system since the beginning is traced in this chapter and the present functional characteristics of the major cities are considered.

The sizes and growth of urban centres were determined by the use of Statistics Canada decennial census publications from 1871 to 1981. An attempt was made to express city size in terms of the geographic urban entity wherever possible, rather than simply the legally incorporated city. The use of this approach is especially important during the post-World War II period of rapid suburbanization. For example, in 1981 the City of Victoria contained 64,380 people, but the geographic city was far larger because continuous urban development extended into Oak Bay, Saanich and Esquimalt, and discontinuous development into a number of other municipalities and unincorporated territories beyond. The census metropolitan area (C.M.A.)[2] and census agglomeration (C.A.)[3] regions based on labor markets are better approximations of geographic cities than the legal city areas. Their drawback is that they often include populations situated at considerable distances away from the urbanized cores because commuting distances have extended so greatly. Among the 12 census agglomerations identified in 1981 all but one were single central cities surrounded by commuter zones. The exception was Terrace which included Kitimat, a separate city of similar size 58 kilometres (36 miles) distant. In this study Terrace and Kitimat were considered as separate urban entities, but the other agglomerations were used to indicate city size.

These more sophisticated urban delineations are available for only a few places and only in the most recent censuses. In order to extend this use of a geographic city concept into the past whenever possible, the population figures in early censuses for incorporated entities that eventually became parts of CMAs or CAs were combined to produce more accurate indications of city size. Thus, the Vancouver urban entity, which today as a CMA includes all municipalities from the north shore of Burrard Inlet south to the international border and east to Langley and Maple Ridge, is considered to include any places listed in earlier censuses that fall within the present CMA area. As a result of this approach, one must consider New Westminster as the major component of the "Vancouver" urban complex in 1881 before the City of Vancouver was founded. And to avoid confusion the names of the eventual major cities of urban agglomerations were used to designate them, rather than using hyphenated names, such as Courtenay instead of Courtenay-Comox and Trail instead of Trail-Rossland.

DEVELOPMENT OF THE URBAN SYSTEM

The past century of urban development is traced through four distinct periods of even length with the use of census data. Owing to the irregularities of urban growth, boundary changes of incorporated places, and the limited number of city listings in the early censuses, the population figures for each city were averaged over the 30-year periods. A city listed in two of the three censuses of the period was included as representative of that period, but if it appeared only once it was omitted.

1871 - 1891

During much of the nineteenth century Victoria was the leading city of British Columbia, performing the chief financial, trade and manufacturing, as well as government, functions for the province (Table 1,16) (Figure 1,16). New Westminster at the mouth of the Fraser River was the main centre of population on the mainland until the rise of Vancouver on Burrard Inlet after the arrival of the Canadian Pacific Railway at that location in 1887. In the 1870 British Columbia census which was published as part of the Census of Canada, 1870-71, Victoria was the only city listed, but figures were given for New Westminster and Nanaimo districts. Figures referring specifically to cities in the 1881 and 1891 censuses were limited to Victoria and New Westminster, appearing in both, and Vancouver, listed for the first time in 1891. A number of other places were listed in 1891 as

TABLE 1,16 AVERAGE POPULATION AND RANK OF CITIES BY PERIODS

Rank	*1870-1891*		*1901-1921*		*1931-1951*		*1961-1981*	
1.	Victoria	11,900	Vancouver	146,000	Vancouver	420,600	Vancouver	1,059,100
2.	Vancouver	8,300	Victoria	37,300	Victoria	79,500	Victoria	195,000
3.	Nanaimo	3,600	Nanaimo	6,200	Trail	13,200	Prince George	45,800
4.			Prince Rupert	5,300	Chilliwack	13,000	Kelowna	44,800
5.			Trail	5,200	Abbotsford	11,900	Kamloops	43,900
6.			Kamloops	4,100	Powell River	7,800	Nanaimo	41,100
7.			Chilliwack	3,300	Kamloops	7,400	Abbotsford	35,400
8.			Vernon	3,200	Prince Rupert	7,200	Chilliwack	33,900
9.			Fernie	3,000	Penticton	7,000	Port Alberni	26,600
10.			Cranbrook	2,900	Port Alberni	6,900	Vernon	21,900
11.			Revelstoke	2,500	Nanaimo	6,900	Courtenay	20,500
12.			Ladysmith	2,200	Nelson	6,300	Trail	19,500
13.			Kelowna	2,100	Kelowna	6,300	Penticton	18,400
14.			Courtenay	1,600	Vernon	6,100	Prince Rupert	16,000
15.			Grand Forks	1,400	Ocean Falls	4,500	Powell River	14,500
16.			Merritt	1,000	Cranbrook	3,100	Terrace	12,000
17.					Prince George	3,100	Cranbrook	12,000
18.					Courtenay	3,000	Dawson Creek	11,900
19.					Mission	2,800	Mission	11,900
20.					Summerland	2,800	Kitimat	10,900
21.					Fernie	2,600	Campbell River	9,900
22.					Revelstoke	2,600	Fort St. John	8,900
23.					Duncan	2,300	Nelson	8,500
24.					Dawson Creek	2,100	Salmon Arm	7,500
25.					Spallumcheen	1,900	Kimberley	7,000
26.					Ladysmith	1,700	Quesnel	6,400
27.					Salmon Arm	1,700	Castlegar	6,100
28.					Grand Forks	1,400	Squamish	6,000
29.					Creston	1,400	Summerland	5,800
30.					Merritt	1,200	Williams Lake	4,900

NOTE: Populations were rounded to the nearest hundred and a city had to be listed in at least two of the three census years to be included.

SOURCES: Censuses of Canada 1871-1981.

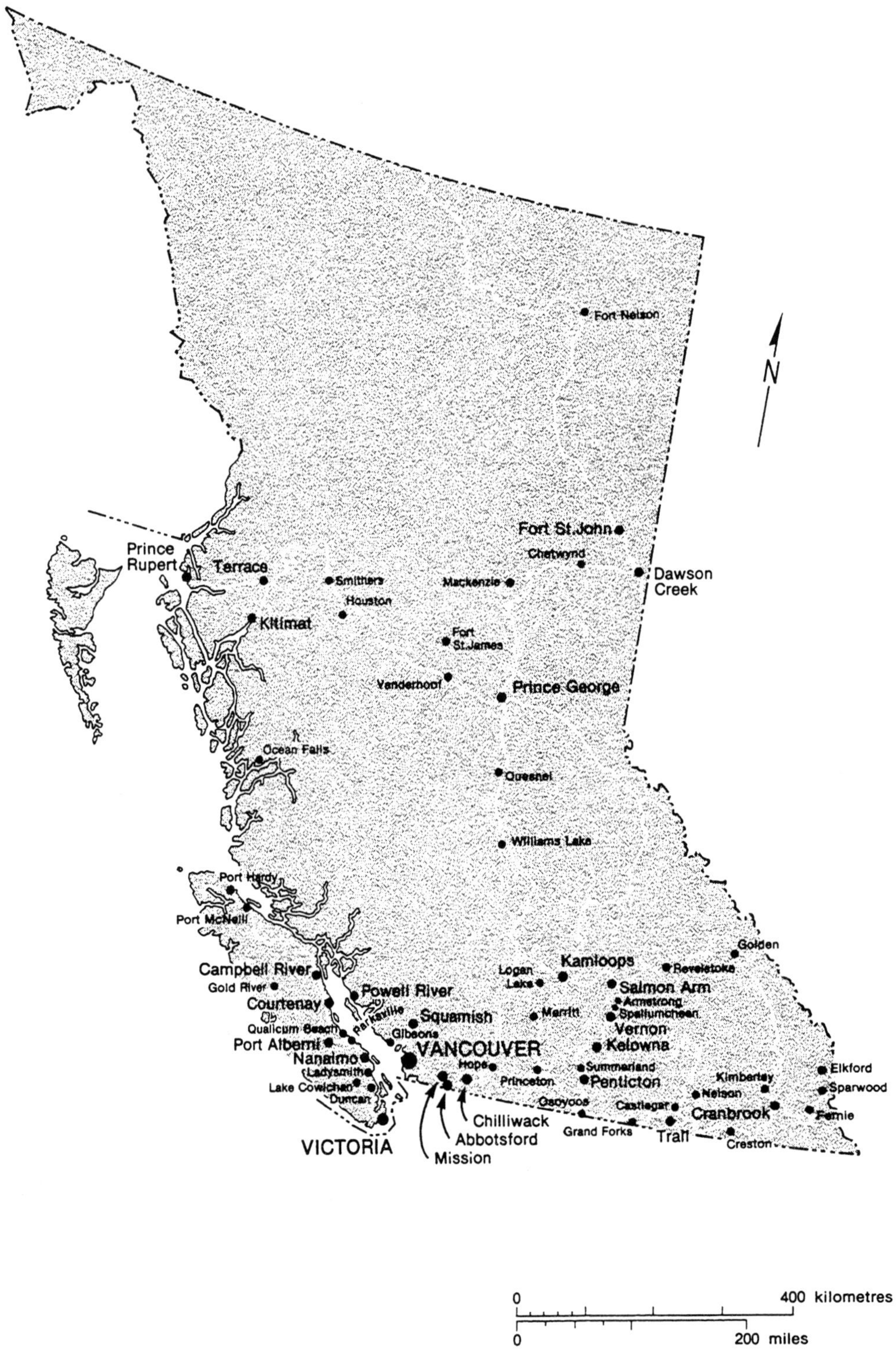

FIGURE 1,16 British Columbia Cities

districts including, in order of population, Nanaimo, Kamloops, Lillooet, Quesnel, Cache Creek, Williams Lake, Clinton and Alberni. However, it is clear from the figures available that Victoria, New Westminster and Nanaimo, a coal mining town on Vancouver Island, were the three key urban centres in the province.

1901 - 1921

By 1901 Vancouver had become the leading transportation centre and had taken over from Victoria the leadership in finance, trade and manufacturing. Its average population during the period was about three times that of Victoria, which retained its governmental role and a second city status in industry, trade and commerce. But Victoria's relative decline is indicated by the fact that Vancouver was four times the size of Victoria in 1921 at the end of this period. Nanaimo still ranked third, but several hinterland centres achieved sizes nearly as large, specifically Prince Rupert, Trail (includes Rossland) and Kamloops. Prince Rupert appeared on the map suddenly during the first decade of the present century when it was selected as the western terminus of the Grand Trunk Pacific Railway (now Canadian National Railway). Rossland (Trail) was the largest of a number of towns in the Kootenay region which experienced a mining boom during the 1890s. Kamloops was one of the older settlements that benefited greatly from the building of the Canadian Pacific Railway and became a service centre for a large part of the southern Interior. The Okanagan cities averaged a little smaller in size, Vernon being significantly larger than Kelowna. Penticton was not listed until 1921, but it is clear that there were three sizable cities in the Okanagan region. The seventh largest city was Chilliwack, the agricultural service centre of the eastern part of the Lower Mainland. The Kootenay mining region was represented by four other cities besides Rossland (Trail): Fernie, Cranbrook, Revelstoke and Grand Forks. The other cities on Vancouver Island, Ladysmith and Courtenay (includes Cumberland), were also connected with mining activities.

1931 - 1951

Rapid population growth, despite the worldwide depression which occurred in this period, produced a much more extensive urban system in the south and a few significant places in the north to join Prince Rupert, the only one appearing in the previous period. However, the overwhelming dominance of Vancouver was further asserted, Vancouver becoming more than five times the size of Victoria. Below Victoria there were no cities that amounted to more than one-sixth of its population and none of these had assumed a broad regional service function. Trail was ranked third, but was chiefly an ore smelting and refining centre. The next ranked, Chilliwack and Abbotsford, were strongly tributary to Vancouver in the heavily populated lower Fraser Valley and Powell River was essentially a pulp and paper manufacturing town near Vancouver. Below these were eight cities of similar size that included Kamloops, the three Okanagan centres, Nanaimo, Port Alberni and two in outlying regions, Prince Rupert and Nelson. Although by 1951 they were in the 7,000 to 10,000 population range, none of them was large enough to attract higher level urban services. Even by 1951, however, the process of suburbanization was underway and in some cases the city boundaries may not have included as much of the urbanized areas as they had in the past, resulting in geographic city population underestimates. One exception is Abbotsford which, perhaps, is overestimated in comparison with the others because it was possible to use the boundaries of the modern municipalities in calculating its population. It is noteworthy that the Kootenay region cities, except for Trail, experienced little growth and dropped substantially in rank. In the north Ocean Falls, Prince George and Dawson Creek were the three largest cities, apart from Prince Rupert, while the other dozen or so cities in the top 30 were spread in a broad band across the more developed southern part of the province from Vancouver Island to the Kootenays.

1961 - 1981

The decades of the 1950s, 1960s and 1970s included in this period were characterized by tremendous urban growth in British Columbia, as well as suburbanization. Both Vancouver and Victoria were classed as CMAs in all three censuses and some of the other cities were classed as CAs in the 1971 and 1981 censuses. Vancouver continued to be a little more than five times the size of Victoria, but there was a striking change in the relationship between the metropolitan centres and those immediately below them in the urban hierarchy. Four major regional centres had emerged, cutting the size gap between Victoria and the other cities. On average during the period Victoria was a little over four times the size of those ranked three to six, down from six times the size of the third largest city in the previous period. Actually, in 1981 Victoria was only three times the size of the third ranked city, Kelowna. The emergence of these four regional capitals was a response to the widespread economic growth of the province and is symptomatic of the degree of maturity that was achieved in the various regions of British Columbia. All of these centres rose from lower positions in the hierarchy, the most striking rise being that of Prince George. Its commanding centrality for serving northern British Columbia and its burgeoning forest industry propelled it to urban dominance in the north. Kelowna asserted its primacy within the Okanagan region rather dramatically, perhaps also partly due to its centrality for regional service functions. Though situated relatively close to Kelowna, Kamloops achieved a regional city status in south central British Columbia, being, like Prince George, a key transport node in the provincial system. Nanaimo benefited greatly from the large-scale development of northern Vancouver Island which occurred during this period: it capitalized on its key ferry transport link with Vancouver, consolidating its role as the service centre for much of the Island.

The urban places ranked seven and eight lie in the path of suburbanization developed along the Fraser River east of Vancouver. At present the CMA includes Maple Ridge and Langley, north and south of the Fraser River, respectively. Abbotsford, composed of the municipalities of Abbotsford and Matsqui, lies immediately east of the CMA boundary on the south side of the river. It is likely that it will meet the criteria for inclusion within the Vancouver CMA by the 1986 or 1991 census. Mission, on the north shore, ranked 19 in Table 1,16, is in the same position as Abbottsford. Chilliwack's location farther inland may protect it from inclusion for some years to come and it likely will remain a viable, independent city. Among the other leading cities there were many changes in ranking and a number of cities appeared in the top 30 that were not there previously. While Prince George shot upward in the urban hierarchy, a number of other hinterland cities dropped. Trail went from rank 3 in the 1931-1951 period to 12, Prince Rupert from 8 to 14 and Nelson from 12 to 23. Port Alberni and Courtenay on Vancouver Island held their own and were joined by a fast growing Campbell River in the recent period. In the Okanagan region Vernon and Penticton experienced rapid growth, but were outstripped by Kelowna. The size relationship among the three is still difficult to determine because even in 1981 Penticton was not classed as a CA, while the other two were. Nevertheless, it is safe to say that Kelowna is now nearly twice the size of either of the others. A total of nine places that were among the top 30 in the 1931-1951 period were replaced by others in the recent period. Most of the replacements were in hinterland locations and more than half lay north of those displaced. Resource development in central and northern British Columbia aided the growth of Williams Lake and Quesnel in the Cariboo, Kitimat and Terrace in the North Coast region and Fort St. John in the Peace River. In fact, Fort St. John replaced Dawson Creek as the largest city in the Peace River area, according to the 1981 census. Dawson Creek situated south of the river was oriented toward servicing the agricultural economy of the region, while Fort St. John to the north became the centre for the oil and gas industry that expanded northward during the 1960s and 1970s. Large scale industrial developments, such as the establishment of an aluminum smelter and refinery at Kitimat and the building of pulp and paper mills at Campbell River, Quesnel and Castlegar, fostered urban expansion. On the other hand, the experience of Ocean Falls illustrates the opposite effect. Ranked 15 in the 1931-1951 period, this isolated coastal pulp and paper manufacturing town was almost completely depopulated when the mill was closed permanently in 1980 and was not even listed as a place in the 1981 census.

A glance at a more complete list of the actual rankings in 1981 affords a more accurate appreciation of the present situation (Table 2,16) (Figure 1,16). It is apparent that the populations of the four regional centres below Victoria, plus the suburban-type cities, Abbottsford and Chilliwack, have diverged to a certain extent, with Kelowna, Prince George and Kamloops most successful in narrowing the size gap below Victoria. Among the 22 cities over 10,000 population below the two metropolitan centres there are five at the top in a group over 50,000 in size, four in the 32,000-42,000 range, five in the 18,000-23,000 range and the remaining eight in the 10,000-15,000 range. These groups are still dominated by cities in the southern part of the province. One must look in the list below 10,000 to find many of the historic northern towns that figured prominently in the early history of British Columbia, such as Quesnel and Fort St. James. Lillooet, the historic gateway to the Cariboo gold rush area, had a population in 1981 of only 1,725. Many cities throughout the province have grown very slowly or languished, while others have burgeoned, and the changing economy inevitably will continue to alter the urban hierarchy in future. Mackenzie, a pulp and paper centre north of Prince George ranked 32, was a new city of the 1960s, Logan Lake, a dormitory community in the copper mining area near Kamloops ranked 53, was a new city of the 1970s, and Tumbler Ridge, southwest of Dawson Creek, is a new city of the 1980s being built to serve the northeast coal mining area.

CHARACTER OF CITIES

The urban system described so far is like a skeleton which must be fleshed out by other information on environmental, population and economic characteristics to be fully understood and appreciated. Comparisons of British Columbia cities with those in other regions of Canada, as well as comparisons of one to another, are helpful in understanding the nature of cities. Detailed comparisons are restricted to the major cities because more

TABLE 2.16 POPULATION AND RANK OF CITIES, 1981

Rank	*City*	*Population*	*Rank*	*City*	*Population*
1.	Vancouver	1,268,183**	31.	Merritt	6,110
2.	Victoria	233,481**	32.	Mackenzie	5,890
3.	Kelowna	77,468*	33.	Revelstoke	5,544
4.	Prince George	67,559*	34.	Fernie	5,444
5.	Kamloops	64,997*	35.	Parksville	5,216
6.	Nanaimo	57,694*	36.	Port Hardy	5,075
7.	Abbotsford	54,746	37.	Smithers	4,570
8.	Vernon	42,158*	38.	Ladysmith	4,558
9.	Chilliwack	41,471*	39.	Duncan	4,228
10.	Courtenay	35,218*	40.	Spallumcheen	4,213
11.	Port Alberni	32,558*	41.	Creston	4,190
12.	Penticton	23,181	42.	Sparwood	4,157
13.	Trail	22,939*	43.	Houston	3,921
14.	Mission	20,056	44.	Fort Nelson	3,724
15.	Powell River	19,364*	45.	Grand Forks	3,486
16.	Prince Rupert	18,402*	46.	Golden	3,476
17.	Cranbrook	15,915	47.	Hope	3,205
18.	Campbell River	15,832	48.	Elkford	3,126
19.	Terrace	15,245	49.	Princeton	3,051
20.	Fort St. John	13,891	50.	Qualicum Beach	2,844
21.	Kitimat	12,814	51.	Osoyoos	2,738
22.	Dawson Creek	11,373	52.	Armstrong	2,683
23.	Salmon Arm	10,780	53.	Logan Lake	2,637
24.	Squamish	10,272	54.	Gibsons	2,594
25.	Nelson	9,143	55.	Chetwynd	2,553
26.	Williams Lake	8,362	56.	Port McNeill	2,474
27.	Quesnel	8,240	57.	Lake Cowichan	2,391
28.	Summerland	7,473	58.	Vanderhoof	2,323
29.	Kimberley	7,375	59.	Fort St. James	2,284
30.	Castlegar	6,902	60.	Gold River	2,225

** Census Metropolitan Area

* Census Agglomeration

SOURCE: Census of Canada, 1981.

elaborate statistical data are available for these in the censuses and other government sources.

Regional Distinctiveness

There is considerable evidence to indicate that British Columbia cities, as a group, are distinctive compared with those of other major regions of Canada. This is due partly to the distinctive, mountainous environment of the province and partly to differences in the character of urban populations that are related to historical development and economic activities. A number of obvious environmental generalizations can be made about British Columbia cities, particularly in comparison with cities of the Prairie Provinces, but also with those of other Canadian regions. British Columbia cities have a Pacific orientation: over half of the cities over 10,000 population, including the two largest, are situated on or near the coast and transportation and communication systems facilitate interaction between coastal and interior centres of the province. The dominance of Vancouver and Victoria within the urban hierarchy assures a high level of interaction with these particular coastal locations in industrial, commercial, trading and governmental activities. Urban settings are in contrast with those of Prairie cities. More often than not the cities are built on sloping terrain and are surrounded by tree-covered hills or mountains, instead of having relatively flat sites surrounded by agricultural land. Perhaps because of this, urban development is less compact in British Columbia cities than on the Prairies where well established agricultural land uses around cities more successfully resist random urban intrusions. Although climates vary considerably, the coastal cities at least have distinctive environmental conditions compared with other Canadian cities. They tend to have milder winters, cooler summers and gradual transitions between seasons, along with substantial rainfall, conditions that produce luxuriant forest growth.

A study of Canadian cities of 10,000 population or over by major regional groupings was carried out based on the 1971 census data.[4] The characteristics of the 19 British Columbia cities were compared with those of Prairie, Ontario, Quebec and Atlantic cities. The British Columbia cities had a number of distinctive characteristics. Their populations were remarkably cosmopolitan and mobile. Ethnic diversity was next highest to that of Prairie cities, religious diversity was the highest and the British Columbia cities were the most strongly Protestant. They also had the highest proportion of males per 100 females of all the groups, a result, perhaps, of the strong orientation to primary forest and mineral processing activities in the urban economies. The percentages of population born in other provinces and outside Canada were by far the highest in British Columbia cities, while the in- and out-migration ratios were also the highest. Conversely, they had the lowest percentage born within the province of all the groups. Educational achievement was remarkably high in British Columbia cities: fully 65 percent of the population had attended secondary school compared with only 55 percent in Prairie and Ontario cities, the next highest. Similarly, eight percent had some university attendance compared with five percent in Prairie and Atlantic cities, the next highest. In general, British Columbia cities are relatively young and fast growing and some housing characteristics are distinctive. Unlike those of eastern Canada, most British Columbia cities lack extensive older areas and the houses are mostly built of wood and stucco instead of brick. Both home ownership and the proportion of single-family dwellings were highest in British Columbia cities, and the percentage of apartments was lowest. British Columbia cities also led the regions in value of houses and in average rents. A study of labor force data by industry groups indicated the dominant functions of cities in the various regions.[5] The British Columbia cities were much less oriented to manufacturing than those of Ontario and Quebec and to retail trade and services than those of the Prairies. Instead, transportation appeared more frequently as a dominant function than in any other region, probably owing to the significance of the port role in many coastal cities. This is borne out by the fact that Atlantic cities were second highest in percentage of cities with transportation as a dominant function. These various measures of differentiation indicate that major British Columbia cities as a group have a surprising number of distinctive features.

City Functions

Although cities perform many different roles, they do not operate in isolation and can, therefore, develop specializations within their urban economies. They are linked into a dynamic "system of cities" by numerous transportation and communication systems which foster interaction. The importance of cities in the urban hierarchy is indicated by the strength of these connecting links, as well as by population size. The maps of railway and highway networks (Figures 1,16 and 2,16) indicate the great importance of Vancouver and some other cities, such as Prince George and Kamloops.

Another indicator of the importance of cities, particularly those in hinterland areas, is scheduled air passenger traffic. The urban system portrayed by air traffic is somewhat different from that indicated by population (Figure 2,16).[6] Air travel focusses on relatively few centres and is most competitive

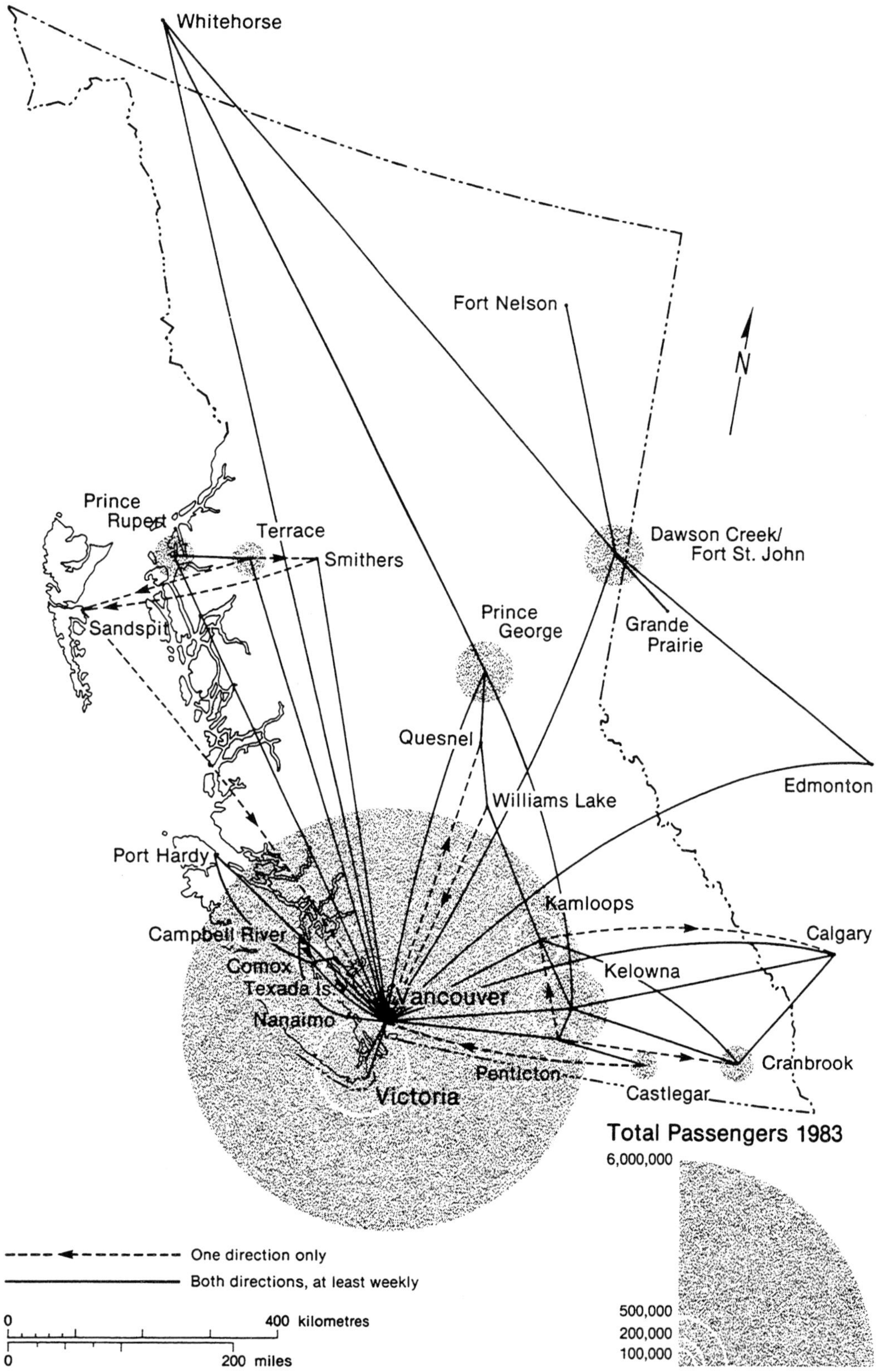

FIGURE 2,16 Scheduled Airline Linkages, Winter 1985 and Passenger Volume, 1983

with other modes of transportation when distances are great. Even sizable cities like Nanaimo that are situated close to a major air traffic centre like Vancouver are unimportant in air travel. Some airports, including Nanaimo's, are indicated in Figure 2,16 as having scheduled service in 1985, but were not listed in the most recent (1983) air passenger statistics. Victoria, although situated close to Vancouver, obviously is large enough to command air service and have the second highest volume of traffic. The next three air traffic generators, Kelowna, Prince George and Kamloops, appeared in the same order as by population in 1981. They were followed by the joint airport destination of the remote northeast, Dawson Creek-Fort St. John. It appears that Cranbrook is the service centre for the East Kootenay and that Castlegar and Terrace are service centres for their neighboring industrial cities of Trail and Kitimat, respectively. Surprisingly, Penticton looms as of sufficient importance to attract substantial traffic, likely because it is on the direct route to the West Kootenay load centre of Castlegar. It is interesting to note an apparent anomaly that the network development of aircraft routing produces numerous one way services.

The strengths and weaknesses of the various aspects of the urban economic base provide a good indication of a city's diversity or specialization and an insight into its function in the urban system. Statistical data on labor force by major industry groups has proven useful in developing profiles of city economies. The significance of each industry group can be determined easily by converting the labor force figures into percentages of the total for each city. In Table 3,16 the percentages of each city's labor force in selected industry groups are shown and the cities are ranked by industry group accordingly. In order to simplify the picture portrayed and to concentrate on the most significant urban functional categories, some of the major industry groups were omitted and two were combined. Extraction or primary industries, including agriculture, mining, fishing and forestry, were omitted because they are typical usually of rural rather than urban areas. Construction was omitted because it varies from one city to another in a rather cyclical fashion, depending upon degree of prosperity and growth at the precise date when the census was taken and, therefore, is not a good indicator of urban function. Finally, community services (for example, schools, health clinics and churches), business services (for example, accountants, lawyers and engineers), and personal services (for example, hotels, restaurants and cleaners) were combined to produce a single service indicator.

It is apparent in Table 3,16 that the percentages of the labor force vary on average from one industry group to another, services, for example, accounting for a much higher proportion than wholesale trade.[8] Also, in some industries, such as manufacturing, the percentages vary

Table 3,16 PERCENTAGE* OF LABOR FORCE AND RANK OF CITIES IN SELECTED INDUSTRY GROUPS, 1981

Rank	*Manufacturing*	*%*	*Transportation, Communications and Other Utilities*	*%*	*Wholesale Trade*	*%*	*Retail Trade*	*%*	*Finance, Insurance and Real Estate*	*%*	*Community, Business and Personal Services*	*%*	*Public Administration and Defence*	*%*
1.	Kitimat (DM)	52	Squamish (DM)	21	Dawson Creek (City)	8	Dawson Creek (City)	19	Vancouver (CMA)	7	Mission (DM)	37	Courtenay (CA)	20
2.	Powell River (CA)	34	Fort St. John (City)	14	Vancouver (CMA)	7	Penticton (City)	17	Mission (DM)	6	Victoria (CMA)	35	Victoria (CMA)	20
3.	Trail (CA)	33	Dawson Creek (City)	14	Cranbrook (City)	7	Nanaimo (CA)	17	Cranbrook (City)	6	Salmon Arm (DM)	34	Chilliwack (CA)	16
4.	Port Alberni (CA)	32	Terrace (DM)	13	Prince George (CA)	6	Cranbrook (City)	17	Victoria (CMA)	6	Vancouver (CMA)	33	Mission (DM)	10
5.	Prince Rupert (CA)	27	Cranbrook (City)	12	Kamloops (CA)	6	Fort St. John (City)	17	Kelowna (CA)	6	Vernon (CA)	32	Prince Rupert (CA)	9
6.	Squamish (DM)	23	Prince George (CA)	12	Salmon Arm (DM)	6	Salmon Arm (DM)	16	Vernon (CA)	5	Kamloops (CA)	32	Terrace (DM)	9
7.	Prince George (CA)	17	Kamloops (CA)	11	Terrace (DM)	6	Vernon (CA)	16	Salmon Arm (DM)	5	Penticton (City)	32	Kamloops (CA)	8
8.	Penticton (City)	16	Vancouver (CMA)	11	Kelowna (CA)	6	Abbotsford**	15	Nanaimo (CA)	5	Courtenay (CA)	31	Dawson Creek (City)	8
9.	Campbell River (DM)	16	Prince Rupert (CA)	11	Fort St. John (City)	6	Kelowna (CA)	15	Courtenay (CA)	5	Kelowna (CA)	31	Penticton (City)	7
10.	Vancouver (CMA)	15	Nanaimo (CA)	9	Vernon (CA)	5	Chilliwack (CA)	15	Kamloops (CA)	5	Dawson Creek (City)	30	Cranbrook (City)	7
11.	Mission (DM)	15	Vernon (CA)	8	Penticton (City)	5	Courtenay (CA)	15	Chilliwack (CA)	5	Nanaimo (CA)	30	Nanaimo (CA)	7
12.	Kelowna (CA)	14	Salmon Arm (DM)	8	Abbotsford**	5	Kamloops (CA)	15	Penticton (City)	5	Cranbrook (City)	30	Abbotsford**	7
13.	Vernon (CA)	14	Campbell River (DM)	8	Chilliwack (CA)	4	Terrace (DM)	14	Dawson Creek (City)	5	Prince George (CA)	28	Prince George (CA)	6
14.	Nanaimo (CA)	14	Abbotsford**	7	Nanaimo (CA)	4	Prince George (CA)	14	Fort St. John (City)	5	Chilliwack (CA)	27	Vancouver (CMA)	6
15.	Terrace (DM)	11	Penticton (City)	6	Campbell River (DM)	3	Campbell River (DM)	13	Abbotsford**	5	Abbotsford**	27	Fort St. John (City)	5
16.	Chilliwack (CA)	10	Powell River (CA)	6	Prince Rupert (CA)	3	Victoria (CMA)	13	Prince George (CA)	5	Campbell River (DM)	27	Vernon (CA)	5
17.	Salmon Arm (DM)	10	Trail (CA)	6	Victoria (CMA)	3	Vancouver (CMA)	13	Campbell River (DM)	4	Fort St. John (City)	26	Port Alberni (CA)	5
18.	Abbotsford**	10	Victoria (CMA)	6	Courtenay (CA)	3	Mission (DM)	12	Prince Rupert (CA)	4	Terrace (DM)	25	Kelowna (CA)	5
19.	Kamloops (CA)	10	Kelowna (CA)	6	Mission (DM)	3	Trail (CA)	12	Trail (CA)	4	Prince Rupert (CA)	24	Trail (CA)	5
20.	Cranbrook (City)	7	Port Alberni (CA)	5	Squamish (DM)	2	Port Alberni (CA)	11	Terrace (DM)	3	Powell River (CA)	24	Kitimat (DM)	4
21.	Victoria (CMA)	7	Courtenay (CA)	5	Kitimat (DM)	2	Powell River (CA)	10	Powell River (CA)	3	Squamish (DM)	22	Salmon Arm (DM)	4
22.	Courtenay (CA)	5	Chilliwack (CA)	4	Trail (CA)	1	Prince Rupert (CA)	9	Port Alberni (CA)	3	Port Alberni (CA)	21	Powell River (CA)	4
23.	Fort St. John (City)	5	Mission (DM)	3	Port Alberni (CA)	1	Squamish (DM)	8	Squamish (DM)	3	Trail (CA)	21	Campbell River (DM)	4
24.	Dawson Creek (City)	4	Kitimat (DM)	3	Powell River (CA)	1	Kitimat (DM)	7	Kitimat (DM)	2	Kitimat (DM)	16	Squamish (DM)	4
	AVERAGE	17	AVERAGE	9	AVERAGE	4	AVERAGE	14	AVERAGE	5	AVERAGE	28	AVERAGE	8

NOTES: *Percentages are rounded to nearest whole number, **Abbotsford includes Abbotsford DM plus Matsqui DM; CMA = census metropolitan area; CA = census agglomeration; DM = district municipality

SOURCE: Census of Canada, 1981.

over a much greater range than in others, such as finance, insurance and real estate. These differences arise owing to the nature of particular industries, thus, manufacturing can account for a large portion of an industrial city's labor force, but no city could be so highly specialized in the financial industry.

At first glance, Table 3,16 seems to present a chaotic picture of urban economic profiles, but there are several distinct patterns that can be traced, along with a number of anomalies. It is noteworthy that a few cities are highly specialized, however the majority are diversified. The most obvious and consistent pattern to appear is the grouping of cities that are high in manufacturing labor force, including Kitimat, Powell River, Trial, Port Alberni, Prince Rupert and Squamish. Conversely, they are near the bottom of the list in the wholesale, retail, finance and services categories. With the exceptions of Prince Rupert and Squamish, they are low also in transportation and public administration. These cities are dominated by pulp and paper or smelting and refining activities. Prince Rupert and Squamish deviate from the consistent profile, owing to the additional importance of their port and railway functions. At the other end of the scale, a group of cities that are low in manufacturing and high in trade may be classed as significant "central places" that serve regional hinterlands. These include Dawson Creek and Fort St. John in the Peace River area, Terrace in the North Coast region, Cranbrook in the Kootenay, Kamloops and Salmon Arm in the Thompson region, Vernon, Kelowna and Penticton in the Okanagan, and Nanaimo on Vancouver Island. For the most part, they are high in services also. Dawson Creek, leading in both wholesale and retail trade and lowest in manufacturing, may be taken as the epitome of the central place.

Several others that are low in manufacturing are special cases. Victoria, Courtenay and Chilliwack have high components of public administration, consisting of military bases in all three and, of course, the provincial capital function in Victoria. Victoria also is very high in services, probably due to its great importance as both a tourist and retirement centre where accommodation and food services catering to visitors, and health services catering to the elderly are well developed. The significance of the finance, insurance and real estate industries in Victoria may be accounted for partly by the high proportion of retired population. The two closest suburbs of metropolitan Vancouver, Abbotsford and Mission, have slightly erratic profiles that must be considered less valid than those of more distant and viable independent cities because so much of their goods and so many of their services are provided by the metropolis, and as a result their labor forces are relatively small for their population sizes. Hence, in view of Mission's location, its high ranking in finance and public administration is not of great significance: it may be a product of underrepresentation of the wholesale and retail trade functions. Vancouver, as the provincial metropolis, is a unique case. Although its economy is highly diversified, it stands out in two key areas that reflect its metropolitan function within the province — finance and wholesale trade.

Another approach to the problem of defining urban character differences is through the use of personal income data. Information on 10 British Columbia cities is available in the taxation statistics published annually which include all personal income declared on tax forms.[9] The study of the period 1966-1978 was carried out using these annual statistics.[10] The 23 different sources of income listed were grouped into three categories to facilitate study: employment, investment, and pension-government-miscellaneous income.[11]

It was discovered that the cities fell into three distinct groups in terms of both employment and non-employment income. The significance of employment income as a proportion of the total in all cities declined during the period, but more rapidly in some cities than in others (Figure 3,16). The overall decline may be accounted for by the increasing affluence during the postwar period, leading to more widespread receipt of income from savings and investments, together with substantially increased transfers of government funds in the form of such things as pensions, welfare payments, family allowances and unemployment insurance benefits. Prince Rupert, Port Alberni and Prince George remain the most dependent on employment sources of income and experienced the lowest decline, while Kelowna, Victoria and Penticton had greater declines, and by 1978 their urban economies were being supported by non-employment income to the extent of one-quarter of the total. The other four cities were grouped half way between these extremes.

The same grouping of cities appears when their positions are plotted in terms of non-employment income sources (Figure 4,16). The two most important non-employment sources were investment and pension income and it was noted that investment income increased more rapidly during the 12-year period than pension income, although there was some variation among cities. Prince Rupert, Port Alberni and Prince George were remarkably similar in having relatively low proportions of both investment and pension income, while Kelowna, Victoria and Penticton exceeded all others in both categories of non-employment income. This clearly differentiates the cities with an industrial-transportation role from those with a retirement-tourism role. Kelowna, Victoria and Penticton had the highest proportion of population 65 and over and are known as

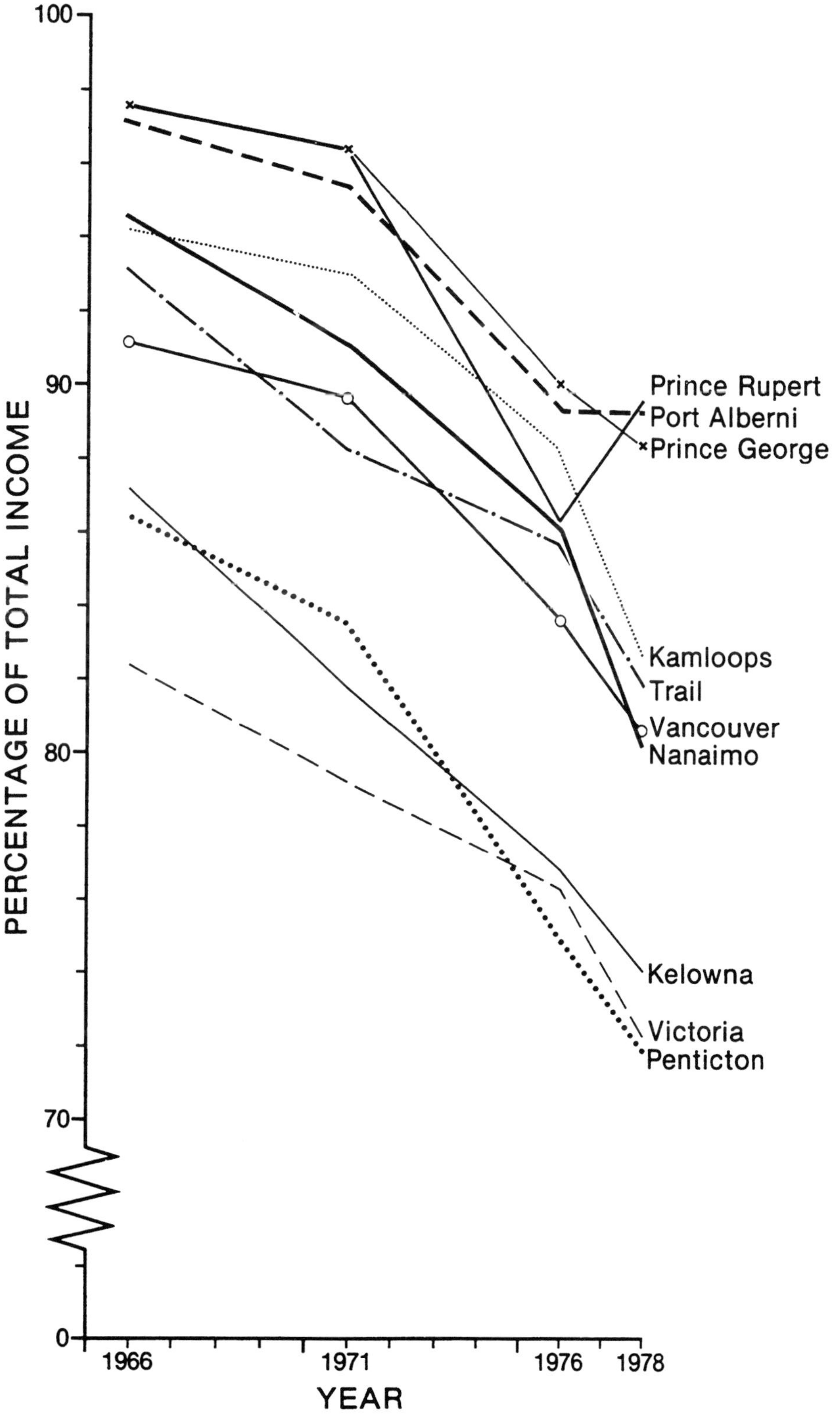

FIGURE 3,16 Employment Income as a Percentage of Total Income in Selected Cities, 1966-1978

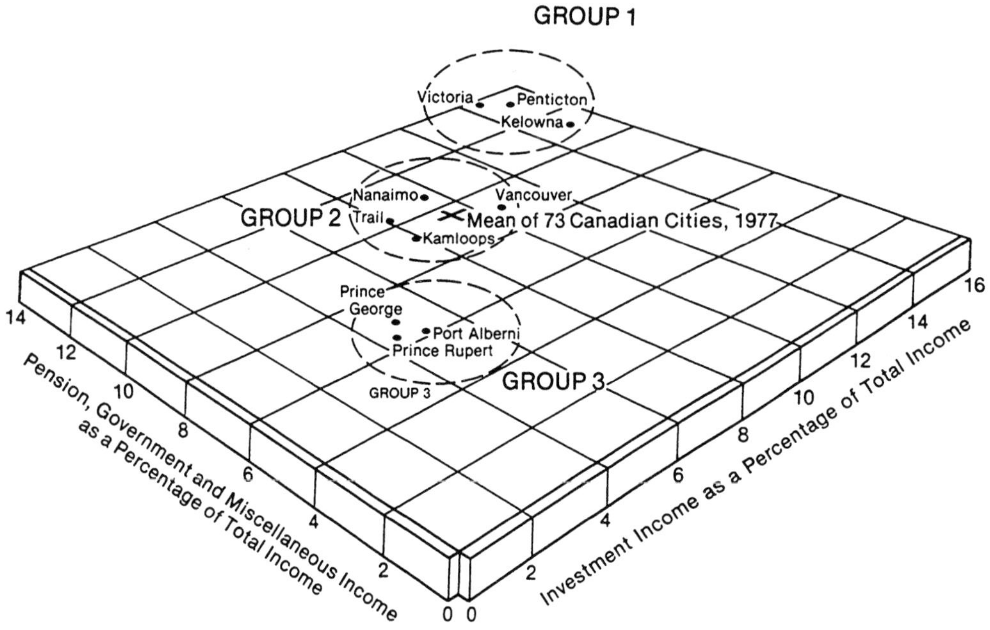

FIGURE 4,16 Non-Employment Income Categories as Percentages of Total Income in Selected Cities, 1978

destinations for retirement, even for those retiring before 65 years of age.[12] It seems likely that the high component of retired people was in large part responsible for the high investment income, as well as the high pension income, received in these cities.[13] Trail (includes Rossland) in group two, although fourth highest in proportion of population 65 and over, ranked next to Victoria in pension income specifically. It falls in group two, however, owing to its much lower percentage of investment income. Unlike the other retirement centres, Trail is a strongly industrial city of slow growth where the population has virtually aged in place due to out-migration of the younger age groups. Vancouver,[14] Kamloops, Trail and Nanaimo constitute an "average" group in terms of income characteristics, being clustered around the mean of all Canadian cities 10,000 and over. It is apparent from this study that the use of labor force as an indicator of urban character differences is more accurate in the case of industrial cities like Prince Rupert and Port Alberni than in the case of retirement-tourism cities like Victoria and Penticton, but both approaches yield valuable insights.

CONCLUSION

The growth and development of a system of cities in British Columbia has been traced through the past century and some elements of the present character of the cities have been identified. It was discovered that the urban system altered considerably as it developed, changing from a simple system focussed almost exclusively on Victoria and Vancouver during much of the period to one more complex that has evolved recently in which four regional centres have arisen as significant intermediaries between the metropolitan centre and the large number of small cities. The change in role and size relationship of Victoria and Vancouver from the nineteenth century when Victoria was dominant to the twentieth century which belonged to Vancouver was striking. Vancouver today is a little more than five times the size of Victoria and this ratio has changed little during the last 20 years. Perhaps a stability in the size relationship has been reached, but if Vancouver manages to increase its western Canadian and national functions in future, at the expense of Toronto and Montreal in particular and the rising Prairie cities in general, it may be able to widen the gap between the two. Kelowna, Prince George, Kamloops and Nanaimo rose from relative obscurity in the prewar period to positions of regional dominance and their future growth seems assured, particularly that of the first three. Being well removed from Vancouver and commanding sizable regions, these three seem well positioned to further narrow the size gap between them and Victoria. Over the long term Prince George probably has the best prospects of becoming a rival of Victoria because of its strategic location and its lack of potential competitors as the key urban centre serving the resource-rich northern half of the province. As the regional centres become larger they will be able to offer more elaborate urban services to the smaller cities within their orbits, thereby decreasing the dependence on Vancouver, except for the highest level of services.

British Columbia cities are varied in character, but when taken as a group they are distinctive compared with those of other regions of Canada. The mountainous environment and Pacific orientation sets them apart and their populations are more cosmopolitan and well educated than those of cities in other regions. Within the province, however, there is great diversity of character among cities. Measures of differentiation utilizing labor force by industry group and personal income sources indicated several distinct types among the major cities for which data were available. A manufacturing employment orientation in industries such as pulp and paper, sawmilling and mineral processing is characteristic of several major cities throughout the province. This "resource town" character is even more common among smaller cities in which primary mining or forestry employment may be as important or moreso than manufacturing. Commercial cities or central places that serve well defined hinterland areas form another group, of which Dawson Creek is a prime example. Many of the smaller cities are of this type also. Falling in a third category are specialized cities, such as Victoria, that are oriented toward government, tourism, retirement or other special roles. Beyond these are other cities that are really quite diversified and do not fit neatly into any of these categories.

REFERENCES

1. The research assistance of Diana Hocking who compiled some of the information for the maps and tables is gratefully acknowledged.

2. The term Census Metropolitan Area refers to the main labor market area of an urbanized core or continuously built-up area having 100,000 or more population. CMAs are comprised of (1) municipalities completely or partly inside the urbanized core; and (2) other municipalities if (a) at least 40 percent of the employed labor force living in the municipality works in the urbanized core, or (b) at least 25 percent of the employed labor force working in the municipality lives in the urbanized core.

3. The term Census Agglomeration refers to the main labor market area of an urbanized core or continuously built-up area having between 10,000 and 99,999 population. CAs are created in exactly the same manner as CMAs, using whole municipalities or census subdivisions.

4. FORWARD, C.N., "The Regional Distinctiveness of Canadian Cities" in BARR, B.M. (ed.), *Studies in Canadian Regional Geography*. Vancouver: Tantalus Research, B.C. Geographical Series, No. 37, Chapter 10, 1984, pp. 150-171.

5. This study was carried out by Shiu-Yeu Li using the minimum requirements method. LI, S.-Y., "Labor Force Statistics and a Functional Classification of Canadian Cities" in RAY, D.M., *et. al.* (eds.), *Canadian Urban Trends*. Toronto: Copp Clark and Ministry of State for Urban Affairs, National Perspective, Vol. 1, 1976, pp. 96-100.

6. In Figure 2,16 the volume of air passenger traffic is based on figures published in: CANADA, *Air Carrier Traffic at Canadian Airports*. Ottawa: Statistics Canada, 1983, while the scheduled routes are derived from airline timetables published in the winter of 1985.

7. *Ibid.*

8. For a detailed analysis of the 1971 labor force data based on the minimum requirements method of determining the urban economic base see LI, S.-Y., *op. cit.*, pp. 58-102, reference 5.

9. CANADA, *Taxation Statistics*. Ottawa: Revenue Canada, annual.

10. FORWARD, C.N., "Variation of Personal Income Sources as an Indicator of Urban Character Differences" in WATERS, N.M. (ed.), *Aspects of Human Geography: The Kelowna Papers, 1981*. Vancouver: Tantalus Research, B.C. Geographical Series, No. 34, 1982, pp. 63-74.

11. Miscellaneous includes a number of minor sources, such as scholarships, prizes and alimony payments.

12. FORWARD, C.N., "Variation of Personal Income Sources as an Indicator of Urban Character Differences", *op. cit.*, pp. 70-72; FORWARD, C.N., "The Elderly Population in Canadian Metropolitan Areas" in BROWN, M.C. and WYNN, G. (eds.), *The Bellingham Collection of Geographical Studies*. Vancouver: Tantalus Research, B.C. Geographical Series, No. 27, 1979, p. 60; FORWARD, C.N., "The Development of Victoria as a Retirement Centre", *Urban History Review*, Vol. 13, No. 2, 1984, pp. 118-119.

13. FORWARD, C.N., "The Importance on Nonemployment Sources of Income in Canadian Metropolitan Areas", *Professional Geographer*, Vol. 34, No. 3, 1982, p. 294.

14. For a detailed comparison of Vancouver and Victoria with other Canadian CMAs see FORWARD, C.N., *Ibid.*, pp. 289-296.

PLATE 24 A small resource based town on Vancouver Island. ▶
Top - Thompson, Berwick & Pratt Photo; Bottom - J.H. Bradbury Photo

PLATE 25 Port Alberni, a resource based community. *B.C. Government Photo* ▶

17 SINGLE ENTERPRISE COMMUNITIES

J. Douglas Porteous
University of Victoria

The most compelling fact about population distribution in British Columbia is its extreme concentration in the southwest corner of the province, where Greater Vancouver and Greater Victoria alone account for almost 60 percent of the provincial population. The corollary is an extremely lightly-populated hinterland which covers the bulk of the provincial land surface. In this peripheral zone the few regional service centres (Nanaimo, Kamloops, and Prince George, for example) are outnumbered by the prevailing form of hinterland settlement, the company town.

British Columbia is a classic example of the centre-periphery relationships conceived by Marxist economists and geographers and generally associated with the Third World. Trans-national mining corporations, with headquarters in Toronto, Montreal, and abroad, locate subsidiary offices in Vancouver. These, in turn, direct the operations of work sites in the hinterland, appropriating peripheral wealth for the benefit of the metropolis, and, ultimately, for corporations based outside British Columbia.[1] Forestry corporations tend to be headquartered in Vancouver, thus exhibiting the centre-periphery relationship on the provincial scale.[2] Dominated by forestry and mining corporations, British Columbia has been termed "the company province".[3]

This chapter is concerned with the nature of the British Columbia company town. The latter term is loosely defined as any town in which at least 80 percent of the employment is provided by a single company and its supporting institutions.[4] Such towns are now generally known as "single-enterprise communities", "one-industry towns" or even "little companies with big industries".[5] These terms are merely euphemisms which mask the raw fact of high levels of dependence upon a single powerful corporation. Originally, a company town would have company housing and a company store. Since the 1960s most of these have been phased out, but in terms of residents' perceptions and the power structure, the towns remain company towns in all but name.

THE NATURE OF THE COMPANY TOWN

Company towns exist in all Canadian provinces. Yet, British Columbia in the 1950s had more than any other province[6] and, according to recent lists, has today almost as many as Ontario or Quebec, which each contain more than twice as many people as British Columbia (Figure 1,17).[7] Except for the southwestern corner, almost the whole of the province is a resource frontier where Canadians hew wood, draw water, and dig holes. Before World War I, all company towns were privately built by the operating companies and were regarded as wholly private property. Intruders, denied "ingress, egress, and regress", were regarded as trespassers by company officials. To correct the many abuses of such total power, the British Columbia legislature enacted in 1919 an Act to make Provision for Access by the Public to Company Towns, thus making the towns accessible to government officials, union organizers and the like. At this time Nanaimo was a typical company town, its layout planned by coal barons in London, its labor relations hostile and violent, and almost its entire work force employed in mining as late as 1921.[8]

Between the wars Canadian company towns enjoyed a period of "holistic planning" with, for the first time, garden city planning ostensibly in the interests of the employees.[9] Prewar bunkhouse "sleep camps" (Bradwin, 1928), occupied largely by single males, were replaced by larger villages with a less ill-balanced population. The most recent stage is the "comprehensive planning" which followed World War II and which is represented in British Columbia by the "instant towns" that were developed following the adoption of the "Instant Town Act" (Municipal Act Amendment Act) in 1965.

It is now appropriate to detail the principal characteristics of British Columbia company towns in 1971, the last census date before most instant towns became fully operational. The chief characteristics, besides dependence upon the company, are isolation and small size.

A mail questionnaire survey conducted in 1971-72 was directed to the managers of companies which were expected: (1) to provide the bulk of employment, and (2) to provide the bulk of the housing in their communities. This, then is the narrower definition of a company town. All plants operating in or near sizable non-company settlements were omitted, for these would be unlikely to operate company housing. Thus, larger pseudo-company towns which are extremely dependent upon a single industrial type (such as forestry in Port Alberni) were omitted. A 51.2 percent response rate provided usable data on 41 townsites.

All 41 towns were operated by forestry or mining companies which were geared for export to the tune of 82 to 91 percent. Whereas mine sites were

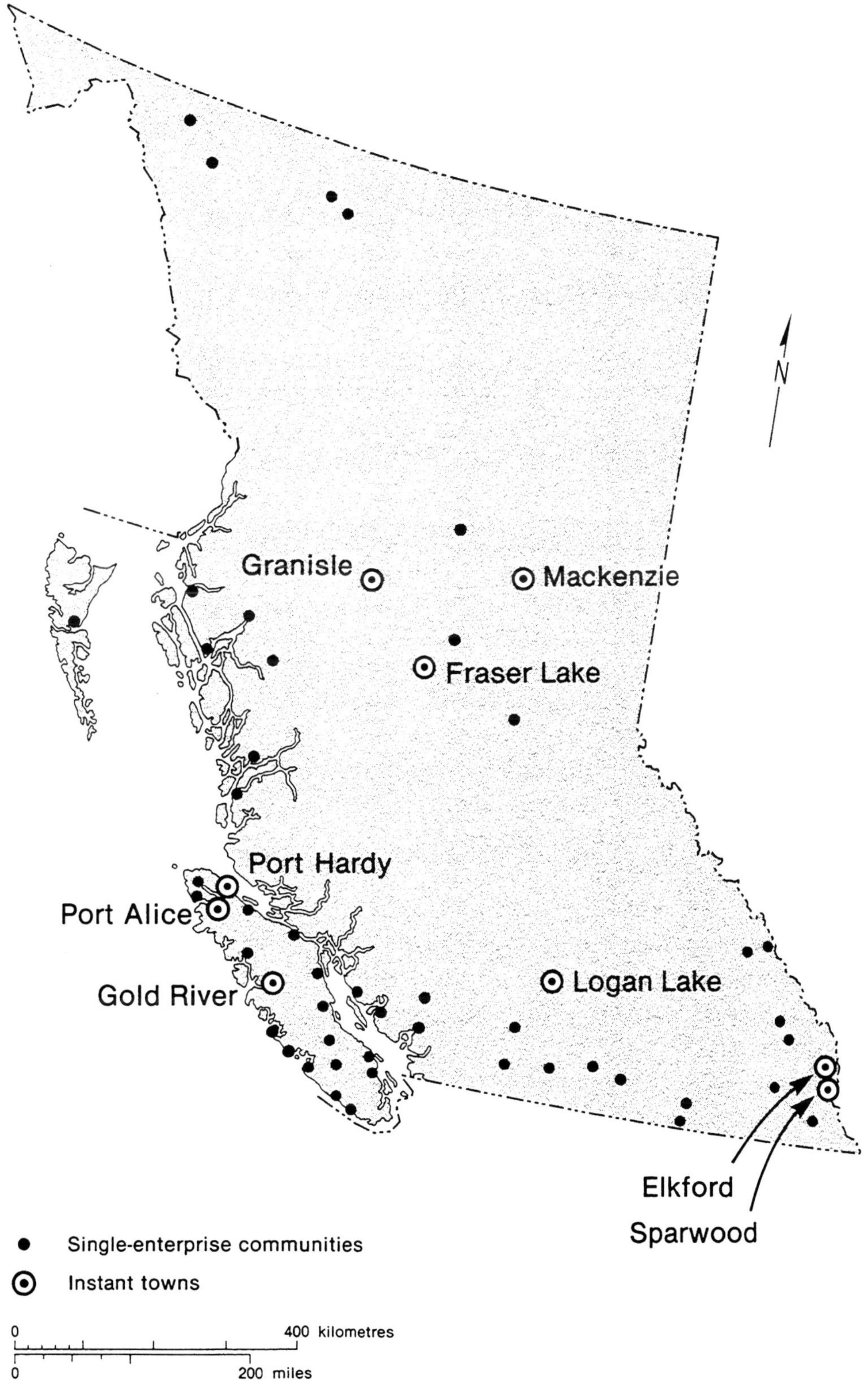

FIGURE 1,17 Single Enterprise and Instant Towns

almost wholly single postwar operations, the vertical integration of the forest industry was matched by several cases of interdependent townsite hierarchies; the Tahsis Company, for example, operates a pulp mill at Gold River (population 2,500), a sawmill at Tahsis (1,400) and several logging operations, including a company camp at Fair Harbor (200). Respondents who did not provide significant amounts of company housing were either very small (under 65 employees) or very large (over 600 employees).

Workforces were almost entirely male. Indeed, the proportion of females employed in these company towns varied from 1.1 percent in logging camps, through 1.5 percent in mining and 1.8 percent in sawmill towns, to 3.9 percent in pulp mill towns. This gradation almost exactly reflected the gradation in mean size between these types of towns. The problem of imbalanced sex ratios in resource frontier towns is common throughout Canada, but reaches its extreme in British Columbia. This Wild West province is, par excellence, Bunge's "macho" or "he-man Canada" (Figure 2,17).[10] This excess of males is, in part, responsible for much of the costly high labor turnover associated with resource towns.

Respondents were asked to indicate whether, in their opinion, their town was isolated. As expected, unincorporated settlements were more likely to be perceived as isolated. Whereas only 21.4 percent of the forestry towns were considered to be isolated, the corresponding figure for mining towns was 80 percent (16 of 20). Forestry towns regarded as isolated averaged only 17.2 kilometres (10.7 miles) from the nearest sizable non-company settlement. In contrast, the range for isolated mining towns was 11 to 209 kilometres (7 to 130 miles), averaging 61.3 (38.1) from the nearest non-company settlement.

Because of this isolated nature of mining towns, a separate study of social infrastructure problems in such towns was completed in 1975.[11] This study derived a measure of the difficult concept of isolation from an index of community services (covering 34 facility types) and from the average mileage to services not available in town. On this basis a 2 × 2 matrix was created, giving four types of towns:

1. A well-appointed town close to a larger centre (examples: Kimberley, Salmo, Sparwood);
2. A poorly-appointed town close to a larger centre (examples: Britannia Beach, Gillis Bay, Logan Lake, Remac);
3. A well-appointed town distant from a larger centre (examples: Fort St. John, Fraser Lake, Port Hardy, Stewart); and
4. A poorly-appointed town distant from a larger centre (examples: Cassiar, Granisle, Hendrix Lake, Elkford).

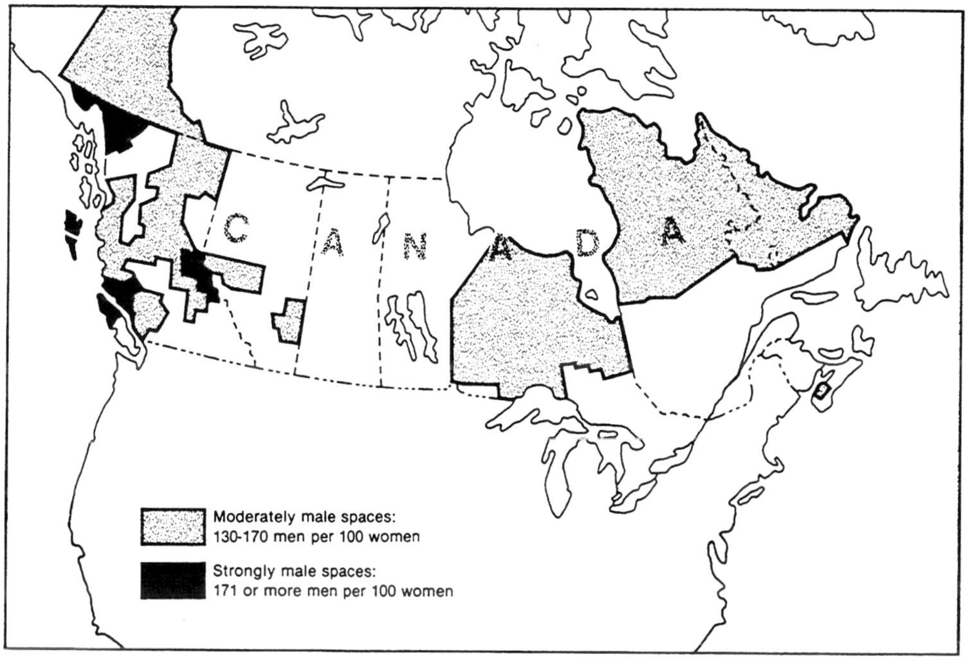

FIGURE 2,17 Male Dominated Areas of Canada

In terms of isolation, type 1 is the least isolated and type 4 the most isolated. Preference for 2 or 3 would depend on personal needs, road conditions, and climate. Whether or not a new resident would be affected by perceived isolation would depend on his or her expectations. The trouble begins when there is more isolation than expected.

Strongly related to isolation was the opportunity to live in a settlement other than the company town. In 85.4 percent of cases sampled in 1971-72 the company provided housing in one townsite only, and in more than one-third of the cases over 90 percent of all workers lived in the company town. Overall, of the 9,923 workers involved 51.7 percent of miners and 62.3 percent of forestry workers used company housing, and as many as 77 percent of loggers.

In terms of problems associated with small size, isolation, and one-company control, unstructured interviewing uncovered significant sources of dissatisfaction with life in British Columbia resource towns, including instant towns. The following points were frequently mentioned: poor alternative opportunities for jobs, recreation, shopping, and education; a feeling of impermanence, the town being expected to decline following resource depletion or corporate retrenchment; inability to reach other centres easily to satisfy needs not met locally; a feeling of company paternalism, despite strenuous attempts by companies to divest themselves of their landlord and service roles; and an overall sense of isolation, expressed in restricted mobility and lack of accessibility to urban services, and poor communications and media reception. Some respondents also felt the lack of a psychic division between work and home, a common phenomenon in situations with inadequate physical separation of these facilities. It is notable that this feeling was less apparent in Granisle, where a short journey by water separates the townsite from the mine site. The individual's solution to these problems is to leave the town. Company towns have registered turnover rates which reach up to 200 percent per year. In such conditions community stability is impossible.

In sum, it appears that the twin ideals of urbanism, choice and control, cannot be fully exercised in the resource town milieu. Indeed, Lucas has suggested that apathy and fatalism are common experiences of company town life.[12] The company town lifestyle may be characterized as a "culture of dependence".

RESIDENT ATTITUDES AND PERCEPTIONS

Canadian research on resource towns since the 1960s has emphasized behavioral approaches and questionnaire methods. In line with recent Australian research,[13] there has also been some investigation of the myths which have developed around the company town.

Folk myth depicts the company town as an oppressive structure wherein the company exercises an excessive degree of influence on social and economic life. With the notable exception of the now defunct Centre for Settlement Studies at the University of Manitoba, most work on residents' attitudes toward and perceptions of company town life has been derived from unstructured interviewing, frequently of male managers. To correct this bias, an attitude survey of the residents of Gold River, Tahsis, and Granisle was undertaken in 1971 to supply data on:

1. motives for migration to the town (money, adventure, inability to cope in cities);
2. social life in isolation (gossip, boredom, age-sex structure);
3. social class problems (segregation, ethnicity);
4. physical and social planning (recreation, community facilities, aesthetics, layout);
5. dependence upon the company (housing, stores, individual responsibility);

6. overall company influence (a general overview); and
7. future prospects (incorporation, impermanence, economic base expansion).

All 34 questions covering these topics were derived from statements in major sources which suggested that certain attitudes were stereotypically held by inhabitants of resource towns.[14]

Respondents were largely Canadian-born (76 - 96 percent), between 20 and 40 years of age (58 - 68 percent), and had moved no more than twice in the last 10 years, an average rate for North America. About half had children in school and half had lived in the settlement less than two years. None of these variables, nor any of the others which were investigated, had any significant statistical relationship with the opinions expressed.

The results of this investigation suggest that generalizations concerning residents' attitudes to company town life should be made only with extreme caution. Of the 34 statements culled from the literature, only nine met with an overall 60 percent agreement or disagreement response in all three towns. Significantly, these nine areas of similar response were almost wholly negative responses in terms of satisfaction with company town life. Though there was a general perception that ethnic discrimination was absent and that the company did not over-organize recreational facilities, there was a general consensus that: population structures were ill-balanced, with a dearth of females; the towns were architecturally monotonous; opportunities for youth were limited; dissatisfaction among women was evident; and incorporation, accompanied by the broadening of the economic base, would prove ameliorative.

Similarly, statistical analysis (multivariate analysis of variance) showed response pattern congruence over all three towns in nine areas only, of which four cases involved a similar sharp dichotomization of opinion. Outside the "future prospects" area there was general agreement only in the belief that migration to the resource town was often motivated by migrants' inability to cope elsewhere, a rather defeatist opinion. Only in the area of future prospects was there consistent agreement among respondents of all three towns over all relevant questions, and this agreement was clearly in favor of change.

The terminal question asked: "And finally, do you consider this town to be a company town?". This was, perhaps, a leading question, but the affirmative response was overwhelming: Tahsis (90 percent); Gold River (82 percent); and Granisle (75 percent). It is significant that at least three-quarters of the respondents in Gold River and Granisle, both instant towns, were of the opinion that they lived in a company town.

TOWARD THE INSTANT TOWN

During the 1970s a wide variety of attempts were made by companies to counteract feelings of isolation. Southeast coal companies helped new employees with removal costs, as did Utah Mines at Port Hardy. The latter, moreover, provided a grant for travelling/shopping expenses before the paved road to Campbell River was completed. Granduc Mines at Stewart recognized its isolation problem to be so acute that it provided not only free air tickets to Vancouver, but also a subsidized company boat to Prince Rupert.[15]

Further, some companies have attempted to relieve the problems of female unemployment and the prevailing lack of single females by permitting females to undertake traditionally "male" jobs. A number of women are employed as coal miners in the southeast, and at Granisle a job-sharing scheme was begun for the wives of employees. Nevertheless, feminist opinion is that the new female employees are generally regarded as an expendable labor source suitable only for unskilled, repetitive, and non-supervisory work.[16] The current development of coal mining in northeast British Columbia is predicated upon enhanced employment opportunities for women, native Indians, and the locally unemployed.[17] It is too early, however, to assess the success of the town of Tumbler Ridge.

The sale of housing to employees has also become common, although house ownership in an isolated town dependent upon a single, non-renewable resource is a dubious proposition, even with buy-back plans. Australian research, indeed, suggests that homeownership in resource towns might have negative effects.[18] Further, the development of innovative housing for extreme climates may be difficult to implement, for resource town residents overwhelmingly desire southern Canadian housing styles and suburban layouts, even when climatically inconvenient, as in Kitimat.[19]

Phasing-out of company housing is proceeding fastest in the less isolated towns. Port Mellon, a pulp mill site founded in 1908, is an example. Until the 1950s the company provided housing for all its employees, as the only means of transportation to and from the site was by water. With the establishment of a road between Port Mellon and Gibsons in the late 1950s, however, employees began to leave the townsite. By 1971 83 percent of company employees were living in the Gibsons and Sechelt areas and the former town had been reduced to a mere 35 units, some of them trailers for single men.

But the most comprehensive planning strategy emerged with the "Instant Town Act" of 1965. This permits the incorporation of a municipality, in conjunction with resource development, upon the petition of five local landowners. Homeownership, incorporation, a wide range of services, and well-regulated planning are techniques of instant town development, which takes place through contractual collaboration between a company and the provincial government.[20] The overall goal is to encourage population stability by providing a well appointed town with at least an appearance of permanence and evidence of both corporate and governmental commitment. The time was ripe for this development, for by the 1960s many local companies had sold out to trans-nationals which no longer wished to continue with the onerous burden of providing and managing resource towns. This burden has now been involuntarily undertaken by the taxpayer and the instant town resident.

In a series of papers, Bradbury has evaluated the success of the nine instant towns created under the aegis of the Social Credit government in the period 1965-1972.[21] The overall assessment is negative. Although physical conditions have indeed been improved, attempts to change community age and sex structures and reduce labor turnover have been less successful. Indeed, the attraction of former city residents to instant towns may have created or exacerbated problems of community and personal instability. Moreover, the emphasis on homeownership has meant that many residents feel locked in by mortgages, high taxes, high living costs, and poor opportunities for profitable resale.

According to Bradbury, class stratification is just as apparent in instant towns as in the old company towns, and women still have few job opportunities. In appearance, there may be little difference between the old company town and new instant town, for in the latter:

> Snob hill is a lawned cul-de-sac, part encased by standing timber, which presides over the main residential area . . . . Below snob hill are spread single family dwellings in descending order of price. Below them, further still and decently hidden from the picture windows of mill managers, is a dirty trailer park.[22]

Furthermore, despite the creation of "instant" municipal governments, company officials still make up the largest percentage of town councillors, while businessmen, sometimes granted monopoly rights by companies, dominate mayoralties. Overall, company workers have occupied only 13 percent of mayoralties and 28 percent of councillor positions. Local control by company officials is matched by the overall control of the mainly trans-national companies. Of the nine instant towns extant in 1974, only one was not ultimately controlled from Eastern Canada, the United States, Europe, or Japan. This position has been ameliorated to some extent in the last decade. Instant towns, therefore, are even more susceptible than the older company towns (some of which were indigenously-controlled) to decisions made in distant metropolitan power centres. "A picture emerges of instant towns with fairly rigidly defined classes, controlled economically from afar by multinational corporations, and still showing a lot of company influence in the new arena of local politics."[23]

THE FUTURE: REGIONAL PLANNING?

Until the northeast coal development townsite at Tumbler Ridge can be evaluated, perhaps in the 1990s, there will be no evidence of successful large-scale regional planning in the British Columbia resource frontier. It is significant that Riffel's overview of major regional planning ventures in the Canadian resource frontier has no entry for British Columbia.[24] Tumbler Ridge is designed on modern, behavioral planning principles,[25] including those of environmental criminology and, unlike most resource towns, will serve two companies' mines. This "megaproject" may be a step towards rational regional planning in the British Columbia resource frontier.

Canada, in general, lacks a basic resource frontier settlement policy. The following is a first step toward such a policy. In regional planning terms should a decision be made to co-ordinate exploitation of a large area of the resource frontier, a series of alternative decision-paths are available (Figure 3,17). The main choices are related to scale and permanence.

If the area is not to be permanently settled a range of alternatives is available, including the strategy most frequently used today, the instant or company town (CT). These include:

1. *The Mobile Town* (MT), a non-permanent settlement, is well adapted for the exploitation of resources with a short working life. Such a town is composed of mobile homes or houses which can be dismantled or demounted, specially designed structures which may each be plugged into a central utilities core. The chief advantage of the MT is that upon resource depletion or corporate withdrawal the whole settlement may be moved to another site by truck, on skids, or by helicopter.
2. *The Long-Range Commuter Town* (LRCT) involves the very minimum of construction at the work site, where only a bunkhouse complex is needed. This is occupied on a week-on/week-off (oil rig) fly-in basis by workers whose permanent homes are in pre-existing southern towns or in a Multi-enterprise Community.

Should a policy of large-scale, long term settlement of the resource frontier be decided upon, the existing scatter of CT's should be replaced as far as possible by:

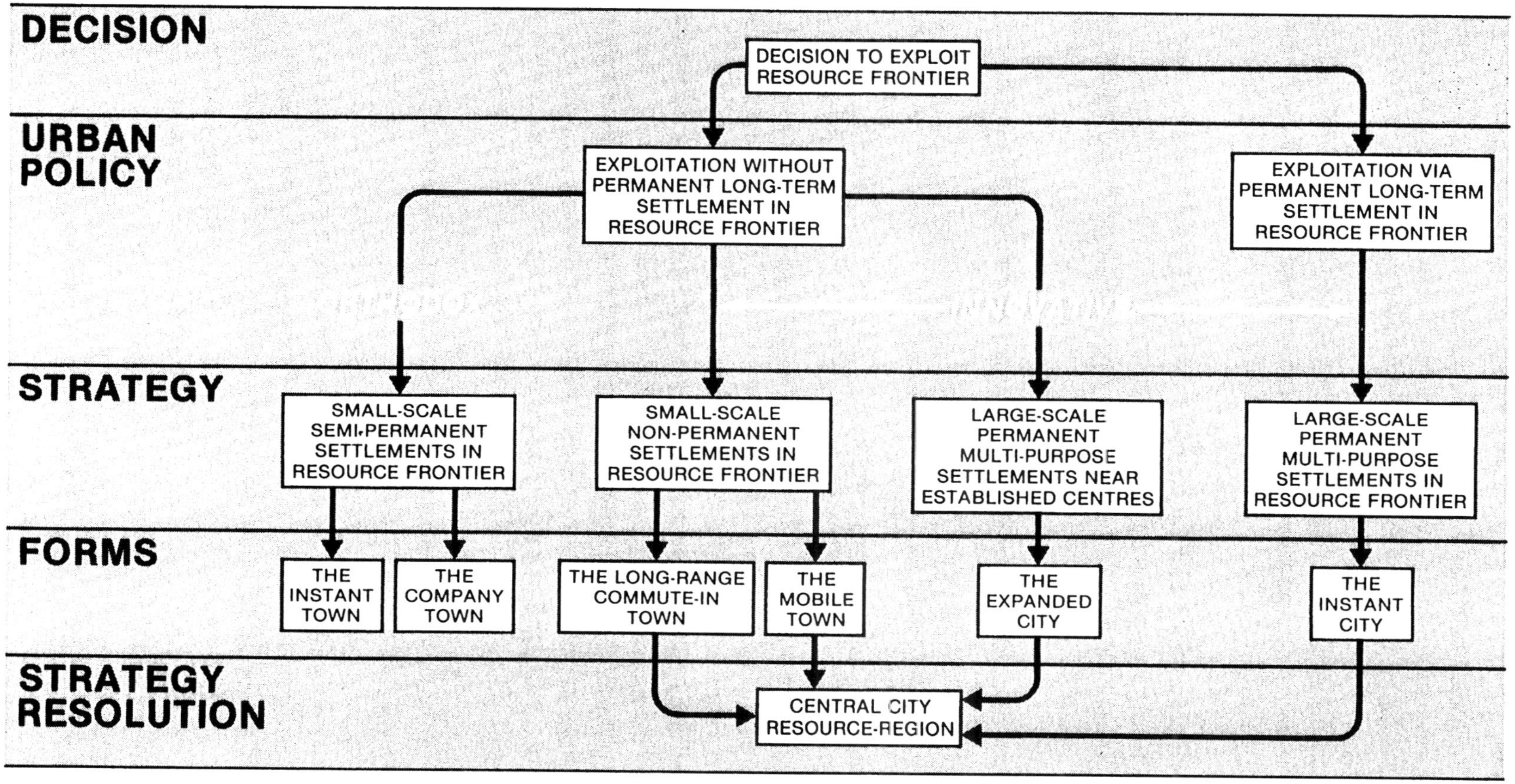

FIGURE 3,17 Alternative Regional Planning Decision-Paths

3. *The Multi-Enterprise Community* (MEC), as a large permanent settlement, avoids many of the problems associated with CT's. The MEC requires a broad economic base, and should, therefore, be strategically located so as to provide basic employment in forestry, mining, fishing, tourism, government, and a wide variety of ancillary and non-basic occupations, all within daily commuting range. The MEC should support a far wider range of services and amenities than even the best equipped CT.

A development of the MEC concept is the Central City Resource Region (CCRR), which involves a regional integration of the MEC with the MT and the LRCT (Figure 4,17). Around a large population centre of considerable size are:

1. a zone of no settlement where work sites are within daily commuting range;
2. a zone of MT's within weekend visiting range of the central city; and
3. a zone of LRCT's beyond this.

This concept is flexible. If well sited the city might need to operate only the daily commuting zone. Alternatively, the zone of mobile towns could be omitted in favor of a daily/weekly commuting dichotomy.

Central cities may be located directly in resource rich regions (instant cities) or in or near pre-existing settlements (expanded cities). Such a regional population complex would require major transportation facilities which would reduce isolation and company town problems. Job opportunities should expand to absorb more school graduates; families and single females should find the CCRR more attractive than the CT; retirement might be encouraged; and the resistance of professionals to taking up northern posts might be lowered. The result should be a more balanced population with, for the first time, adequate numbers of women, children, and the elderly; reduced socio-psychological problems; reduced need to travel south; a healthier economic base; and a sound basis for mature local government. In the CCRR resource depletions may result in structural employment shifts, rather than economic uncertainty or the death of the towns, as frequently happens with the CT. Above all, some degree of individual choice and control would be possible. Life in the central city of the CCRR might still resemble that in a resource town, but it would in no way be a company town.

Planning a CCRR system would require co-ordinated co-operation between three levels of government, many corporations, native peoples, and both existing and potential residents. A public development corporation

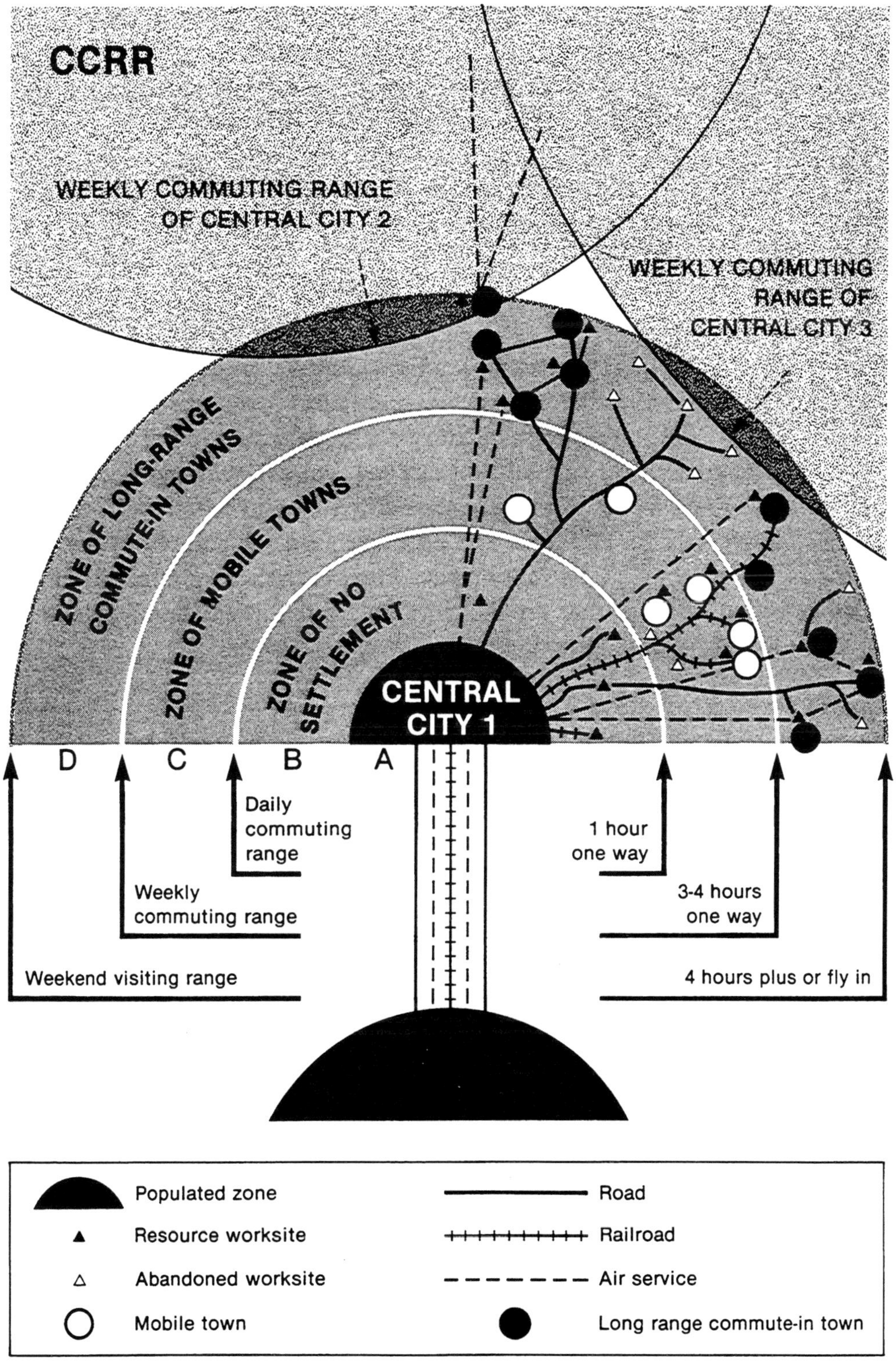

FIGURE 4,17 The Central City Resource Region

would be required. Location is crucial: northern British Columbia is not the isometric plain beloved of central place theorists. Above all, implementation would require co-ordinated, long range resource development planning, rather than the fragmented, short range planning which prevails today.

CONCLUSION

About two percent of British Columbia's population lives in true company towns where the majority of the population occupies rented company housing. A much larger population occupies pseudo-company towns where company housing is of little importance, but where a single company still provides most of the employment. All the instant towns fall into this latter category and, to all intents and purposes, are still company towns in terms of social structure, isolation, size, local company control, ultimate alien control, and persistent problems. Only innovative regional planning procedures are likely to permanently solve this problem.

REFERENCES

1. BRADBURY, J.H., "The Instant Towns of British Columbia: A Settlement Response to the Metropolitan Call on the Productive Base" in EVENDEN, L.J. (ed.), *Vancouver: Western Metropolis*. Victoria: University of Victoria, Department of Geography, Western Geographical Series, No. 16, 1978, pp. 117-133.

2. HAYTER, R., "Forestry in British Columbia: A Resource Basis of Vancouver's Dominance", in EVENDEN, L.J., *op. cit.*, pp. 95-115.

3. ROBIN, M., *Pillars of Profit: The Company Province 1934-1972*. Toronto: McClelland and Stewart, 1973.

4. LUCAS, R.A., *Minetown, Milltown, Railtown: Life in Canadian Communities of Single Enterprise*. Toronto: University of Toronto Press, 1982.

5. BOWLES, R.T., *Little Communities and Big Industries*. Toronto: Butterworth, 1982.

6. INSTITUTE OF LOCAL GOVERNMENT, *Single-enterprise Communities in Canada*. Kingston: Queen's University, 1953.

7. PRESSMAN, N.E.P., *Planning New Communities in Canada*. Ottawa: Ministry of State for Urban Affairs, 1975; RIFFEL, J.A., *Quality of Life in Resource Towns*. Ottawa: Ministry of State for Urban Affairs, 1975.

8. GIDNEY, N., "From Coal to Forest Products: The Changing Resource Base of Nanaimo, B.C.", cited in STELTER, G.A. and ARTIBISE, A.F.J., "Canadian Resource Towns in Historical Perspective", *Plan Canada*, 18, 1978, pp. 7-16.

9. STELTER, G.A. and ARTIBISE, A.F.J., *op. cit.*, reference 8.

10. BUNGE, W.W. and BORDESSA, R., *The Canadian Alternative: Survival, Expeditions, and Urban Change*. Toronto: York University Geographical Monographs, No. 2, 1975.

11. BANCROFT, C.G., BANCROFT, K.B., and PORTEOUS, J.D., *Mining Communities in British Columbia: A Social Infrastructure Analysis*. Victoria: Ministry of Mines and Petroleum Resources, 1975.

12. LUCAS, R.A., *op. cit.*

13. BREALEY, T.B., "Popular Myths and Folklore Concerning Planning in New Mining Towns in Australia", *Man-Environment Systems*, 10, 1980, pp. 146-152.

14. INSTITUTE OF LOCAL GOVERNMENT, *op. cit.*; LUCAS, R.A., *op. cit.*; ROBINSON, I., *New Industrial Towns on Canada's Resource Frontier*. Chicago: University of Chicago, Department of Geography, Research Monograph, No. 73, 1962.

15. BANCROFT, C.G., *et. al.*, *op. cit.*

16. VEIT, S. and Associates, *Women in Mining*. Victoria: Ministry of Economic Development, 1976.

17. BRITISH COLUMBIA, *Report of the B.C. Manpower Sub-committee on N.E. Coal Development*. Victoria: Ministry of Economic Development, 1976.

18. NEILL, C.C. and BREALEY, T.B., "Home Ownership in Resource Towns: Will it Change or Reinforce Existing Social Trends?", *Human Resource Management*, 20, 1982, pp. 38-44.

19. PORTEOUS, J.D., "Resource Towns in B.C. Opt for Wilderness Suburbs", *Habitat*, 17, 1974, pp. 2-7.

20. PORTEOUS, J.D., "Gold River: An Instant Town in British Columbia", *Geography*, 55, 1971, pp. 317-322.

21. BRADBURY, J.H., *op. cit.*; BRADBURY, J.H., "Class Structures and Class Conflicts in 'Instant' Resource Towns in British Columbia, 1965-1972", *B.C. Studies*, 37, 1978, pp. 3-18; BRADBURY, J.H., "Instant Resource Towns Policy in British Columbia: 1965-1972", *Plan Canada*, 20, 1980, pp. 19-38.

22. *Vancouver Province*, August 18, 1971, p. 16.

23. BRADBURY, J.H., "Class Structures . . .", *op. cit.*, reference 21, p. 12.

24. RIFFEL, J.A., *op. cit.*, reference 7.

25. PORTEOUS, J.D., *Environment and Behavior: Planning and Everyday Urban Life*. Reading, Massachusetts: Addison-Wesley, 1977.

PLATE 26 Gastown in Vancouver. *B.C. Government Photo*▶

CAMBIE ST.
100
307
305

18 TOURISM

Peter E. Murphy
University of Victoria

INTRODUCTION

The year 1984 was declared "The Year of Tourism" in Canada, in an attempt to increase public awareness of this industry and its potential for our country. This chapter has a similar goal. It will describe the tourism industry in British Columbia, examine its problems and prospects, and comment upon strategies that have been proposed to guide its development over the next 20 years. The perspective taken in this chapter is that tourism is a resource industry, much like forestry and fisheries, but with some important differences.

First, although tourism depends heavily on the quality of the natural environment for its appeal, these resources are not the complete product it sells. Along with attractive scenery, one must consider British Columbia's cultural heritage, level of hospitality and competitive value. Therefore, tourism's prospects and future are linked to many facets of our daily lives.

Second, it is a unique resource industry in that the customer travels to use the resources, rather than awaits the import of these resources to his locality. Since nearly every community in British Columbia has some form of touristic resource, be it scenery, heritage, or facilities, it can expect a certain number of visitors, and in some cases a great number. Furthermore, since public goods often form the major attractions (beaches, parks, museums, heritage buildings) and the facilities are used by both visitors and residents alike (restaurants, theatres, museums), almost everyone comes into contact with this industry and its customers. Therefore, whether or not individual residents and businesses feel they are part of a community's tourism industry, they are part of tourism's local product.

Third, although tourism has been described as a renewable resource industry, with visitors expected to view rather than consume the product, leaving the same resources for future visitors and generations, this cannot be taken for granted. The personal nature of the tourism experience means that under outwardly satisfactory conditions individual perceptions of crowding, degradation or commercialization can mar a visitor's experience. When a destination area becomes popular the associated volume of visitors cannot fail to alter its physical and social environment; while there may be little actual consumption, there may be considerable modification of the local ecosystem.

From these brief observations it is apparent that tourism is a complex and multi-faceted resource industry. There has been considerable debate on what constitutes a tourist and how to distinguish between recreation and tourism. A more detailed discussion of these questions is presented in a previous volume of this series,[1] but as far as this chapter is concerned tourism is defined as:

> . . . the travel, the activities, and services used by any British Columbia resident beyond a 40 kilometre (25 miles) radius from home for the purpose of personal enjoyment and the travel, the activities and services used by non-residents who enter the province for any reason other than work. It is recognized that elements of tourism are often involved when people travel for business purposes.[2]

This comprehensive definition includes regular vacations, weekend and recreational travel that takes someone beyond his local community boundaries, and the tourist activities associated with business travel — particularly conventions. It includes facilities and services that may be designed primarily for resident use, yet are used by tourists during their stay, as well as the more obvious accommodation and attraction facilities associated with the industry.

SIGNIFICANCE OF TOURISM

British Columbia's tourism has grown in conjunction with tourism's worldwide expansion since World War II. By 1982 some 280 million international tourist trips were being taken annually, but the bulk of tourism still consisted of travel within one's own country (domestic tourism) which, in 1981, accounted for 2.3 billion trips and generated 813 billion dollars.[3] By 1984 tourism was the third largest revenue earner in British Columbia, with a revenue of 2.325 billion dollars, providing an estimated 100,000 person years of employment in both full and part-time positions combined.[4] In 1984 the single largest source of visitors and revenue was British Columbia itself, and when other Canadian visitors were counted the

Table 1,18 REVENUES AND VISITOR VOLUMES BY MARKET AREAS IN 1984

	Revenue (millions of dollars)	*Percentage*	*Visitors* (thousand person trips)	*Percentage*
British Columbia Residents	995	43.0	5,600	46.0
Other Canadians	665	29.0	3,160	26.0
United States Visitors	470	20.0	2,970	24.0
Overseas Visitors	195	8.0	470	4.0
TOTAL	2,325	100.0	12,200	100.0

SOURCE: British Columbia, *Tourism Highlights, 1984*, Ministry of Tourism, 1985, p. 2.

domestic tourism component accounted for 72 percent of the revenue and visitors (Table 1,18).

The major attractions of this province, and the country in general,[5] are its scenic landscapes, open spaces and outdoor recreational opportunities — all of which have a strong appeal in an urbanized world. British Columbia's size, terrain and coastal location create a variety of vegetation types that support diverse wildlife species and offer a wide range of land and water based recreational activities. These include skiing, sailing, mountaineering, river rafting, fishing and hunting.[6] But most visitors seem content with general sightseeing through the car windows. A 1979 out-of-province visitor survey, for example, revealed that sightseeing was the most important activity for 26 percent of the respondents, and passing through for another 11 percent.[7] In comparison only 12 percent rated outdoor activities as their most important activity.[8]

Despite the province's "super natural" appeal and the widespread availability of such attractions, most tourist traffic focusses on four of the province's nine tourist regions (Table 2,18). This imbalance comes about because of the varying degrees of accessibility and opportunity for multipurpose trips within the province. Since most tourists still travel by car, good highway connections tend to channel visitors into certain regions.[9] The effect of this can be seen in Table 2,18 and Figure 1,18 where three of the top regions straddle the Trans-Canada Highway. Towns and cities also have great appeal because they not only offer a wide range of facilities and entertainment, but also a greater chance to visit with friends and relatives.

Table 2.18 REGIONAL MARKET SPACE OF TOURIST REGIONS IN 1984

	Region	*Person Trips* (thousands)	*Percentage*	*Revenue* (mill. of dollars)	*Percentage*
A.	Vancouver Island	2,465	20.2	505	21.7
B.	Southwestern B.C.	3,595	29.5	765	32.9
C.	Okanagan-Similkameen	1,540	12.6	275	11.8
D.	Kootenay Country	550	4.5	95	4.1
E.	High Country	1,445	11.8	210	9.0
F.	Cariboo	495	4.1	90	3.9
G.	North by Northwest	825	6.8	185	8.0
H.	Peace River-Alaska Highway	300	2.5	60	2.6
I.	B.C. Rocky Mountain	985	8.1	140	6.0
	TOTAL	12,200	100.1	2,325	100.0

SOURCE: British Columbia, *Tourism Highlights, 1984*, Ministry of Tourism, 1985, p. 6.

This last fact was the most important visitation motive for out-of-province visitors in the 1979 survey, being cited as the major reason for visiting by 28 percent of them.[10] Another reason for the visitor distortions is the overwhelming attraction of water. Vancouver Island and southwest British Columbia offer seascapes, boating, salmon fishing and beach activities, while the Okanagan-Similkameen and the High Country offer lake activities and river rafting.

This province's natural and cultural resources combine to create a tremendous potential for the tourist industry — one which both individuals and government are eager to grasp. Tourism has not only been one of the strongest growth industries since the 1950s, but also it has shown unexpected resilience during the recent recession. While some of British Columbia's major resource industries have experienced significant business downturns during the early 1980s, tourism's visitor volumes declines by only two percent and its direct revenue, measured in current terms, increased by 29 percent between 1980 and 1984.[11] Hence, tourism and tourism related activities have begun to receive greater attention and investment.

Examples of this renewed interest in tourism's potential to pick up the slack resulting from declining primary industries can be seen in Kimberley and Port Alberni. The City of Kimberley has adopted a tourism theme

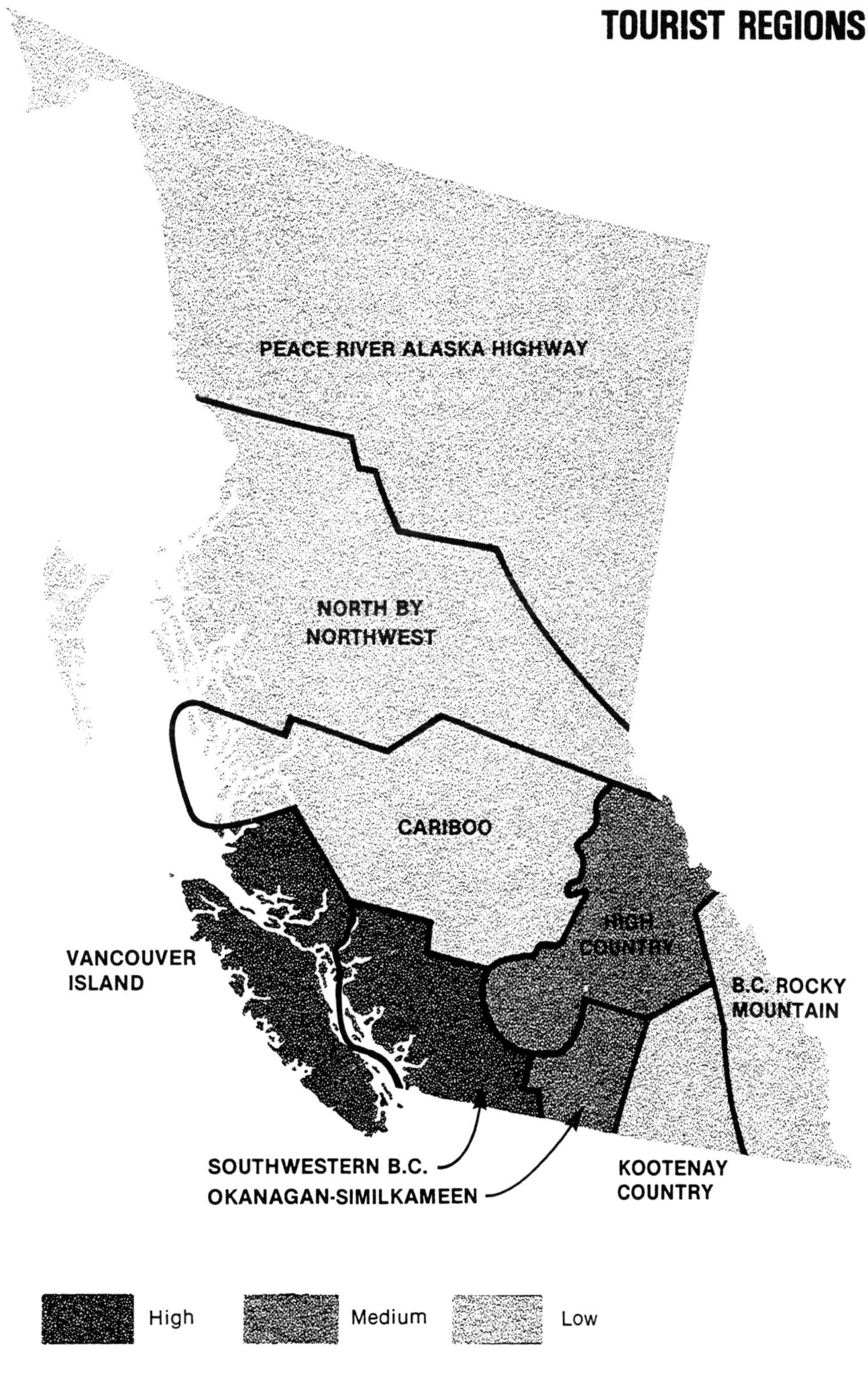

FIGURE 1,18 Ranking of Tourist Regions by Person Trips and Revenue, 1984

which it hopes to develop into a year-round business to replace its dependency on a local mine. It faces the closure of its Cominco mine by the year 2000 and has decided to develop its small ski operation into a year-round resort destination, using a "Bavaria of the Rockies" theme to unite the town and its various tourism components.[12] Port Alberni has experienced high unemployment with the recent declines in salmon and lumber prices, so it has cleared and developed part of its unused industrial waterfront as a tourist attraction, in the hopes of diversifying its local economy.[13]

Tourism's other major attraction in these hard economic times is that it is a labor-intensive service industry. While other British Columbia industries are resorting to further automation in an attempt to remain competitive in world markets, tourism needs workers in order to provide a world class product and experience. Man-years of employment now form an important consideration in any development proposal, for as James Callaghan noted when he was Prime Minister of Britain:

> Now new plants in manufacturing and new investment and new methods bring about greater efficiency but it does not necessarily mean more jobs, and for this reason, we need to look at the service industries, of which tourism is a notable example, as an additional and important source of income and work.[14]

Much of the employment generated by tourism will be for the young and less skilled workers, the largest unemployment group in British Columbia. While many of the positions will pay the minimum wage, there are also numerous avenues for advancement to more responsible positions and the ever present opportunity to start a related small business — the prime source of new investment and employment in the present economy.

These are some of the reasons behind the provincial government's mega-project of Expo 86, with its associated B.C. Place, Canada Place and light rapid transit system. A major investment of 1.5 billion dollars has gone into Expo 86, another 137 million into Canada Place, and 126 million to build B.C. Place stadium. These developments will replace derelict and dying industrial areas surrounding Vancouver's core with service-related functions that will complement its commercial activities. While there is some debate over Expo's direct revenue and cost picture,[15] it will undoubtedly leave a tangible legacy behind that will enhance the existing hospitality and retail sector of downtown Vancouver. The fact that such an investment is being undertaken in these difficult times reflects the province's faith in the industry, and the fact that this gamble has generated so much debate reflects not only the New Orleans World Fair problems, but also the contentious nature of tourism development in British Columbia.

ISSUES

Examination of the tourist industry in the late 1970s revealed several issues which needed to be addressed if the province were to maintain its market position, even before it considered expanding this resource industry in a growing, competitive market. These concerns were expressed in the provincial government's tourism discussion paper and can be condensed into four major groups.[16]

1. Conservation — the need to retain attractive environmental and social resources;
2. Investment — the need to attract investment and improve the quality of the "tourism plant";
3. Administration — the need to co-ordinate the industry's efforts and integrate the goals of various jurisdictional levels; and
4. Planning — the need for long range development planning that could adjust to changing markets and circumstances.

Each of these issues will be examined in the light of British Columbia's experience and the proposed tourism strategy for the province.

Conservation

Industries must have a saleable product in order to prosper and tourism is no exception. A prime concern, therefore, is to conserve and protect the integrity of British Columbia's scenic resources and outdoor recreation opportunities, since these have proven to be its major tourist attractions. With many claims being made on the province's landscape and resources by other industries, some form of integrated resource management would appear desirable. Two areas where this is particularly relevant concern the visual repugnance of clear cutting in the forest industry and the squabble over declining salmon stocks between the sports fishermen and commercial fishing interests.

Clear cut harvesting has been championed as the most economical way of harvesting our timber resources, but the visual aftermath of stumps, debris and bare soil contradicts the province's "super natural" promotions. This contradiction has not been missed by travellers and members of the public, as various letters to the editor attest:

> I toured your beautiful province last month and, even as the ads say, it was Super Natural. Except for one thing: everywhere we went, those magnificent mountain slopes had been

devastated by clear-cutting. A nuclear blast couldn't have wrought greater or more unsightly destruction.[17]

After reading for months about the dispute over logging Meares Island, I have come to the conclusion that no matter how important and valuable the land is to the Indians, tourists and residents of the area, for its natural beauty and historic value, money is the most important thing to B.C. . . . (Serena Meares, a direct descendant of Capt. John Meares).[18]

Nor has the contradiction been ignored by the Ministry of Forests. It has produced a report that recommends a landscaping approach to tree harvesting in aesthetically sensitive areas.[19] It recommends that forest companies leave a buffer zone of uncut timber between the clear cut areas and prime tourist routes and attractions. This suggestion calls for some financial sacrifice on the part of forestry companies, namely the loss of accessible timber, but in some cases such a financial loss in one industry can be recouped by larger returns in another.

Unfortunately, the presentation of this landscaping report and concept has had little impact on the forest industry, where various companies are committed to exercising their timber licence rights to the full. A recent example of this unwillingness to acknowledge the aesthetic and commercial value of a scenic forest landscape occurred on Meares Island. This island lies opposite Tofino, a service centre for the Pacific Rim National Park, and is visible from various lookouts in the park itself. At present it offers a scenic backdrop to a growing tourist destination, but despite the pleas of local residents and business groups (many tourist related) Macmillan Bloedel has exercised its tree licence rights and announced plans to log the island, including parts that are visible from Tofino and the park. A similar conflict is emerging over the future of the South Moresby archipelago in the Queen Charlotte Islands, where ecological and tourism-recreation interests are lining up against the forestry and mining industries.[20]

There is discord over the future and best use of British Columbia's salmon fishing resources. There is little doubt that the stock needs to be conserved and rebuilt, but at the same time two groups are vying for a bigger share of the harvest. Commercial fishermen feel that sports fishing is contributing to their demise, while sports fishing interests feel that their share of the catch should be increased.

A strong advocate for the recreational interests is Alec Merriman. He maintains that the commercial fishing industry not only takes too many fish, but that its harvesting methods waste the resource by killing thousands of sub-legal size grilse in seiner drums as they pull their nets aboard.

He advocates that more of the catch be allocated to sports fishing because it produces a higher dollar return on each fish caught, when one considers the money spent on guides, boats, gear, accommodation and food by the visitors or local fishermen. Merriman suggests that a more balanced use of this resource can be achieved along the lines developed by the Great Lakes states, where sports fishing has been built into a "one billion dollar world class recreational fishery".[21]

Evidence that the province needs to focus on the quality and variety of its resource attractions can be found in its declining relative competitiveness. The province offers nothing in its product mix that cannot be found elsewhere in North America, and closer to the major tourist generating areas of the East. In addition, its northern location makes summer weather unreliable and winter untenable for most forms of recreation and many constitutions. Thus, it comes as no surprise that a basic indicator of relative competitiveness, the travel budget balance, has long recorded a deficit. For 1984 it was estimated that British Columbians spent over 200 million dollars more outside the province than visitors brought into the province.[22]

Investment

To make the province appealing so that more visitors come and more residents stay at home will require more than conservation; it will take investment in facilities and the development of world class attractions. Investment will be needed to update the existing plant, create new attractions and facilities, and to overcome the major problem of seasonality. The cost of such investment can be considerable and frequently requires government assistance, especially when major projects or infrastructure developments are needed.

An example of government investment support for major projects tapping new tourist resources in the province is the 1978 Travel Industry Development Subsidiary Agreement (TIDSA) between the federal and provincial governments.[23] The agreement had five objectives, or subprograms, and of these the three investment programs took the lion's share of the 50 million dollar budget (Table 3,18). A major beneficiary of the skiing development funds was Whistler Mountain and Resort, where over 10 million dollars was spent to create a world class ski resort intended to compete with such well known ski areas as Vail and Aspen, Colorado, and Mount Baker, Washington.

Similarly, investment to extend the tourist season has become an area for government assistance. Private investors are unable to justify major investment in slack periods, but if the costs were spread over various tourist

Table 3,18 FIVE SUB-PROGRAMS OF BRITISH COLUMBIA'S TRAVEL INDUSTRY DEVELOPMENT SUBSIDIARY AGREEMENT

	Program	*Projected Costs in Dollars**
1.	Studies and Planning	1,500,000
2.	Travel Industry Organization	4,500,000
3.	Travel Industry Upgrade	13,000,000
4.	Travel Generators	15,000,000
5.	Skiing and Related Facilities	16,000,000

*Costs shared equally by federal and provincial governments.

SOURCE: Canada and British Columbia, *Canada-British Columbia Subsidiary Agreement on Travel Industry Development*, Ottawa and Victoria: Department of Regional Economic Expansion and Ministry of Economic Development, October 17, 1978, p. 21.

businesses and non-tourist users then such investment could be justified. This is the basis of government involvement in convention centre construction. It is felt that the high cost of these facilities, which attract off-season business, can be justified because of the indirect revenue and taxes generated over a wide range of businesses, and the opportunity to provide multi-purpose public facilities. For example, the federal government's Trade and Convention Centre Program committed 69 million dollars worth of grants toward the construction of convention and trade centres between 1981 and 1985.[24] British Columbia's portion of this government investment is Canada Place on the Vancouver waterfront. It will house the Canadian pavillion during Expo 86 and revert to being a cruise ship terminal and conference-trade centre after the exhibition closes.

Administration

A major problem facing the province in the conservation of British Columbia's resources and the development of a world class tourist industry that can compete with other destinations is the numerous competing interests and jurisdictions involved, plus the fragmented nature of the industry itself. For tourism development to be effective there is a need for multiple use management so that the needs of various resource industries can be met. To achieve this requires collaboration between the various industrial sectors to a greater extent than has happened in the past. The overall fragmentation of the industry makes it easier for conservation and investment conflicts to arise, and as Gunn observes:

> . . . the evidence of fragmentation of tourism development is mounting. As the volume of tourism expands and as development increases, many segments of tourism are running on collision courses with growing frequency. These conflicts appear to be expressed in reduced satisfactions to visitors, reduced rewards to owner-managers, and erosion of basic resource assets.[25]

One attempt to bring about greater co-ordination and administrative efficiency in British Columbia's tourist industry was made by the TIDSA program. The second sub-program was designed to correct for the lack of organizational and institutional structures required to enunciate the needs of the industry and co-ordinate its development. As a result, TIDSA proposed to spend 4,500,000 dollars on creating and subsidizing an organizational structure for the province's tourist industry. This was to include the creation of a Travel Industry Advisory Board to liaise with the provincial government and reflect the industry's concerns and aspirations; the development of regional and community travel industry organizations to interact with local governments; and the creation of industrial boards or commissions to police the industry and assure that basic standards of operation were maintained.[26]

However, as Montgomery and Murphy discovered in their examination of TIDSA's implementation process, it is one thing to set objectives and goals for this industry; it is another to pull all the elements together and achieve co-ordinated development.[27] They noted that there was considerable doubt concerning the public sector's role in tourism development. Tourism was seen as a bastion for small business and some involved in the industry were unsure whether or not the government was acting as a "banker, political patron, project co-ordinator or prospective competitor"[28] in the TIDSA program. Furthermore, the intra-sector and intra-group rivalry and suspicion frustrated several attempts to stimulate joint or inter-regional projects. The appearance of such problems was not unexpected, given Gunn's observations, but Montgomery and Murphy felt an important step had been taken by TIDSA in that it established forums and procedures for consultation where few had previously existed.

Planning

The process through which the above issues can be addressed and linked is planning, but large scale and comprehensive planning for tourism has been slow to emerge. During the heady days of tourism expansion in the 1950s and 1960s most government interest focussed on promotion and marketing.

> Governments as well as business take tourism promotion seriously, spending millions of dollars directly for its support. Every state in the U.S., all provinces of Canada and many nations spend enormous sums on colorful, stimulating and informative material.[29]

But as growth slowed during the recessionary 1970s and 1980s, and more negative environmental and social impacts were linked to the unplanned growth of this earlier era, there was a move toward planning as well as promotion. This is not to suggest that planning is a recent innovation, only that it has taken on a more central and relevant role.

Planning in tourism has evolved with the industry. It started as an isolated and site-specific process, with little or no concern about possible externalities or linkages. A classic example of this type of planning was the work of the Stanford Research Institute in 1952 for Disneyland. But as the interrelated nature and complexities of tourism development became better appreciated there was the development of more comprehensive master plans and flexible systems planning.[30]

The monitoring, feedback and flexibility of systems planning is evident in the more recent moves to incorporate management into the planning process. "Effective management of tourism will involve a range of techniques including marketing and publicity, information and interpretation as well as planning controls and traffic management."[31] This acknowledges that tourism involves people, as well as places, and that planning for facilities must proceed hand-in-hand with marketing and a general process. Such is the comprehensive approach put forward in the proposed tourism strategy plan for British Columbia.

PROPOSED BRITISH COLUMBIA TOURISM STRATEGY

After examining the industry's evolution in this province, its current market and weaknesses, and its future prospects, a group of consultants recommended that the best strategy for British Columbia would be:

> . . . to identify and encourage the development of *high quality tourism* resource opportunities and destination areas on a *nodal-corridor* concept that will attract international, U.S., and domestic markets to complement existing provincial and regional markets with specific attention to *changing tourism leisure trends* and the *world energy* situation.[32]
> (This author's emphases)

While this report has been accepted by the Ministry of Tourism as fulfilling its terms of reference and certain of its recommendations acted upon, such as the adaptation to the proposed nine regions, it has not been declared policy. This leaves room for further analysis and discussion as the province continues to deliberate on what strategy to adopt.

The strategy's overall objective is to build upon British Columbia's present base of magnificent scenery and outdoor recreation opportunities, with the type of facilities that appeal to up-market segments of the emerging leisure trends. These trends include the activity-oriented vacations for the "yuppies" and the attraction of world class resorts. The strategy suggests that a nodal-corridor approach would be the most feasible manner for British Columbia to take advantage of these trends, due to its existing linear development and relative distance from major tourist generating regions. The concept consists of identifying gateway cities to the province, including Calgary, Edmonton, Spokane and Seattle, in addition to Vancouver; developing major corridor routes for travellers who wish to tour the province, and nodal attractions to lure them along those routes; and creating resort destinations to function as nodal attractions or complete vacation experiences in their own right. The emphasis on quality tourism development is seen as a way of tempting international visitors to travel the great distances involved and as a way of keeping more British Columbians at home, for there have been few world class facilities to complement our world class resources.

The proposed strategy has four stages, each set within a desired time frame. The *first stage* is an immediate endorsement of the province's regional development approach, due to the diverse social and biophysical nature of British Columbia, and some suggested boundary changes so that the nine tourism administration regions would coincide more with resource management region boundaries. It was hoped that such standardization would make integrated resource planning more feasible.

The *second stage* is to develop planning sub-strategies to direct the regions toward the overall strategy objective. Four sub-strategies are proposed: a market orientation, a resource orientation, an economic orientation, and a conserver orientation. It was intended that these should be viewed as complementary approaches and that each region would select the mix that best suited its situation. This second stage was seen as a transitional short-term (five year) adjustment from the current ad hoc development approach to one designed to fit in with the provincial objective, based on the region's market potential, resource base, economic structure and conservation needs.

The market orientation attempts to identify potential market segments that could be attracted to a region and justify the development of nodes and corridor routes in that area. The resource orientation identifies the natural, cultural and historic resources of a region and integrates them into a regional product or combines them with a neighboring region's attractions to make an overall appealing destination. The economic orientation is to use tourism to stimulate or broaden the regional economy, particularly in the more peripheral areas where primary industries are in difficulty and more amenities would be welcomed by resident and tourist alike. Finally, the conserver orientation recognizes that the carrying capacities and tolerance levels of certain areas within the regions may be at their threshold limits. Under those conditions a conserver orientation is recommended, with its emphasis on retaining resource (physical and social) quality and integrity.

The relative importance of each sub-strategy is expected to vary in accordance with the situation and potential of each tourism region, but in general terms the suggested strategies for the nine regions are those outlined in Table 4,18. Within these general emphases there are areal variations as can be seen in Figure 2,18. Vancouver Island, for example, has been recommended for a resource orientation at the regional level. But within that general emphasis on the need to identify and develop new resources, it is suggested that a conserver orientation be adopted for Victoria and an economic development strategy for the area around Port Hardy. It is apparent that a sub-strategy identity leaves plenty of scope for local variation, taking into account the seasonal pressures experienced in Victoria and the untapped tourism resources in the primary economy of the northern part of Vancouver Island.[33]

The *third stage* involves implementation of the sub-strategies, and once again four approaches have been identified. They are to maximize the economic impact of tourism development by increasing local economic linkages; to optimize future benefits by adopting a conserver approach whenever and wherever it becomes necessary; to facilitate growth by identifying development opportunities and supplying start-up funding; and to encourage free enterprise initiatives. Figure 3,18 indicates the various implementation strategies suggested by the consultants, and using Vancouver Island as an example, again one can see considerable variation within this region, based on varying local needs and priorities.

Table 4,18 RECOMMENDED SUB-STRATEGY EMPHASES FOR BRITISH COLUMBIA'S TOURISM REGIONS

Sub-Strategy	*Tourism Region*
RESOURCE ORIENTATION	Vancouver Island (A)
	Okanagan-Similkameen (C)
	B.C. Rocky Mountain (I)
	Peace River/Alaska Highway (H)
MARKET ORIENTATION	Southwestern B.C. (B)
	Cariboo/Chilcotin (F)
CONSERVER ORIENTATION	Kootenay Boundary (D)*
	Thompson/Shuswap (E)*
ECONOMIC ORIENTATION	Yellowhead 16 (G)*

* Since 1984 Region D has been renamed "Kootenay Country", Region E "High Country" and Region G "North by Northwest".

SOURCE: Marshall, Macklin, Monaghan Ltd., *Technical Report of the Province of British Columbia Tourism Development Strategy*, Ottawa and Victoria: Department of Regional Economic Expansion and Ministry of Industry and Small Business Development, 1979, pp. 406-407.

The implementation stage is seen as the medium term (five to ten years) of the development strategy. It involves a continued movement away from the province's dependency on the conventional automobile market, with increased emphasis on package tours and activity vacations in specific destination resorts.

If all went well with the implementation stage a long-term planning strategy would take over, and in a 10 to 20 year period British Columbia would be developed into an international destination area. Combining its scenic and recreation resources with high quality facilities is expected to draw more visitors over longer distances. Included in this phase is more emphasis on the establishment of northern destination areas and travel loops, so that this area of the province can become more involved with the industry and the socio-economic benefits it brings.[34]

While it is the task of planning experts to chart future directions for the industry, it is the responsibility of their customers, the province and its residents, to determine whether or not these suggestions meet the needs which have been identified. The 555-page technical document[35] represents a gallant attempt to bring some structure and direction to a very diverse and

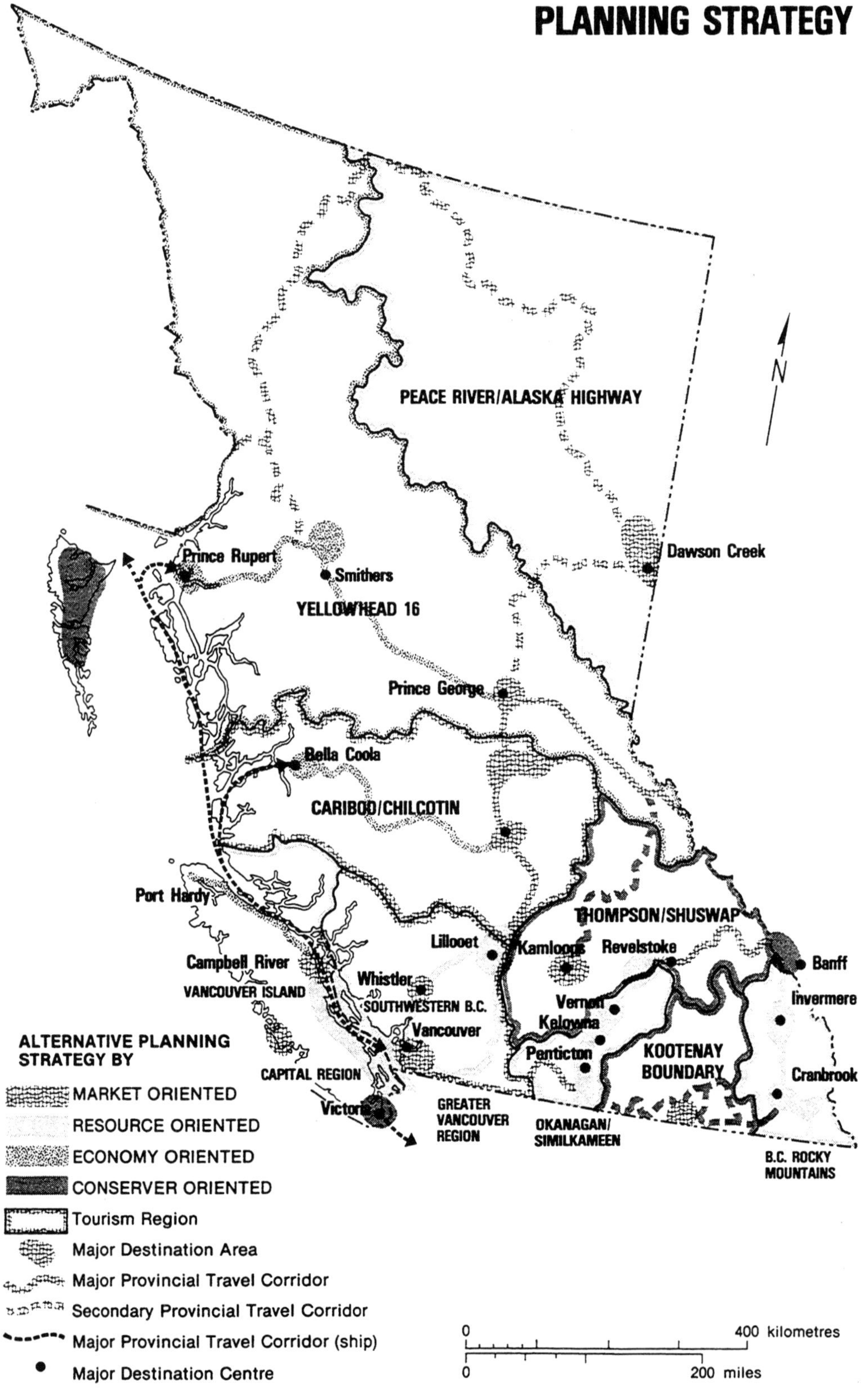

FIGURE 2,18 Tourism Planning Strategy and Sub-Strategies

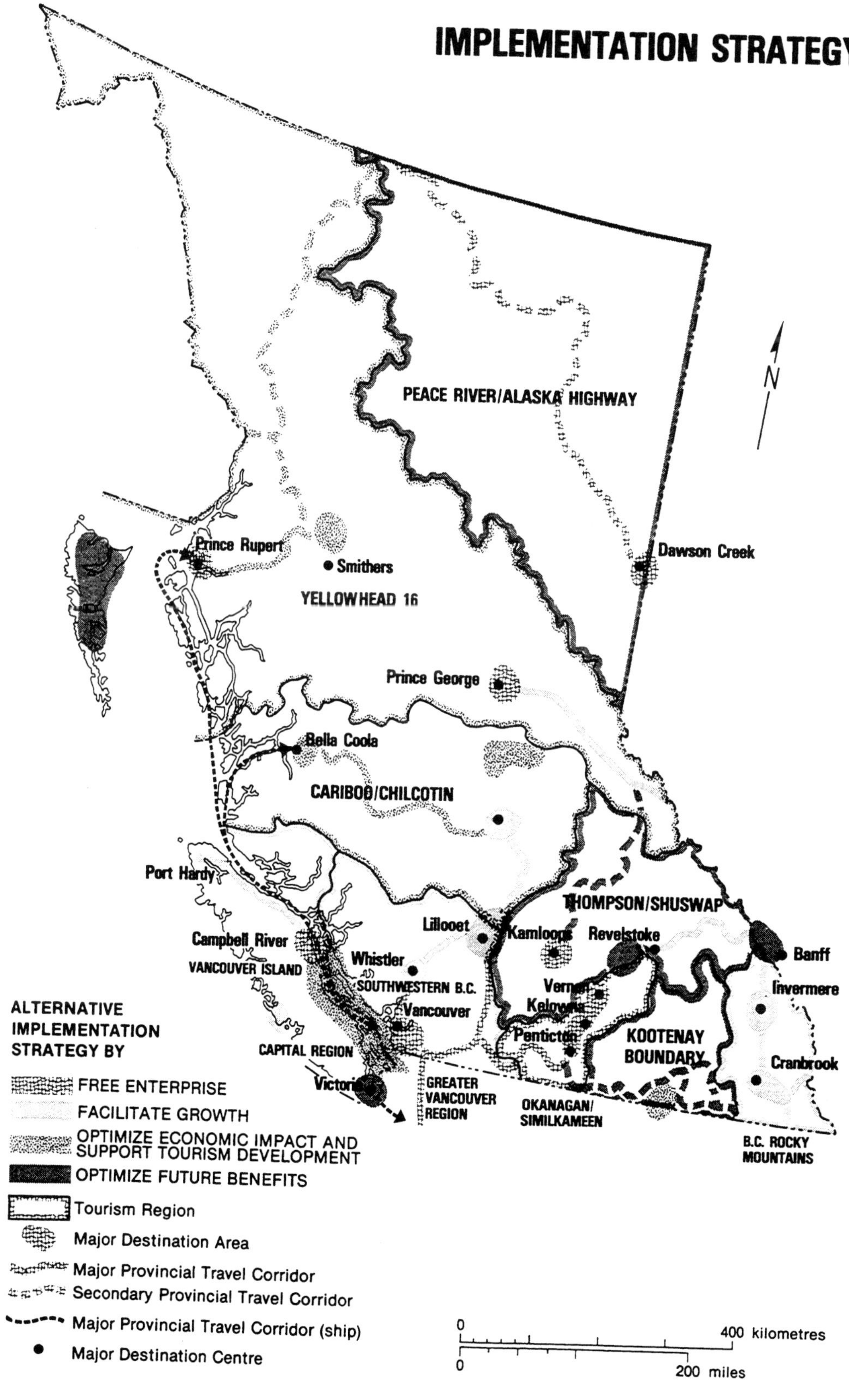

FIGURE 3,18 Tourism Implementation Strategy

unco-ordinated industry, but does it address the four needs identified in the Discussion Paper[36] on British Columbia's tourism? To answer that question an assessment of the proposed strategy in terms of British Columbia's conservation issue, investment potential, administrative structure and planning process is needed.

COMPARISON OF STRATEGY AND NEEDS

Conservation

The conservation issues would appear to have been addressed if one is influenced by its prominent labelling as a sub-strategy concern. The conservation sub-strategy was "to ensure retention of resource quality and integrity and to prevent degradation or misuse".[37] It is to be applied wherever physical or social tolerance levels have been exceeded or are in danger of becoming so, and has been recommended for the southern Kootenay Boundary region on physical grounds and the Victoria area on social grounds.

A major problem with this concept is one of application. It is difficult and time consuming to determine when the natural and social carrying capacity threshold of an area has been reached, for the signs may be site or time specific and the cause may be indistinguishable.[38] The industry and its planners do not want to be spooked into action by occasional incidents, but there is also the danger that inaction can lead to irreversible damage, not just to the local tourist product, but also to the destination area's image. An example of where reluctance to act led to more serious consequences and a tarnished image can be seen in the industry's slow reaction to the milfoil infestation of Okanagan Lake. Dearden reported that by failing to acknowledge the problem and tackling it in its early stages the industry was left with a major deterrent, one which now requires draconian remedies, such as the application of chemical herbicides that may in themselves deter future visitors.[39]

It is apparent that the conserver sub-strategy, as envisioned in the tourism strategy proposal, is a reactive form of planning. A crisis must be at hand before the area qualifies, but by then the pressures and damage may be excessive. Tourism is built upon imagery, and if a destination develops a poor reputation for quality or hospitality it takes a long time to recover from such publicity. It would be better to integrate the conservation ethic into all sub-strategies so that it becomes a permanent benchmark by which to assess the viability and consequences of the other three. Once a resource, whether it be physical or human, is damaged the marketing, development and economic sub-strategies will all be compromised.

Investment

The proposed strategy gave substantial attention to economic and investment concerns, noting that tourism can "stimulate or broaden the economy of selected urban centres or rural areas".[40] Such concepts are not unexpected, given that the TIDSA umbrella for this strategy proposal was already committed to spreading the investment and benefits of tourism development to areas outside of the two major metropolitan regions.[41] This goal was rigorously pursued, with millions of dollars being spent on large and small scale projects, designed to upgrade existing facilities and develop new attractions in the peripheral regions of the province.

However, experience elsewhere has revealed difficulties in stimulating economic growth via tourism. Brownrigg and Greig's examination of tourism's contribution to economic development in rural Scotland found that most of the cash injected into the countryside leaked out of local economies because of their high dependence on outside goods and services.[42] Low multipliers were the rule for small rural communities, the result being that "it involves a lot of noise and activity but, at the end of the day, locals have surprisingly little to show for it".[43] Diamond has pointed out yet another problem: unknown and untested peripheral destinations are placed at a disadvantage in attempting to break into established market patterns.[44] They must place themselves in the hands of travel developers and agencies, and thanks to the nearly ubiquitous nature of tourism resources there is nothing to stop an intervening area from cutting into their market, once it becomes established.

Some of the problems recognized elsewhere have been noted in the early stages of TIDSA's development program in British Columbia. Montgomery and Murphy reported that despite the intent to spread investment throughout the province most of it took place in areas surrounding the current magnets of Vancouver and Victoria.[45] The reason is not hard to find. The development and revenue potential for new projects is always greater near existing attractions and tourist routes, thus the officials who were responsible for the program were only being cautious and conservative with public funds when they endorsed the more viable projects. Furthermore, the project proposals that were most successful were those put forward by large companies or organizations. These were the organizations which had the staff, or could afford to hire the expertise, to fill out the forms and make the presentations needed to convince the government. Thus the program which had the good intention of spreading investment to the more peripheral regions of the province often was frustrated by operational considerations.

One problem that the investment program failed to address sufficiently was the seasonality of the existing tourism industry. One may view the Whistler ski development as an attack on seasonality, bringing visitors to the province during the off-season and providing residents with an alternative to American ski areas. Although the long term goal is to create a four-season resort at Whistler, its present usage is still seasonal, albeit with a winter peak in this case. Apart from the ski attraction investment, there was little attempt to alleviate this serious constraint to British Columbia's tourism development. There were no funds for convention and trade centres, beyond those for the Whistler centre, and there were only limited funds for small-scale local, all-weather attractions such as museums. Without more high profile, all-weather attractions the industry will continue to depend on a two month season, with all its problems of inefficiency and congestion.

Administration

A prime concern for British Columbia and all administrations involved in tourism development is co-ordination. Co-ordination is recognized as the key to meaningful planning and marketing, yet the proposed strategy suggested little beyond the need to create a Ministry of Tourism with sufficient funding and staff to stand alongside other major industries.[46] Its focus is on centralized leadership and planning, yet it fails to indicate how the tourism concerns and strategies will become co-ordinated with the priorities of other ministries and individual regions around the province.

Although the strategy document calls for a more integrated approach to resource development, no suggestions are provided as to how this may be done. At present the Ministry of Tourism sends representatives to public hearings or committee debates in other ministries, but lacks a political presence in many of the key decision-making areas. One recent suggestion to remedy this was put forward by the Outdoor Recreation Council of British Columbia in its position paper on tourism.[47] It noted, quite correctly, that it does not make much sense to plan land use in consultation with only two of the three top revenue-generating industries in the province. Consequently, it called for the minister of tourism to take his rightful place on the cabinet's Environment and Land Use Committee, which is the last court for all major land use decisions in British Columbia, along with the ministers of highways, energy and mines, environment, industry and small business, municipal affairs, agriculture and forests.

Another administrative problem is the integration of marketing and product development. Traditionally, the former has been the responsibility of the Ministry of Tourism, while the Ministry of Industry and Small Business

Development has strongly influenced the latter through its development and grant programs. The separation of these two functions creates an unnatural division in the tourism development process, yet it is difficult to resolve because tourism is but one of the province's industries and it is certainly dominated by small business. A recent attempt to improve communication and co-ordination between the two ministries has been the establishment of a development section within the Ministry of Tourism, which operates very closely with the Ministry of Industry and Small Business Development.

Despite the strategy's call for more co-ordinated marketing and development within and between the nine tourist regions, little practical advice was given regarding ways to overcome this fragmentation. In an attempt to foster greater marketing co-ordination, the Ministry of Tourism now offers tempting 50-50 cost share partnerships for approved regional marketing strategies. But to develop and refine a tourist product that supports such regional marketing strategies requires the investment and co-operation of regional and municipal districts. Again, the divisions between marketing and development reappear, now magnified and confused by the ambitions and hopes of 26 regional districts and numerous municipalities operating within the nine tourist regions. One attempt to solve this problem is the creation of program managers within the Ministry of Tourism's development department. These people assist local authorities and businessmen to develop tourist products that fit in with the regional strategy, helping them with the paperwork and arranging financial assistance where necessary.

Planning

The consultants state bluntly that tourism planning in British Columbia "has generally been lacking and often, where it has existed, it has been fragmented and not related to other planning efforts".[48] This is not an unusual situation for tourism, being a relatively new industry and one with seemingly few problems as it moved from one record-breaking season to another. But by the early 1980s its increasing size and demand for resources placed it in competition for physical resources, land and investment funds with the more established provincial industries. Its growing external impacts, some negative, some positive, could no longer be ignored by regional districts and municipalities which formed the jurisdictions of major tourist destinations.

The suggested response to this need and fragmentation is a "decision-tree approach" to British Columbia's tourism development:

> It allows various ministries, branches and sections of the government to clearly identify their particular roles and involvement in each sub-strategy and the anticipated results of each decision. Further, the decision tree allows government resource planners to clearly map the implications of a particular strategy prior to implementation.[49]

The links of this decision-tree approach can be seen in Figure 4,18. Two points stand out in this approach. First, the lack of interconnecting links between the various alternatives suggests the consultants' statement that the four sub-strategies can be combined may be wishful thinking. Figure 4,18 indicates a linear development, rather than a reinforcing combined approach, once a sub-strategy has been selected. Second, as one would expect with a development mandate, there is no room for a "no growth" option. The conservation orientation still involves a focus on "developing resource potential" and the "status quo" option emphasizes increased efficiency of the plant and "no change to the government's approach to the planning, development and use of resources".

Several problems emerge from the planning approach outlined in the suggested development strategy. Although the consultants have made a case for the need for tourism planning, their argument has apparently received only limited support. The TIDSA planning program was undertaken concurrently with the other four, not first as the program ranking may suggest to some. Consequently, major investment decisions were being made as researchers examined the regional tourist pictures and developed their planning recommendations.

Perhaps an even bigger problem is the scale and origin of the planning recommendations. To many communities and individuals the planning has come from Victoria, based on the province's needs and wishes after only fleeting surveys and investigation of local needs and priorities. Getz refers to this form of planning as a top-down approach, and after reviewing Canada's general experience concludes:

> Perhaps the sad experience of numerous mega-projects in Canada is gradually encouraging a shift from big plans and top-down policies to more local and small scaled tourism planning. This change in emphasis will have to come about in any event if the federal and provincial schemes are to be implementable.[50]

The motivation to implement master plans must come from destination areas themselves, and to generate such motivation requires more input and

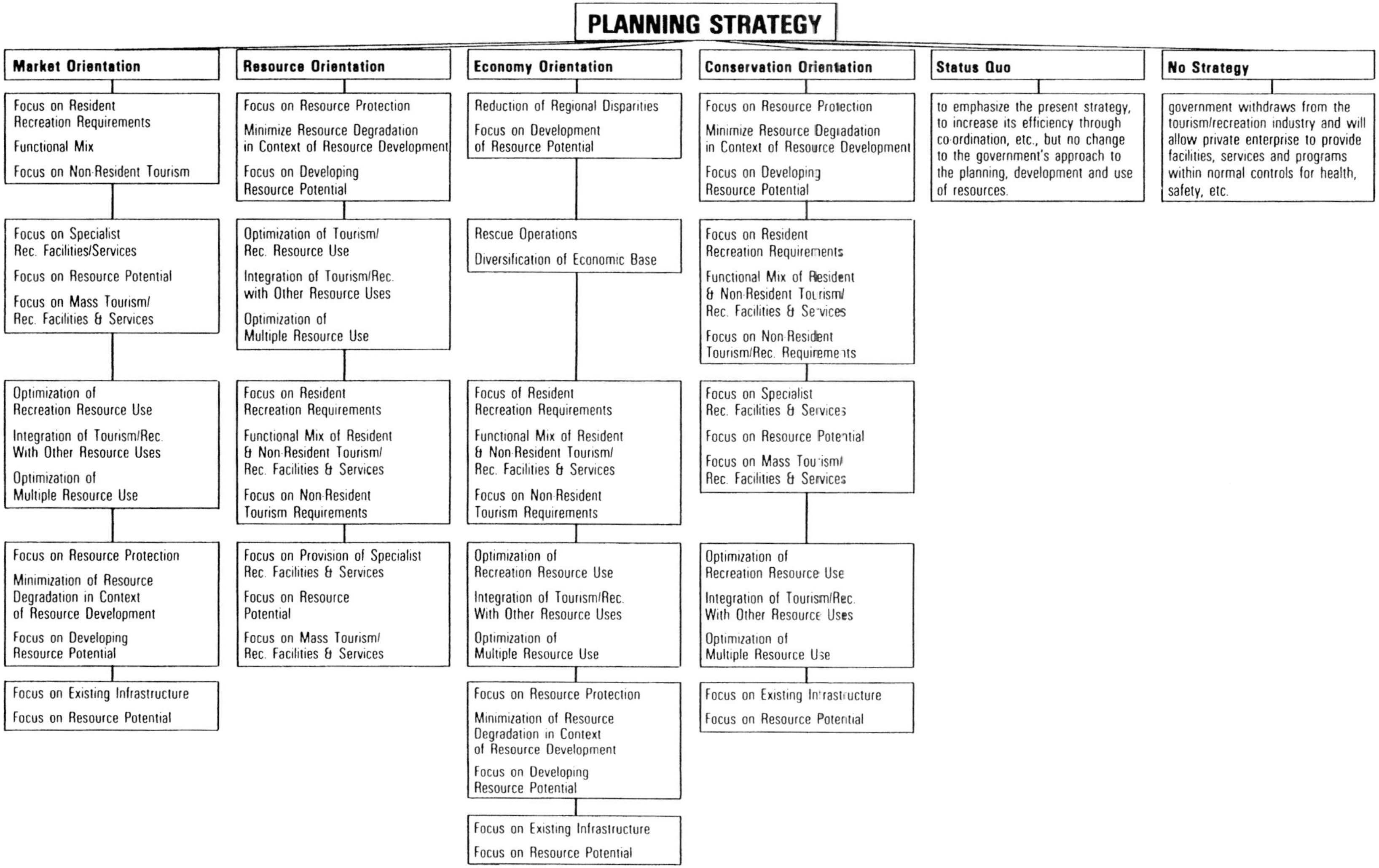

Figure 4,18 BRITISH COLUMBIA TOURISM STRATEGY

control at the local levels of municipalities and regional districts. Tourism, in fact, should become part of municipal and regional district planning. It should be viewed as a tool for economic development or conservation, as a supporter of local culture and heritage, and as a developer of amenities for both resident and visitor. As such, more of the stimulus should originate at local levels, to be supported and co-ordinated by the province when it coincides with provincial aims and objectives.

SUMMARY

Many observers consider that British Columbia is going through an economic adjustment, and that its traditional primary industries will no longer be the mainstay of our economy and employment, as in the past. One of the activities which can pick up the slack over the short term and become an important contributor to our future economy is tourism. Tourism, however, has some obstacles to overcome because of the province's peripheral location in North America; the seasonal nature of our current product and its complex, fragmented character. The TIDSA program of 1978 and proposed strategy of November 1979 provided a necessary stimulant and analysis, but still left some serious questions unanswered.

There is a need to build upon the results of TIDSA and to tackle some of the problems that still remain. Since we sell ourselves as Super Natural British Columbia, we must ensure that we conserve more of our precious landscape vistas and heritage. This means more integrated resource management with other ministries and more conservation of our urban-industrial heritage, since the province was settled at an earlier date than its western neighbors.

To reap a more acceptable return on our investment we must extend the season. A few destinations can be developed into four-season resorts, but most will remain dependent on the summer business. These areas should stress the shoulder season opportunities of the spring and fall and some could develop regional trade and convention markets to take advantage of the upcoming information age. The prime conference centre will be Vancouver because of its modern facilities and international air links.

Both planning and administration need to become more integrated if efforts to modernize the industry are to succeed. Tourism is so diffused throughout our economy and province that no single agency or jurisdiction can plan and implement on an efficient basis in this competitive business.

There should be a partnership between provincial ministries and between the province and local jurisdictions, rather than the fiefdoms that currently control development and destinies. The Tourism Ministry should head an umbrella organization that has components in related ministries, and representatives in the nine tourist regions to confer and co-ordinate with local jurisdictions. The ideal would be a partnership where local initiative and character were encouraged to develop within a provincial framework. In this way individual communities could benefit from the amenities and revenues of tourism, while the province benefited from the increased drawing power of a co-ordinated, yet varied, product.

REFERENCES

1. MURPHY, P.E., "Tourism: Canada's other resource industry", in MURPHY, P.E., (ed.), *Tourism in Canada: Selected Issues and Options*. Victoria: University of Victoria, Department of Geography, Western Geographical Series, Vol. 21, 1983, pp. 3-23.

2. MARSHALL, MACKLIN, MONAGHAN LIMITED, *et. al.*, *Executive Summary of the Province of British Columbia Tourism Development Strategy*. Ottawa and Victoria: Department of Regional Economic Expansion and Ministry of Industry and Small Business Development, 1979, p. 1.

3. MAUNDRELL, P., "Review of Tourism Trends and Planning", a paper presented at the University of Victoria, February 27, 1984.

4. BRITISH COLUMBIA, *Tourism Highlights, 1984*. Victoria: Ministry of Tourism, 1985, pp. 1 and 16.

5. TAYLOR, G., "Canada's tourism trends for the 1980's", in MURPHY, P.E. (ed.), *op. cit.*, reference 1, pp. 29-55.

6. O'BOYLE, P. and WHITE, L., *A Study of British Columbia Ski Areas and their Market Potential*. Victoria: Ministry of Industry and Small Business Development, 1979; CANADA, *The Economic Impact of Recreational Boating in British Columbia*. Ottawa: Environment Canada, 1977; THOMPSON, BERWICK, PRATT and PARTNERS, *British Columbia River Rafting Study*. Ottawa and Victoria: Department of Regional Economic Expansion and Ministry of Industry and Small Business Development, 1979; PEARSE BOWDEN ECONOMIC CONSULTANTS LIMITED, *The Value of Fresh Water Fishing in British Columbia: An Economic Analysis*. Vancouver: Fish and Wildlife Branch, Ministry of Environment, 1979; BRITISH COLUMBIA, *Proposed Wildlife Management Plan for British Columbia*. Victoria: Ministry of Environment, 1979.

7. B.C. RESEARCH, *Visitors '79*, Ottawa and Victoria: Department of Regional Economic Expansion and Ministry of Industry and Small Business Development, 1980, p. 2.

8. *Ibid.*

9. Car and recreation vehicle traffic represented 76 percent of British Columbia domestic travel, and 74 percent of United States visitors staying overnight, according to Marshal, Macklin, Monaghan, Limited, *Technical Report of the Province of British Columbia Tourism Development Strategy*. Ottawa and Victoria: Department of Regional Economic Expansion and Ministry of Industry and Small Business Development, 1979, pp. 226 and 249.

10. B.C. RESEARCH, *op. cit.*, p. 2.

11. BRITISH COLUMBIA, *op. cit.*, pp. 3 and 4.

12. D'AMORE, L.J., "Guidelines to Planning in Harmony with the Host Community", in MURPHY, P.E. (ed.), *op. cit.*, reference 1, pp. 135-159.

13. The Port Alberni development received a 1,185,500 dollar TIDSA loan to develop its waterfront, market, observation tower, pier and passenger facilities for the MV "Lady Rose" ferry. *B.C. Economic Bulletin*, Vol. 2, No. 4, December-January 1984, p. 10.

14. ENGLISH TOURIST BOARD, "A 5 Billion Pound Industry", *Tourism in England*, No. 23, 1977, p. 5.

15. BLACKORBY, C., DONALDSON, G., and SLADE, M., *Expo 86: An Economic Impact Analysis*. Vancouver: University of British Columbia, B.C. Economic Policy Institute, Paper No. P-84-11, 1984, p. 15; MARKET FACTS OF CANADA, *Expo 86: Tracking Study, The March 1984 Market*. Vancouver: Market Facts of Canada Limited, 1984, p. iii; *Sun*, "Expo Cuts Its Gate Forecasts", December 21, 1984, p. A-3; *Times-Colonist*, "Expo 86 Loss Seen Topping Half-billion", October 11, 1984, p. A-1.

16. BRITISH COLUMBIA, *British Columbia Tourism Development Strategy: A Discussion Paper*. Victoria: Ministry of Tourism, n.d., pp. 9 and 10.

17. HOOPER, A., "Our Ugly Slopes", *Times-Colonist*, October 12, 1981, p. 5.

18. MEARES, S., "Historical Value Too", *Times-Colonist*, January 2, 1985, p. A-4.

19. BRITISH COLUMBIA, *Forest Landscape Handbook*, Victoria: Ministry of Forests, 1981.

20. "Give Moresby Top Protection", *Times-Colonist*, August 22, 1984, p. B-1.

21. MERRIMAN, A., "Recreational Fishing", *Times-Colonist*, August 15, 1984, p. B-7.

22. Communication with a representative of British Columbia, Ministry of Tourism, January 10, 1985.

23. CANADA and BRITISH COLUMBIA, *Canada-British Columbia Subsidiary Agreement on Travel Industry Development*, Ottawa and Victoria: Department of Regional Economic Expansion and Ministry of Economic Development, October 17, 1978.

24. "Convention Centres Confident of Success", *Globe and Mail*, November 19, 1984, pp. B-1 and B-14.

25. GUNN, C.A., "Industry Pragmatism vs. Tourism Planning", *Leisure Sciences*, Vol. 1, 1977, p. 92.

26. CANADA and BRITISH COLUMBIA, *op. cit.*, pp. 18 and 19.

27. MONTGOMERY, G. and MURPHY, P.E., "Government Involvement in Tourism Development: A Case Study of TIDSA Implementation in British Columbia", in MURPHY, P.E. (ed.), *op. cit.*, reference 1, pp. 183-209.

28. *Ibid.*, p. 205.

29. GUNN, C.A., "Tourism-Recreation-Conservation Synergism", *Contact*, Vol. 8, 1976, p. 130.

30. GRAVEL, J-P., "Tourism and Recreational Planning: A Methodological Approach to the Valuation and Calibration of Tourism Activities", in PERKS, W.T. and ROBINSON, T.M., (eds.), *Urban and Regional Planning in a Federal State: The Canadian Experience*. Stroudsburg, Penn.: Dowden, Hutchinson and Ross, 1979, pp. 122-134.

31. ENGLISH TOURIST BOARD, *Planning for Tourism in England*. London: English Tourist Board, 1981, p. 82.

32. MARSHAL, MACKLIN, MONAGHAN LIMITED, *Technical Report of the Province of British Columbia Tourism Development Strategy*, *op. cit.*, reference 9, p. 405.

33. MURPHY, P.E., "Development and Potential of Tourism", in FORWARD, C.N. (ed.), *Vancouver Island, Land of Contrasts*. Victoria: University of Victoria, Department of Geography, Western Geographical Series, Vol. 17, 1979, pp. 289-307.

34. MURPHY, P.E., "Prospects and Problems of Tourism Development in the Middle North Region of British Columbia and the Yukon", in BARR, B.M., (ed.), *Geographical Perspective on Western Canada*. Vancouver: Tantalus Press, B.C. Geographical Series, No. 28, 1979, pp. 33-47.

35. MARSHAL, MACKLIN, MONAGHAN LIMITED, *Technical Report of the Province of British Columbia Tourism Development Strategy*, *op. cit.*, reference 9.

36. BRITISH COLUMBIA, *British Columbia Tourism Development Strategy: A Discussion Paper*, *op. cit.*

37. MARSHAL, MACKLIN, MONAGHAN LIMITED, *Technical Report of the Province of British Columbia Tourism Development Strategy*, *op. cit.*, reference 9, p. 376.

38. MITCHELL, B., *Geography and Resource Analysis*. New York: Longman, 1979, pp. 176-199; D'AMORE, L.J., "Guidelines to Planning in Harmony With the Host Community", in MURPHY, P.E. (ed.), *op. cit.*, reference 1, pp. 135-159.

39. DEARDEN, P., "Tourism and the Resource Base", in MURPHY, P.E. (ed.), *op. cit.*, reference 1, pp. 75-93.

40. MARSHAL, MACKLIN, MONAGHAN LIMITED, *Technical Report of the Province of British Columbia Tourism Development Strategy*, *op. cit.*, reference 9, p. 375.

41. MONTGOMERY and MURPHY, *op. cit.*, pp. 188-189.

42. BROWNRIGG, M. and GRIEG, M.A., *Tourism and Regional Development.* Glasgow: University of Strathclyde, The Fraser of Allander Institute Speculative Papers, No. 5, 1976.

43. *Ibid.*, p. 7.

44. DIAMOND, J., "Tourism's Role in Economic Development: The Case Re-examined", *Economic Development and Cultural Change*, Vol. 25, 1977, pp. 539-553.

45. MONTGOMERY and MURPHY, *op. cit.*, pp. 202-203.

46. MARSHAL, MACKLIN, MONAGHAN LIMITED, *Technical Report of the Province of British Columbia Tourism Development Strategy, op. cit.*, reference 9, pp. 413-415.

47. LANG, S., "Tourism Paper, Food for Thought", *Times-Colonist*, August, 24, 1984, p. B-7.

48. MARSHAL, MACKLIN, MONAGHAN LIMITED, *Technical Report of the Province of British Columbia Tourism Development Strategy, op. cit.*, reference 9, p. 43.

49. *Ibid.*, p. 367.

50. GETZ, D., "A Research Agenda for Municipal and Community-based Tourism in Canada", a paper presented at the Travel and Tourism Research Conference, Banff, June 1983.

PLATE 27 Grain elevators at Dawson Creek. *B.C. Government Photo*▶

ALBERTA
WHEAT POOL
DAWSON CREEK

LIST OF CONTRIBUTORS

PHILIP DEARDEN, *B.A. (Birmingham), M.Sc. (Memorial), Ph.D. (Victoria),*
Assistant Professor, Department of Geography, University of Victoria

Dr. Dearden is a biogeographer who has specialized in recreational geography, landscape evaluation and landscape aesthetics.

BRUCE K. DOWNIE, *B.A. (Winnipeg), M.A. (Waterloo),*
Consultant, Parks: Research and Planning Incorporated

Mr. Downie is a recreational geographer who was in the Department of Geography at the University of Victoria as Co-operative Education Co-ordinator and as a lecturer from 1982 to 1984.

MICHAEL C.R. EDGELL, *B.A. (Birmingham), Conservation Diploma (London), Ph.D. (Birmingham),*
Associate Professor, Department of Geography, University of Victoria

Dr. Edgell is a biogeographer whose research areas include the forest industry, landscape evaluation and waterfowl habitats.

CHARLES N. FORWARD, *B.A., M.A. (British Columbia), Ph.D. (Clark),*
Professor, Department of Geography, University of Victoria

Dr. Forward is an urban geographer who has specialized in ports and shipping, as well as urban historical geography.

HAROLD D. FOSTER, *B.Sc., Ph.D. (London),*
Professor, Department of Geography, University of Victoria

Dr. Foster is a geomorphologist whose research areas include natural hazards, disaster planning and renewable energy.

PATRICIA P. GILMARTIN, *B.Sc., M.A. (Georgia State), Ph.D. (Kansas),*
Assistant Professor, Department of Geography, University of South Carolina

A specialist in cartography, Dr. Gilmartin was a faculty member in the Department of Geography at the University of Victoria from 1980 to 1984.

DAVID CHUENYAN LAI, *B.A., M.A. (Hong Kong), Ph.D. (London),*
Associate Professor, Department of Geography, University of Victoria

Specializing in China, Hong Kong and overseas Chinese communities, Dr. Lai also has research interests in urban, cultural and industrial geography.

PETER E. MURPHY, *B.Sc., Diploma in Education (London), M.A., Ph.D. (Ohio State)*
Associate Professor, Department of Geography, University of Victoria

An urban geographer, Dr. Murphy has specialized particularly in tourism management and community planning.

J. DOUGLAS PORTEOUS, *B.A., M.A. (Oxford), Ph.D. (Hull),*
Professor, Department of Geography, University of Victoria

Dr. Porteous is an urban geographer whose research specialties include urban planning, behavioral geography and environmental aesthetics.

WILLIAM M. ROSS, *B.Ed. (British Columbia), M.A. (Toronto), Ph.D. (Washington),*
Associate Professor, Department of Geography, University of Victoria

Dr. Ross is a resources geographer and his specialties include coastal management and fisheries, as well as political and legal aspects of resource development.

W.R. DERRICK SEWELL, *B.Sc. (London), M.A., Ph.D. (Washington),*
Professor, Department of Geography, University of Victoria

A resources geographer, Dr. Sewell has specialized in economic, social and behavioral aspects of planning and policy-making in resources management, with particular reference to water and energy resources.

STANTON E. TULLER, *B.A. (Oregon), M.A., Ph.D. (California at Los Angeles),*
Associate Professor, Department of Geography, University of Victoria

Dr. Tuller is a climatologist whose research specialties include the earth's heat balance and applied aspects of micro-climatology.

COLIN J.B. WOOD, *B.A. (Wales), M.A., Ph.D. (McMaster),*
Associate Professor, Department of Geography, University of Victoria

A resources geographer, Dr. Wood has been engaged in research on conflict resolution, land ownership patterns and agriculture.